HIGH-PERFORMANCE LIQUID CHROMATOGRAPHY

Advances and Perspectives

Volume 2

HIGH-PERFORMANCE LIQUID CHROMATOGRAPHY

Advances and Perspectives

Volume 2

Edited by

Csaba Horváth

Department of Engineering and Applied Science
Yale University
New Haven, Connecticut

1980

ACADEMIC PRESS
A Subsidiary of Harcourt Brace Jovanovich, Publishers
New York   London   Toronto   Sydney   San Francisco

ACADEMIC PRESS, INC.
111 Fifth Avenue, New York, New York 10003

United Kingdom Edition published by
ACADEMIC PRESS, INC. (LONDON) LTD.
24/28 Oval Road, London NW1 7DX

ISSN 0270-8531

ISBN 0-12-312202-3

PRINTED IN THE UNITED STATES OF AMERICA

82 83   9 8 7 6 5 4 3 2

CONTENTS

LIST OF CONTRIBUTORS   vii
CONSPECTUS.   ix
PREFACE   xi
CONTENTS OF VOLUME 1   xiii

Optimization in Liquid Chromatography
Georges Guiochon

     I. Introduction.   1
    II. Resolution and Efficiency.   2
   III. Analysis Time, Flowrate, and Pressure.   6
   IV. Flow Velocity and Column Efficiency   7
    V. What to Optimize in Liquid Chromatography   15
   VI. Equations Used for Optimization   17
  VII. Ways to Achieve 5000 Plates in 5 Minutes   19
 VIII. Column Performance at Minimum Inlet Pressure   29
   IX. Critical Minimum Pressure   33
    X. Achievement of Fastest Analyses   35
   XI. Practical Considerations   39
  XII. Equipment Specifications.   43
 XIII. Conclusions.   52
      References   54

Liquid Chromatography on Silica and Alumina as Stationary Phases
Heinz Engelhardt and Helmut Elgass

     I. Introduction.   57
    II. Stationary Phases   60
   III. Role of the Eluent   70
   IV. Effect of Sample Structure on Retention   90
    V. Anisocratic Analysis   92
   VI. Theory of Adsorption Chromatography   101
  VII. Selected Applications   104
 VIII. Prospects   106
      References   108

Reversed-Phase Chromatography
Wayne R. Melander and Csaba Horváth

     I. Introduction.   114
    II. The Stationary Phase   123
   III. The Mobile Phase   165
   IV. Effect of Temperature   192
    V. Theory of Reversed-Phase Chromatography   201
   VI. Modulation of Selectivity by Secondary Equilibria   229

Contents

VII. Physicochemical Measurements 266
VIII. Selected Analytical Applications 279
    Notation 299
    References 303

Index 321

Erratum 341

LIST OF CONTRIBUTORS

Numbers in parentheses indicate the pages on which the authors' contributions begin.

Helmut Elgass[*] (57), Angewandte Physikalische Chemie, Universität des Saarlandes, 6600 Saarbrücken, Federal Republic of Germany

Heinz Engelhardt (57), Angewandte Physikalische Chemie, Universität des Saarlandes, 6600 Saarbrücken, Federal Republic of Germany

Georges Guiochon (1), Laboratoire de Chimie Analytique Physique, École Polytechnique, 91128 Palaiseau Cedex, France

Csaba Horváth (113), Department of Engineering and Applied Science, Yale University, New Haven, Connecticut 06520

Wayne R. Melander (113), Department of Engineering and Applied Science, Yale University, New Haven, Connecticut 06520

[*] Present address: Hewlett-Packard GmbH., 7517 Waldbronn 2, Federal Republic of Germany.

CONSPECTUS

The analytical technique known in most languages as HPLC, the acronym for *high-performance liquid chromatography*, may very well represent the climax of the development started when the Italo-Russian botanist Michael S. Tswett coined the name chromatography and recognized the potential of his method for separating plant pigments almost 80 years ago. In order to distinguish HPLC from conventional column chromatography, the names high-pressure or high-speed liquid chromatography are also used occasionally. It is a separation method of unsurpassed versatility and a microanalytical tool *par excellence*. Like gas chromatography, HPLC is characterized by a linear elution mode and by the use of a sophisticated instrument, high-efficiency columns, and sensitive detectors.

Recent advances in instrumentation, column engineering, and theory have considerably broadened the field of application of HPLC, which now finds employment in virtually all branches of science and technology. Yet, we may have witnessed only the beginning of a long growth period in which HPLC will become the preeminent method of chemical analysis.

The goal of this serial publication is to provide up-to-date accounts of various topics in HPLC. The individual chapters will cover subjects of particular interest in this rapidly growing field. Throughout the successive volumes, the coverage of applications, instrumentation, and theory will be balanced, although the contents of some volumes may focus on one or the other of these subjects. As the field evolves and the horizon of HPLC expands, future volumes are expected to present full accounts of the advances in HPLC and to unfold the perspective required for exploiting its full potential.

Selection of topics and the level of treatment, at least in the early volumes, are planned to offer useful reading both for the novice and the seasoned chromatographer. Thus the contents will reflect not only the individuality of expression of the contributors, but also the diversity and broad scope characteristic of HPLC.

New Haven, Connecticut   CSABA HORVÁTH

PREFACE

Volumes 1 and 2 of this serial publication deal mainly with fundamental aspects of high-performance liquid chromatography, a technique generally referred to by the acronym HPLC. As the preparation of the first volume coincided with the seventy-fifth anniversary of Tswett's discovery, it was felt that the opening chapter should pay tribute to the history of chromatography. The detailed treatment of the subject by Ettre presents not only a panoramic view of the evolution of liquid chromatography, but also new findings from the author's recent research into chromatographic history. Subsequently the use of bonded phases in HPLC is discussed by Majors, who played a pioneering role in the development of microparticulate bonded phases. The effect of chemical phenomena such as ionization and complex formation on retention and selectivity in reversed-phase chromatography is critically examined in the chapter on secondary chemical equilibria by Karger, LePage, and Tanaka. Definitive treatment of gradient elution is given by Snyder, perhaps the greatest living authority on this subject.

Volume 2 contains three chapters. Guiochon discusses optimization of the column and the operating conditions. His chapter will enable the reader to appreciate the factors determining speed and efficiency of separation in HPLC and to develop a notion about the optimization of the chromatographic system. The use of polar adsorbents and nonpolar eluents is authoritatively presented by Engelhardt and Elgass, whereas reversed-phase chromatography, the principal branch of HPLC, receives a comprehensive treatment by Melander and Horváth.

The contents of the first two volumes of this continuing publication are somewhat theoretically oriented thanks to recent advances in HPLC that make possible a discussion of fundamentals with some rigor. It is expected that focusing on basics at the beginning will facilitate both the treatment and comprehension of new topics dealing with application and instrumentation in the ensuing volumes.

I am indebted to Wayne R. Melander for advice and help in many aspects of the preparation of the first two volumes and for assistance in compiling the index.

Cs. H.

CONTENTS OF VOLUME 1

Evolution of Liquid Chromatography: A Historical Overview

    L. S. Ettre

Practical Operation of Bonded-Phase Columns in High-Performance Liquid Chromatography

    R. E. Majors

Secondary Chemical Equilibria in High-Performance Liquid Chromatography

    Barry L. Karger, James N. LePage, and Nobuo Tanaka

Gradient Elution

    Lloyd R. Snyder

Index

HIGH-PERFORMANCE LIQUID CHROMATOGRAPHY

Advances and Perspectives

Volume 2

OPTIMIZATION IN LIQUID CHROMATOGRAPHY

Georges Guiochon

Laboratoire de Chimie Analytique Physique
École Polytechnique
Palaiseau, France

I.	Introduction	1
II.	Resolution and Efficiency	2
III.	Analysis Time, Flowrate, and Pressure	6
IV.	Flow Velocity and Column Efficiency	7
V.	What to Optimize in Liquid Chromatography	15
VI.	Equations Used for Optimization	17
VII.	Ways to Achieve 5000 Plates in 5 Minutes	19
VIII.	Column Performance at Minimum Inlet Pressure	29
IX.	Critical Minimum Pressure	33
X.	Achievement of Fastest Analyses	35
XI.	Practical Considerations	39
XII.	Equipment Specifications	43
	A. Detector Time Constant	44
	B. Detector Cell Volume and Connecting Tubes	46
	C. Injection	48
	D. Detection Limits and Maximum Peak Number	51
XIII.	Conclusions	52
	References	54

I. INTRODUCTION

Column liquid chromatography has evolved through three active periods, separated by latent induction periods during which almost nothing new apparently happened. The "discovery time," at the beginning of this century (*1*), was followed by a "renaissance" period in the early 1930s (*2*), and finally the "modern times" which started in the 1960s (*3*). The last period began with the "technological age" during which instrumental problems were recognized and solved while few analyses were actually made. The present time is characterized by an exponential

1

growth of the instrument market and practical applications. Research on retention mechanisms is still very active and will remain so for many years before the thermodynamics of most chromatographic systems will be well enough understood. On the other hand, problems concerning instrumentation, the selection of optimum analytical conditions, the preparation of columns, and their operations are now pretty well solved, so that, in a sense, this field of science is closing. Whereas much technological work still remains to be done, the conceptual framework of column chromatography is not likely to change appreciably in the time to come.

In this chapter, we discuss the basic principles of modern liquid chromatography, which have been slowly and painfully recognized as valid over the past ten years. After presenting the main features of the optimization model together with the relevant equations, we shall point out the practical consequences of this approach and the way the analyst can use the optimization scheme in practice (3–26).

II. RESOLUTION AND EFFICIENCY

The goal of the analyst is to separate the components of a mixture under such convenient conditions that a satisfactory quantitative analysis can be obtained at a reasonable cost, or that the compounds can be easily identified. Although everyone would agree on such a broad definition, troubles begin when a more precise definition is sought. We usually do not strive for the fastest possible analysis. Analysis times less than a few minutes, with base peak width less than about 10 sec, raise very difficult technological problems (cf. Section XII). The specifications on sampling systems, detectors, and data acquisition systems become so stringent that full computerization of the chromatograph is necessary. This can be costly, and often that money can be saved by turning to systems having lower performance with respect to speed of analysis.

Modern column liquid chromatography is usually called HPLC. This stands either for *high-performance* or *high-pressure liquid chromatography*. It is curious that what is appealing in this new technique (the high performance), came much later to the forefront than the disagreeable price one must pay, in terms of high pressure. The pressure is needed to achieve a flow velocity of the mobile phase at which the column performances are satisfactory. As we shall show later, a trade-off between pressure and analysis time is often possible. In such cases we can find a compromise which yields an analysis time somewhat longer than the minimum possible with a given column, but at a much reduced pressure. The choice will of course depend on the analyst's needs and means.

Be that as it may, the first requirement is to achieve a minimum degree of resolution between all the compounds of a given mixture. Let us emphasize here that optimization can seriously be carried out only on known mixtures, where all components are identified. If the analyst has an unknown mixture to analyze, he can only make a crude estimate of the resolving power of the column he wishes to use, assuming that compounds that are not resolved by this column are not worth separating, at least in a first trial. This decision is made, consciously or not, while weighing the advantages of separating most or all components against the troubles of using excessive resolving power that means wasting time. For such qualitative optimization some useful estimates can be found in Section XI.

It is convenient to assume that all band profiles have a Gaussian shape. The study of the degree of resolution between non-Gaussian peaks is very difficult, if one wants to be rigorous, and the equations become very complicated. Fortunately, Kirkland (27) has shown that if the peak asymmetry is moderate, conventional equations for Gaussian peaks can be used. It should be kept in mind, however, that the resolution necessary under such conditions to accomplish a certain degree of separation of asymmetric peaks is higher than that calculated for Gaussian peaks.

The resolution between two Gaussian peaks is given by Eq. (1)

$$R = 2(t_{R_2} - t_{R_1})/(W_1 + W_2) \tag{1}$$

where $t_{R,i}$ and W_i are the retention time and base width of peak i. Note that the baseline width for a Gaussian distribution is equal to four standard deviations, σ. If the two peaks are close, the resolution is given by Eq. (2):

$$R = \frac{\sqrt{N}}{4} \cdot \frac{\alpha - 1}{\alpha} \cdot \frac{k_2'}{1 + k_2'} \tag{2}$$

The plate number of the column, N, is evaluated [Eq. (3)]

$$N = 16(t_R/W)^2 = (t_R/\sigma_t)^2 \tag{3}$$

where σ_t is the standard deviation of the peak in time units. The relative retention of the two compounds, α, is obtained from the relationship (4)

$$\alpha = \frac{t_{R_2} - t_0}{t_{R_1} - t_0} = \frac{t_{R_2}'}{t_{R_1}'} = \frac{k_2'}{k_1'} \tag{4}$$

where t_0 is the retention time of an inert, nonsorbed solute, $t_{R,i}'$ is the adjusted retention time and k' is the capacity factor of the column evaluated from the chromatogram by Eq. (5):

$$k = (t_R - t_0)/t_0 = t_R'/t_0 \tag{5}$$

It is seen that k is the dimensionless retention time of a peak given by the adjusted retention time divided by the elution time of an retained solute. The importance of k comes from the fact that it is proportional to the equilibrium constant for the equilibrium phase distribution underlying the retention process. By these definitions of the chromatographic parameters, it is assumed that no extracolumn effects from the equipment contribute to the retention of the compounds and/or the broadening of their bands. In practice this assumption often does not hold, and the problem is discussed further in Section XII. The choice of the inert tracer is sometimes very difficult in liquid chromatography as discussed, among others, by Knox (28) and Horváth (55).

Equation (2) shows that the resolution is a function of three different factors: (1) the resolving power of the column as measured by the plate number that expresses the relative width of bands; (2) the relative retention of the two compounds that measures how far apart the bands are from each other; and (3) the magnitude of retention, as separation is a result of retention. The relative influence of these factors has been discussed by Snyder (12, 13) in a form very easy to use in practice.

One of these three factors, the plate number, is largely controlled by the column parameters, construction, and operating conditions such as flow velocity. In some cases, however, the efficiency can fall sharply for some type of compounds having an unusually small diffusion coefficient or relatively slow equilibrium kinetics. The magnitude of retention can be adjusted to some degree by changing the phase ratio by varying the amount of the liquid stationary phase or the specific surface area of the absorbent in a given column or the eluting strength of the mobile phase (12, 13). In this last case, α will probably change. Thus, finding a suitable chromatographic system, which provides large enough values of α for all the pairs of components of the mixture, is a prerequisite to any optimization. Separations at small values of α require very large plate numbers and consequently long columns. As a result, the time of analysis is long and the column inlet pressure is high. Considerable saving can be achieved, therefore, by finding a stationary–mobile phase system that has high selectivity. This is illustrated by the relationship (29) between molar free energy changes, ΔG_1 and ΔG_2, associated with the equilibrium distribution of the solutes 1 and 2 between mobile and stationary phases [Eq. (6)]:

$$\Delta(\Delta G) = \Delta G_2 - \Delta G_1 = RT \ln \alpha \qquad (6)$$

As this quantity is small we can write that

$$\alpha - 1 = \Delta(\Delta G)/RT \qquad (7)$$

where α is the relative retention of the two compounds.

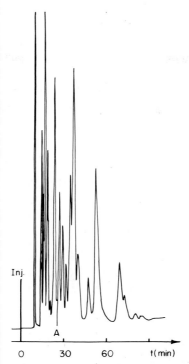

FIG. 1. Chromatogram of a complex mixture of polychlorobiphenyls.

A small change in the free energies, which are usually of the order of several kilocalories per mole, can result in a marked relative change in $\Delta(\Delta G)$. At ambient temperature ($T \simeq 300°K$), a change in $\Delta(\Delta G)$ of 18 cal/mol from 18 to 36 cal/mol would convert a difficult analysis ($N = 32,000$; $\alpha = 1.03$) into an easy separation problem ($N = 8800$; $\alpha = 1.06$) assuming that $k' = 3$ and $R = 1$.

The search for a "good" system is certainly worthwhile for simple mixtures, such as those containing a few closely related isomers. For complex mixtures, such as the one shown on Fig. 1, however, only an improvement in column efficiency can effect a complete resolution of the sample components.

Once the chromatographic system has been selected, the problem reduces to the design of a column having the required resolution power while minimizing analysis time and possibly inlet pressure. Before discussing this problem we need to know the relationships between analysis time, flow velocity, and pressure, and between resolving power and flow velocity.

III. ANALYSIS TIME, FLOWRATE, AND PRESSURE

From the definition of k, the retention time can be explained by Eq. (8)

$$t_R = t_0 (1 + k') \tag{8}$$

where k is proportional to the thermodynamic equilibrium constant for distribution of the solute between the stationary and the mobile phases. The proportionality constant is a function of the amount of solvent in the column, i.e., the total porosity of the packing, and the amount of stationary phase, i.e., the liquid loading of the support in liquid–liquid chromatography (LLC) or the surface area of the solid in liquid–solid chromatography (LSC). Generally, k is defined once the chromatographic system is chosen.

The transit time, t_0, is given by Eq. (9)

$$t_0 = L/u \tag{9}$$

where L is the column length and u the mobile phase velocity. As liquids have a very low compressibility, for all practical purposes, the velocity of the mobile phase is constant along the column in HPLC (30). It is related to the column parameters by Eq. (10)

$$u = \frac{k_0 d_p^2}{\eta} \cdot \frac{\Delta P}{L} \tag{10}$$

where ΔP is the difference between inlet and outlet pressures—the latter is practically always atmospheric pressure. As gauges measure pressures by reference to the atmospheric pressure, ΔP is the reading on the inlet pressure gauge. In Eq. (10), η is the solvent viscosity and d_p is the particle diameter; k_0 is a dimensionless coefficient having a value usually about 1×10^{-3} (23, 31). It is a function of the external porosity, ϵ, given by that fraction of the column volume which is not occupied by the particles, and is therefore available to the solvent flowing around the particles. It is noted that although the particles are porous, the solvent does not flow through them, only around them. The value of k_0 can be evaluated from the Kozeny–Carman equation given by Eq. (11):

$$k_0 = \epsilon^3/180(1 - \epsilon)^2 \tag{11}$$

Equation (11) shows that k_0 increases rapidly with increasing porosity (31). For columns used in HPLC, ϵ is usually about 0.40 and seems not to depend on the material and the packing technique used for the preparation of the column (22). For $\epsilon = 0.40$ the equation predicts that $k_0 = 1 \times 10^{-3}$ in agreement with experimental observations. In order to achieve the very homogeneous packing necessary for a good column efficiency, the methods used tend to give a very dense packing (23). The importance of the parameter k_0 has been stressed by Bristow and Knox (50), who prefer to use $\phi = 1/k_0$.

The product $k_0 d_p^2$ is the column permeability. It is in practice entirely determined by the particle diameter. In reality, the packing material is made of particles of different sizes. Sieving, or various elutriation or sedimentation techniques, have a rather small efficiency and the distribution of particle diameter is usually quite large. Different techniques are used to determine the particle size distribution (51). We should be cautious because these different methods give different distributions. The meanings of number average and mass average of the size distributions are quite different, just like the diameters derived from the experimental determination of the average diameter, the average mass, or the average surface. The experimental determination of the permeability from flow velocity gives the hydraulic radius of the particles which is quite different from the mean particle radius obtained from optical measurements if the size distribution is wide or skewed.

Be that as it may, Eq. (10) allows the calculation of the pressure necessary to achieve a given flow velocity in a column of known permeability. As we shall see in the next section, the choice of the velocity of the solvent is determined by the characteristics of the column as well as the desired efficiency or speed of analysis (18–21, 26).

IV. FLOW VELOCITY AND COLUMN EFFICIENCY

We have seen in Eq. (2) that, once the chromatographic system is chosen, the resolution of the components of the mixture is possible only if the plate number of the column exceeds the value N_{reg} given by Eq. (12)

$$N_{reg} = 16R^2 \left(\frac{\alpha}{\alpha - 1}\right)^2 \left(\frac{1 + k'}{k'}\right)^2 \tag{12}$$

where R is the required resolution. To be able to make any prediction, we need to know how the plate number is affected by changing the column parameters, such as length, particle diameter, and flow velocity.

It has long been recognized that the measured plate number of the column is proportional to its length in a large range, with certain exceptions. In the case of short columns, the peak profile is often perturbed by a contribution of the equipment to bandwidth. The extracolumn effect is always difficult to separate from the band spreading occurring in the column quantitatively and the calculation does not do any good to the analyst whose separation is just ruined anyway.

The magnitude of relative band spreading in the column is often measured by the height equivalent to a theoretical plate (HETP) which is denoted by H and calculated from the expression [Eq. (13)]

$$H = L/N \tag{13}$$

If the bandwidth at the column outlet is expressed in length unit, we have an expression for the plate number, which is similar to Eq. (3), and given by Eq. (14)

$$N = (L/\sigma_l)^2 \tag{14}$$

where σ_l is the standard deviation of the Gaussian peak. Consequently, the plate height can be expressed as [Eq. (15)]

$$H = \sigma_l^2/L \tag{15}$$

Equation (15) shows that H is proportional to the variance of the solute band at column outlet. If several independent phenomena contribute to band spreading, the sum of their respective variances, σ_i^2, determines the overall band spreading measured by σ_l^2 as [Eq. (16)]

$$\sigma_l^2 = \sum \sigma_i^2 \tag{16}$$

Three main independent contributions to band spreading inside the column have been identified (*32*) as longitudinal molecular diffusion, the unevenness of flow through the nonhomogeneous packing, and the resistances to mass transfer in the mobile and stationary phases.

The contribution of molecular diffusion is due to the fact that during the chromatographic run, the solute molecules diffuse along the column axis in the opposite direction to the concentration gradient, that is, away from the mass-center of the zone. The variance contribution arising from this phenomenon is proportional to the time and to the diffusion coefficient. We must, however, distinguish between longitudinal diffusion in the mobile and stationary phases. All solutes spend the same time, t_0, in the mobile phase; accordingly the variance contribution of axial diffusion in the mobile phase $\sigma_{D,m}$ is given by Eq. (17)

$$\sigma_{D,m}^2 = 2\gamma_m D_m L/u \tag{17}$$

where γ, the tortuosity coefficient, accounts for the perturbation to the molecular diffusion which arises from the geometrical effect of the packing structure (*32*). In fact, γ is a complex factor which should also take into account the fact that axial diffusion in the mobile phase occurs both in the interstitial space of the column and inside the particles. Since the geometrical structures of the two media are different, two different tortuosity factors should be used.

There is axial diffusion also in the stationary phase (*32*). In gas chromatography this effect is rightly neglected because the diffusion coefficient in the stationary phase is 10^4 times smaller than that in the mobile phase. There is no reason to neglect it in LC where the diffusion coefficients in the two phases are of the same order of magnitude. However, their respective contributions to zone broadening are by no means equal. Although the diffusion coefficient in the stationary phase is probably smaller than that in the mobile phase, the tortuosity coefficient is certainly smaller, because of the disconnected character of that phase (*32, 33*). The contribution of axial diffusion in the stationary phase $\sigma_{D,s}^2$ can thus be written as [Eq. (18)]

$$\sigma_{D,s}^2 = (2\gamma_s D_s L/u) \cdot k \qquad (18)$$

since the time spent in the stationary phase is $k't_0$ (*32*).

The contribution of the unevenness of the flow pattern is the result of the irregularities of the packing and of the nonuniform particle size distribution. The streamlines have different length and the average velocity in the channels of the packing is different. The particles and the stagnant region between them can be reached only by diffusion. A molecule which travels in the axis of a wide stream has less chance to diffuse into a particle than a molecule traveling through a narrow channel. Since the average velocity is larger in wide channels than in narrow ones, the effects of the distributions of channel length and average velocity, as well as that of the rate of radial mass transfer across channels, are not independent (*32*).

It has been shown (*34*) that this contribution to the variance $\sigma_{D,e}^2$ can be satisfactorily approximated by the relationship (19)

$$\sigma_{D,e}^2 = aLu^{1/3} \qquad (19)$$

where a is a function of the particle diameter and of the diffusion coefficient as shown below.

The solute molecules can enter and leave the particles only by diffusion. The particles are porous, like sponges, and most of the absorbent surface area is given by the surface of the inner pores. In LLC most of the stationary liquid phase is also inside the particles. Only porous layer beads which have been used in the past and may find some applications in the future represent an exception. The diffusion through the particles takes some time. The average time necessary for a molecule to diffuse across a distance d_p is $d_p^2/2D_m$.

Accordingly it has been shown (*32*) that the contribution of the mass-transfer resistance to the variance, $\sigma_{tr,r}^2$, is given by Eq. (20)

$$\sigma_{tr,r}^2 = \frac{CLd_p^2 u}{D_m} \qquad (20)$$

where C is a function of solute retention. The kinetics of adsorption–desorption can also contribute to band broadening. As this contribution is also proportional to $Ld_p^2 \, u/D_m$, its proportionality constant may be included in parameter C. Consequently a different dependence of C on k' is observed when mass-transfer or kinetic resistances predominate. Combining Eqs. (15)–(20) gives Eq. (21):

$$H = \frac{2}{u} [\gamma_m D_m + k' \gamma_s D_s] + au^{1/3} + \frac{Cd_p^2 u}{D_m} \tag{21}$$

Experiments have shown that columns of different length prepared with the same batch of stationary phase and operated under the same conditions yield the same H vs. u plot, and this finding validates this concept (7, 26, 34).

It was difficult, on the other hand, to measure accurate H values over a sufficiently wide velocity range with columns packed with relatively large diameter particles in the 50- to 100-μm range, which were first used in HPLC. In such a case, the velocity at which H, according to Eq. (21), is minimum is so low that experiments are unpractical (see Section VI). Therefore experiments were carried out at flow velocities 10 to 100 times higher than the optimum velocity and quasi-linear plots of H vs. u were obtained. Snyder (13) has shown that the velocity range in which H is evaluated determines the choice of a two-parameter empirical equation. The results of Majors (58) show how excellent experimental data can lead to erroneous conclusions when fitted on empirical equations, even though the data are in good agreement with Eqs. (10) and (21).

In order to compare data obtained with otherwise similar chromatographic systems in which only the particle size of the column packing and solute diffusivity may vary, Eq. (21) should be written in dimensionless form. Using an approach taken from chemical engineering, Knox (7, 34) has shown that a corresponding reduced plate height equation is given by Eq. (22)

$$h = (B/\nu) + A\nu^{1/3} + C\nu \tag{22}$$

where h is the reduced plate height (Eq. 23)

$$h = H/d_p \tag{23}$$

and ν is the reduced velocity given by Eq. (24):

$$\nu = ud_p/D_m \tag{24}$$

Plots of reduced plate height against reduced velocity should give a single curve for columns packed with different size fractions of adsorbents prepared in an identical fashion. These curves vary with the solvent used, es-

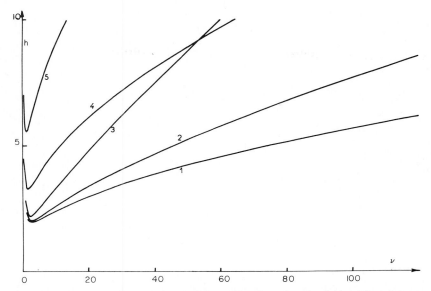

FIG. 2. Plots of the reduced plate height against the reduced velocity. The parameters are as follows: (1) $A = 1$, $C = 0.01$; (2) $A = 1$, $C = 0.03$; (3) $A = 1$, $C = 0.1$; (4) $A = 2$, $C = 0.03$; (5) $A = 4$, $C = 0.03$. In all cases, $B = 1.5$.

pecially for retained compounds, as C is a function of k and other characteristics of the solvent system. Typical plots are given in Figs. 2–4. Figure 2 shows how h vs. v plots change with the parameters of Eq. (22). Figures 3 and 4 show plots of $\log h$ vs. $\log v$ which are easier to use. Minima of the h vs. v curves for Figs. 3 and 4 are given in Tables I and II.

The coefficient A in Eq. (22) characterizes the regularity of the packing. It is shown below that the value of A critically determines the quality of a column. Values of A significantly larger than unity indicate that the column is poorly packed, because either the packing technique used is inappropriate or the particular batch of stationary phase contains nonuniform particles. The latter case occurs, for instance, when the particle size distribution is too broad or bimodal, or the particles have very irregular shapes. Values of A about unity have been reported for excellent columns. In principle, A is independent of the nature of the solvent and solute and of the retention. In practice, experimental values of A for a given column depend slightly on these factors. Combination of Eqs. (22), (23), and (24) yields Eq. (25):

$$H = \frac{BD_\mathrm{m}}{u} + \frac{Ad_\mathrm{p}^{4/3}}{D_\mathrm{m}^{1/3}} \cdot u^{1/3} + \frac{Cd_\mathrm{p}^2 u}{D_\mathrm{m}} \tag{25}$$

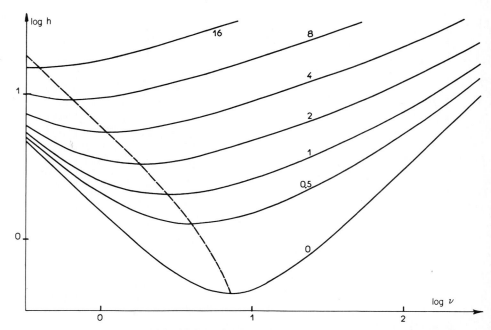

FIG. 3.   Double logarithmic plots of the reduced plate height against the reduced velocity. For all curves C is constant and equal to 0.03. A is stated for each curve on the graph. The dotted line connects the loci of the minima for the individual curves, cf. Table I.

TABLE I

Minima of the h vs. ν Curves Shown in Fig. 3[a]

A	h_{m}	ν_0	$h_{\mathrm{m}}\nu_0$
0	0.42	7.2	3.0
0.5	1.26	4.1	5.0
1	2.04	2.69	5.5
2	3.31	1.78	5.89
4	5.50	1.07	5.89
8	9.33	0.65	6.03
16	15.5	0.38	5.89

[a] The value of the parameter C is fixed at 0.03.

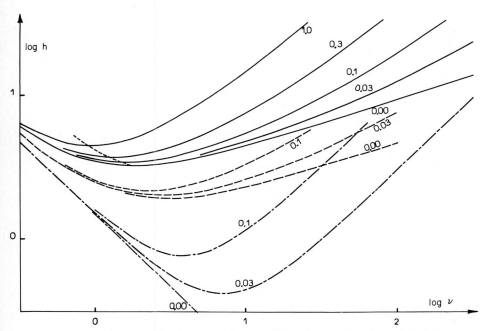

FIG. 4. Double logarithmic plots of the reduced plate height against the reduced velocity. The values of the parameter A are 2, 1, and 0 for the data shown by solid, dotted, and dash–dot lines, respectively. Parameter C is shown for each curve on the graph. The coordinates of the minima are given in Table II.

TABLE II

Minima of the h vs. ν Curves Shown in Fig. 4

A	C	h_m	ν_0	$h_m\nu_0$
0	0.03	0.42	7.2	3.0
	0.1	0.78	3.89	3.0
1	0	1.94	3.09	6.0
	0.03	2.02	2.75	5.6
	0.10	2.20	2.24	4.9
2	0	3.24	1.82	5.89
	0.03	3.31	1.78	5.89
	0.10	3.39	1.58	5.37
	0.30	3.72	1.32	4.90
	1.0	4.47	0.93	4.17

Equations (21) and (25) permit the calculation of a and B. B is a function of k whenever the contribution of axial diffusion in the stationary phase is not negligible [cf. Eq. (21)]. Otherwise B is equal to $2\gamma_m$ and usually close to $1.5-1.6$.

The value of the coefficient C is usually between 0.01 and 1. Values larger than 0.1, however, are unusual and stationary phases where C is so large should be avoided as the efficiency of the column rapidly deteriorates with increasing velocity above the optimum.

A word of caution is necessary regarding the precision of the measurements of H as a function of velocity. Unless they are carried out at very large values of ν, the unavoidable scatter of experimental HETP values makes the determination of small values of C very imprecise (56). Packing techniques and column test conditions have been reviewed (24, 50). The importance of measuring the reduced parameters and [Eqs. (23), (24)] has been stressed by Bristow and Knox (50).

Three important features of the H vs. u curves should be emphasized here, as they are relevant to the problem of optimization of experimental conditions (19, 21). *First,* there is an optimum flow velocity at which the plate height is minimum. The optimum reduced velocity is usually in the range from 2 to 4, depending on the exact values of A and C. The optimum flowrate and the corresponding column inlet pressure are lower than those employed in practice of HPLC. Working at the optimum flowrate permits the best use of the pressure drop, although it does not necessarily provide the smallest possible analysis time. *Second,* column efficiency is inversely proportional to the particle diameter of the packing. Consequently a shorter column suffices to perform a given separation when smaller particles are used. *Third,* the flow velocity at which column efficiency is maximum is also inversely proportional to the particle diameter.

As a consequence of the last two relationships, by reducing the particle size the time of analysis can also be reduced because short columns and relatively high solvent velocities can be employed. This is the basis for the trend toward the use of finer particles. Since the early 1970s, when typical particle size ranged from 50 to 100 μm, the particle size range in columns used in HPLC has been reduced to $5-10$ μm. Today most companies offer their phases in various sizes ranging from 5 μm to about 20 μm. A further reduction in the particle size, however, does not seem attractive in most cases. Particles finer than 5 μm are difficult to produce in narrow size distribution, and they are difficult to handle and to pack because they tend to agglomerate. Moreover, the pressure drop over the column becomes very important and the bandwidth so narrow that it is very difficult to design equipment which does not contribute to extracolumn broadening of such bands. This problem will be elaborated in the ensuing sections.

V. WHAT TO OPTIMIZE IN LIQUID CHROMATOGRAPHY

Due to the availability of computers and powerful pocket calculators it has become easy to carry out extensive numerical calculations in order to optimize quite complicated sets of experimental conditions by using sophisticated models. Thus the difficulties rest not with the calculations but with choosing a suitable model that best represents the problem.

Many different sets of values can be selected for the experimental parameters to solve a given analytical problem. Some sets are more practical than others or give better performances. Hence in the process of optimization, we can search for the shortest analysis time, the conditions of most sensitive analysis, the shortest column length, the lowest pressure drop, etc., depending on what the analyst considers most important.

The equations which relate the parameters of the system are summarized in Table III. Some of these parameters are determined by the choice of the chromatographic system: the solvent viscosity η, the diffu-

TABLE III

Most Important Equations Used for Optimization of Experimental Conditions in Liquid Chromatography

Equations	Equation number
A. Fundamental	
$R = \dfrac{\sqrt{N}}{4} \cdot \dfrac{\alpha - 1}{\alpha} \cdot \dfrac{k'}{1 + k'}$	2
$t_R = (L/u)(1 + k')$	8,9
$u = \dfrac{k_0 d_p^2}{\eta} \cdot \dfrac{\Delta P}{L}$	10
$L = NH$	13
$H = \dfrac{BD_m}{u} + \dfrac{A d_p^{4/3}}{D_m^{1/3}} u^{1/3} + \dfrac{C d_p^2}{D_m} \cdot u$	25
$C_{max} = m\sqrt{N}/V_R\sqrt{2\pi}$	26^a
B. Practical	
$L = Nhd_p$	27
$t_R = \dfrac{N d_p^2}{D_m} \cdot \dfrac{h}{\nu} \cdot (1 + k')$	28
$\Delta P = N \cdot \dfrac{h\nu}{d_p^2} \cdot \dfrac{\eta D_m}{k_0}$	31

^{a} V_R is the retention volume [$V_R = (1 + k)V_0$, where V_0 is the mobile phase hold-up of the column] and m is the mass of sample introduced.

sion coefficients D_m and D_s, the relative retention α, the column capacity factor k, the specific column permeability k_0, the internal and external porosity of the packing, and the coefficients of the reduced plate height equation, γ, A, and C. The sample size and column diameter also have to be considered in optimizing the loading capacity of the column in preparative work.

The three most important characteristics of the analysis are the resolution, the analysis time, and the maximum concentration of the band. The three most important column and operational parameters are the column length, the particle diameter, and the inlet pressure. Since the velocity is given by Eq. (10), the HETP by Eq. (25), and the plate number by Eq. (13), we have three equations to relate the three characteristics of the analysis and the three major column and operational parameters. We thus have three unknowns and can select one of the above characteristics and parameters (six possibilities) and optimize it as a function of three of the remaining five ($C_5^3 = 10$ possibilities). This approach offers a total of 60 possible optimization theories; most of them, however, are devoid of any practical interest.

In practice the resolution will always be specified, because our goal is to separate a given mixture, and this requires a given minimum degree of resolution between each band. Thus, the value of the required number of theoretical plates is given by the choice of the chromatographic system.

Furthermore we do not consider the maximum solute concentration in the effluent as important in most practical cases except in trace analysis (16). As shown by Eq. (26) (see also Table III),

$$C = m\sqrt{N}/V_R\sqrt{2\pi} \tag{26}$$

the maximum concentration, C_{max}, is in fact entirely determined by the mass of the corresponding compound in the injected sample m, once the chromatographic system, which determines the value of N and V_{R_1} has been selected. Optimization for high sensitivity has been discussed and it has been shown that in many cases a much larger sample than that employed usually could be introduced (16).

We thus are left with four main parameters: the analysis time, the inlet pressure, the column length, and the particle diameter. The first two are measures for the "cost" of each analysis in terms of time and pressure (cost of operation) whereas the choice of the other two determines the "cost" of the column (investment cost). In most applications, such as routine analyses, the latter will be small in comparison to the cost of operation and little attention is paid to the cost of the column, as long as it is possible to prepare it without additional research and investment. On the other hand, in research work often a few separations are performed with the same system only, and much less attention is paid to analysis time and inlet pressure. In such cases a column which is available or readily prepared, will often be preferred. In either situation we can choose one of the parameters and optimize it as a function of the other three; since we have three unknowns again there are only four optimization theories.

As shown by Eq. (13), the shortest column has to yield the smallest possible HETP when the number of theoretical plates N is given. This is accomplished when the column is operated at the optimum flowrate and the smallest particles available are used. There are limits, of course, set either by the maximum pressure which can be delivered by the equipment or by the contribution of the equipment to the peak variance. The latter is always finite and may significantly broaden the peaks eluted from a short column (20). We may as well set a value for the analysis time and calculate the minimum column length and the corresponding values of the pressure and particle size by using the equations developed in the next section.

Alternatively we may want to keep the inlet pressure of the column low (19). This approach is interesting in terms of safety and equipment lifetime. Furthermore the use of columns packed with fine particles is most promising at relatively low inlet pressures so that the pressure does not exceed the practical limit of the instrument when moderately viscous solvents are used. Finally, pumps capable of working at moderate pressures up to 70 atm are less expensive than those currently used in HPLC. It should be noted, however, that a separation is accomplished at the lowest inlet pressure when the column is packed with the largest particles available. In this case very long columns and extremely long analysis times are required.

Therefore we shall optimize the experimental conditions by looking for the minimum pressure at constant analysis time and efficiency for a given solute pair. It has been shown that this goal is accomplished when the column is operated at the optimum flowrate at which the plate height is minimum (19). The particle size and column length then depend on the plate number and the required analysis time.

In general, optimization calculations are considerably easier when they are carried out at the maximum efficiency or when the reduced plate height equation is used.

VI. EQUATIONS USED FOR OPTIMIZATION

In evaluating the parameters of Eq. (22) from experimental data we find that carefully packed columns have A values between 1 and 2. Most phases used now have C values below 0.1, although the measurements are quite imprecise and the exact value depends on the magnitude of retention and the nature of the solvent and solute. In the cases shown in Fig. 2, the minimum value of the reduced plate height is between 2 and 3 and the corresponding optimum value of the reduced velocity ranges from 1.5 to 3. This is a rather narrow range of variation.

The column length can be expressed by using the definition of the reduced plate height as [Eq. (27)]

$$L = Nhd_p \tag{27}$$

Combination with Eqs. (8), (9), and (23) gives the retention time as [Eq. (28)]

$$t_R = \frac{Nd_p^2}{D_m} \cdot \frac{h}{\nu} \cdot (1 + k') \tag{28}$$

which can be rearranged to obtain Eq. (29):

$$\frac{N}{t_R} = \frac{D_m}{d_p^2} \cdot \frac{\nu}{h} \cdot \frac{1}{1 + k'} \tag{29}$$

Combination of Eqs. (10) and (24) gives Eq. (30):

$$\frac{\Delta P}{L} = \frac{\eta D_m}{k_0} \cdot \frac{\nu}{d_p^3} \tag{30}$$

The pressure drop per theoretical plate can be expressed by combining Eqs. (27) and (30) as Eq. (31):

$$\frac{\Delta P}{N} = \frac{\eta D_m}{k_0} \cdot \frac{h\nu}{d_p^2} \tag{31}$$

As we want to generate a given number of plates in a given time, using a column operated at a given reduced velocity so that its reduced plate height is defined, Eq. (28) permits the calculation of the particle diameter. The column length is given by (27) and the pressure drop is evaluated by using either Eq. (30) or Eq. (31). We have already seen that the required pressure drop or column length are minimum when the flow velocity is optimum, i.e., plate height is minimum, therefore it is sufficient to introduce in the equations the values of minimum h and optimum ν in order to obtain the optimum particle diameter and thereafter calculate the necessary column length and pressure.

Alternatively, one can derive from Eq. (29) the necessary particle diameter to achieve the requested performance in terms of N and t_R, at a given reduced velocity that determines h through Eq. (22). Subsequently the column length and the inlet pressure can be calculated by using Eqs. (27)

and (31), respectively. Both equations are very simple and the use of a scientific pocket calculator permits their application to many practical problems. It is often useful to calculate the velocity ν_0, corresponding to the minimum plate height h_m. Differentiation of Eq. (22) shows that ν_0 is the root of the following equation

$$-3B + A\nu^{4/3} + 3C\nu^2 = 0 \tag{32}$$

The analytical solution of Eq. (32) is so complicated that it is useless in practice; therefore numerical procedures are preferred to evaluate ν_0. Combining Eqs. (22) and (32) we obtain Eq. (33):

$$h_m\nu_0 = 4B - 2C\nu_0^2 \tag{33}$$

In practice the product, $h_m\nu_0$, can be considered constant when $C < 0.1$.

VII. WAYS TO ACHIEVE 5000 PLATES IN 5 MINUTES

Plots of reduced plate height equations are depicted in Figs. 2 to 4. Different approaches to accomplish a separation requesting 5000 plates in 5 min are illustrated in Fig. 5. The characteristics of the curves on this figure are given in Table IV. The variation of the pressure drop with the particle diameter for columns having different packing quality is demonstrated. Curves 3 and 4 correspond to very well-packed columns whereas curves 1, 2, and 6 show results expected with correctly packed columns. A poorly packed column would yield results shown by curve 5. The quality of stationary phase particles is either good (curves 1 and 3), average (curves 2, 4, and 5), or poor (curve 6). In all cases the viscosity is 0.5 cP, corresponding to n-heptane or acetonitrile, the retention is moderate ($k' = 3$) and the diffusion coefficient is assumed to be relatively large (3×10^{-5} cm²/sec) like that of benzene in n-heptane. The effect of these parameters will be investigated later. It is seen in Fig. 5 that in order to achieve the requested performances (5000 plates in 5 min) the column has to be operated at a reduced velocity which increases with increasing d_p.

Several features of the curves shown in Fig. 5 are striking. *First,* the minimum pressure drop is much smaller for a well-packed column than for a poorly packed one. With A equal to 1, 2, and 4 the minimum pressures are 8.1, 19.5, and 51 atm, respectively. The gain in performance is remarkable and the importance of the packing will be stressed later when we discuss the achievement of high efficiencies in a reasonable time.

Second, the quality of the particle as measured by the value of the term C, is of much less importance. When C increases from 0.03 to 0.3 (this last

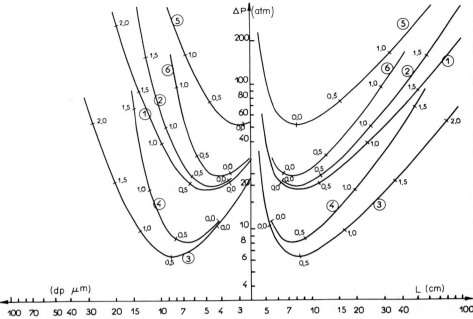

FIG. 5.   Graphical illustration of the dependence of the pressure, which is required to obtain 5000 plates in 5 min, on the particle size (left scale) and on the column length (right scale). The parameters are $k = 3$, $\eta = 0.5$ cP, $D_m = 3 \times 10^{-5}$ cm²/sec. Curves 1, 2, and 6: well-packed column ($A = 2$). Curves 3 and 4: excellent column ($A = 1$). Curve 5: poorly packed column ($A = 4$). Values of parameter C and the coordinates of the minima are given in Table IV. The numbers given on the graph for each curve state the corresponding logarithm of the reduced velocity, cf. Figs. 3 and 4.

value representing a quite poor stationary phase) the minimum pressure increases from 18.5 to 23 atm for a short, typical column.

This is because, in the neighborhood of the minimum plate height, the packing term, A, plays a much more important role in determining efficiency than the mass-transfer term, C. The effect can also be seen by comparing Figs. 3 and 4. Thus, the analyst can choose a stationary phase mainly on the basis of its thermodynamic properties which determine the values of α and k provided it is possible to pack the column material uniformly and efficiently. The results strongly suggest that it is most desirable to develop packing techniques that yield for the column A values below 2. The effect of C on efficiency is greater when the column is well packed. Figure 5 shows that increasing C from 0.01 to 0.10 for a well-packed column ($A = 1$) results in an increase of the minimum pressure from 6.5 to 8.1 atm.

TABLE IV

Parameters and Minima of the Curves Shown in Fig. 5

Curve number	A	C	ΔP_m (atm)	d_{p0} (μm)	L_0 (cm)
1	2	0.03	18.5	5	8.5
2	2	0.10	19.5	4.5	8
3	1	0.01	6.5	4	8.5
4	1	0.10	8.1	7	8
5	4	0.10	51.9	8.5	8
6	2	0.30	23.0	3	7

Third, the column length, which corresponds to the minimum pressure, is a weak function of A and C. For the whole range of parameters examined in Fig. 3 (A from 1 to 4 and C from 0.01 to 0.3) L_0 varies between 7.25 and 8.3 cm. This is due to the fact that when t_R and N are fixed the required particle diameter is given by Eq. (34):

$$d_p = \left[\frac{t_R}{N} \cdot \frac{D_m}{1 + k'} \cdot \frac{\nu}{h}\right]^{1/2} \tag{34}$$

Combination of Eqs. (34) and (27) gives for the column length the expression (35)

$$L = \left[Nt_R \cdot \frac{D_m}{1 + k'} \cdot h\nu\right]^{1/2} \tag{35}$$

so that L is proportional to $\sqrt{h\nu}$. As discussed above [cf. Eq. (33)] $h_m\nu_0$ varies only slightly in the range of practical C and ν_0 values.

Finally, the most efficient columns permit the use of larger particles to achieve the required performance and therefore the necessary pressure drop is smaller. In our case the maximum particle diameter is 9μm for the best and only 3 μm for the worst column. In practice it is much easier to pack an efficient column with 9 μm rather than with 3-μm particles, and this is a good example of a vicious circle encountered with such problems.

The effect of the performance requirements for a given separation is illustrated in Fig. 6, by using parameters which correspond to a correctly packed column with reasonable, but not outstanding qualities, such as represented by $A = 2$, $C = 0.03$, $\nu_0 = 1.8$, and $h_m = 3.3$ (cf. Fig. 2). The parameters of the curves on Fig. 6 are given in Table V. Curves 1, 2, and 3 correspond to increasing analysis time, 5, 10, and 15 min, respectively, to achieve 5000 plates. Curves 4, 1, and 5 represent conditions under which 2000, 5000, and 10,000 plates can be obtained, respectively, in a separation time fixed at 5 min. Note that when $k' = 3$ the elution time of the inert peak is four times smaller than the analysis time.

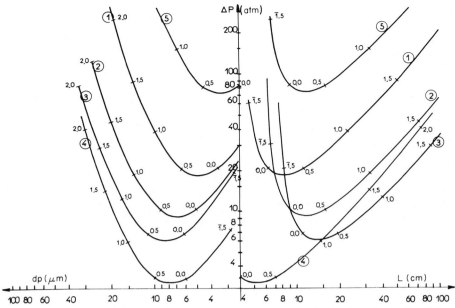

FIG. 6. Graphical illustration of the relationship between the particle size (left scale), column length (right scale), and the pressure drop (vertical scale) required to obtain 5000 plates in 5 min (curve 1), 10 min (curve 2), and 15 min (curve 3), as well as to obtain both 2000 plates (curve 4) and 10,000 plates (curve 5) in 5 min. Coordinates of the minima are given in Table V.

With increasing analysis time, 5, 10, and 15 min, there is a marked decrease in the inlet pressure, 18.2, 9.1, and 6.2 atm, respectively (see Table V). Both the necessary particle diameter, 5, 7, and 8.5 μm, respectively, and the column length, 8.5, 11.5, and 14 cm, respectively, increase with the analysis time. On the other hand, an increase in efficiency at constant analysis time is accompanied by a slight increase in column length, a

TABLE V

Parameters and Minima of the Curves Shown in Fig. 6

Curve number	t_R (sec)	N	ΔP_m (atm)	d_{po} (μm)	L_0 (cm)
1	300	5000	18.2	5	8.5
2	600	5000	9.1	7	11.5
3	900	5000	6.2	8.5	14
4	300	2000	3.0	8	5
5	300	10000	72.5	3.5	12

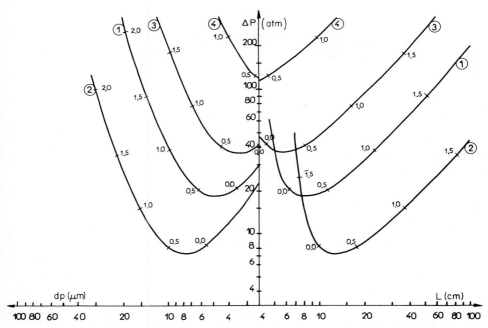

FIG. 7. Graph illustrating the effect of particle size (left scale) and column length (right scale) on the pressure drop (vertical scale) required for obtaining 5000 plates in 5 min with eluents having different viscosities as shown in Table VI.

marked decrease in particle diameter, and a considerable increase in pressure.

It is remarkable that all the analytical results discussed here can be obtained with approximately 10-cm-long columns packed with particles of about 5 μm in diameter. This means, among other things, that the results will be largely influenced by the exact column length and particle diameter. Whereas the column length is easy to measure within a millimeter or even more accurately, if necessary, the particle diameter is usually not too well defined. The range of the size distribution may exceed 50% of the mean diameter, especially for small particles. Furthermore the various methods used to evaluate the average particle diameter do not yield the same values of $\bar{d}_p$. The wider the distribution, the larger is the difference between the mean particle diameter obtained by different methods. In the case of a wide particle size distribution the pressure necessary to achieve given performances will be higher than that predicted by this simple approach.

The effect of the solvent viscosity is illustrated in Fig. 7. The parame-

TABLE VI

Parameters and Minima of the Curves Shown in Fig. 7

Curve	η (cP)	D_M ($\times 10^{-5}$ cm²/sec)	ΔP_m (atm)	d_p (μm)	L (cm)	k'^a
1	0.5	3	18.2	5	8.5	3
2	0.2	7.5	7.2	7.5	13	0.6
3	1.0	1.5	36	3.5	5.8	7
4	3.0	0.5	110	2.0	3.3	23

^{a} Corresponding to a value of 3×10^{-5} cm²/sec for the diffusion coefficient, as explained in the text.

ters of the curves on this figure are given in Table VI. In agreement with the empirical equations on diffusion in liquid we can assume that the diffusion coefficient is inversely proportional to the viscosity. Therefore the product ηD_m in Eqs. (30) and (31) remains constant. The changing viscosity causes a variation of D_m in Eqs. (28) and (29) and the effect is obviously very important. The use of a more viscous solvent requires a larger column inlet pressure to maintain a given flow velocity. Because of the concomitant decrease in the diffusion coefficient, a given value of reduced velocity is obtained at a lower value of the actual flow velocity [cf. Eq. (24)]. In order to compensate for this effect and achieve the analysis within the required time, finer particles have to be used to pack the column. On the basis of this consideration columns used with more viscous solvents should be shorter and packed with finer particles. Since the solvent is more viscous the pressure drop across the column will be considerably higher. As seen from the results presented in Fig. 7, it is not possible to meet our specifications (5000 plates in 5 min) with the most viscous solvent studied because it is beyond our present capability to pack a 3.3-cm-long column with 2-μm particles. The column inlet pressure would be 110 atm and at such a steep axial pressure gradient (33 atm/cm) untoward thermal effects would occur and probably reduce markedly the efficiency of the column.

We see in Fig. 7 that our performance goal, 5000 plates in 5 min, can be rather easily met by using heptane or acetonitrite ($\eta = 0.4$ cP). With more viscous eluents such as CCl_4 ($\eta = 0.94$ cP), water, ($\eta = 1.01$ cP), or ethanol ($\eta = 1.1$ cP), it becomes difficult and the accomplishment of our goal is practically impossible with more viscous solvents such as propanol ($\eta = 2$ cP) or butanol ($\eta = 3$ cP). The latter solvents, however, are rarely used as eluents in HPLC, but water/ethanol and water/methanol mixtures have viscosities in the 0.5–3 cP range.

We can examine the effect of the diffusion coefficient in a similar fashion. Assuming a constant viscosity of 0.5 cP, we could study the effect in two ways. In normal phase LC comprising a polar stationary phase and an eluent of low polarity there is very poor correlation between molecular weight and retention. So we can assume that k' is constant and let D_m vary from 3×10^{-6} to 5×10^{-5} cm^2/sec as shown in Fig. 8. In reversed phase LC by using a polar eluent and a nonpolar stationary phase, the retention increases, whereas the diffusion coefficient decreases with the size of solute molecule. We will use empirical relationship between the capacity factor and the molecular weight (MW) of the solute [Eq. (36)]:

$$\log k' = -0.32 + 0.01 \quad (\text{MW}) \tag{36}$$

Equation (36) reflects a typical increase of 0.14 log k' units for the addition of a methylene group to the molecule and a k' value for 3 for a solute having a molecular weight of 80.

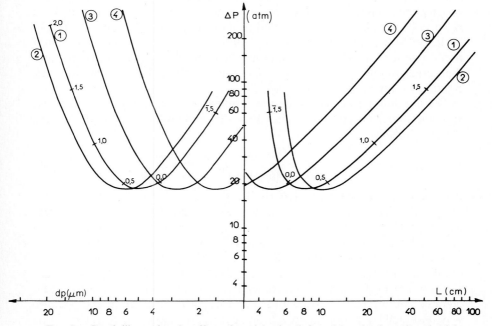

FIG. 8. Graph illustrating the effect of particle size (left scale) and column length (right scale) on the pressure drop (vertical scale) required for obtaining 5000 plates in 5 min with solutes having different molecular diffusion coefficients. (1) $D_m = 3 \times 10^{-5}$ cm^2/sec; (2) $D_m = 5 \times 10^{-5}$ cm^2/sec; (3) $D_m = 1 \times 10^{-5}$ cm^2/sec; (4) $D_m = 3 \times 10^{-6}$ cm^2/sec.

The dependence of diffusivity on the molecular weight is conveniently estimated from the empirical approximation (37)

$$D_m = 4.16 \times 10^{-4} \, (MW)^{-0.6} \tag{37}$$

and the dimensions of D_m are given in square centimeters per second. This relationship is based on the assumption that the molecular volume is proportional to the molecular weight and that D_m equals 3×10^{-5} cm^2/sec for a substance having a molecular weight of 80. If we increase MW from 80 to 160, which corresponds to a large range of k', from 3 to 19, respectively, according to Eq. (36), the variation in D_m is small, it decreases from 3 to 2×10^{-5} cm^2/sec, respectively, in view of Eq. (37). The result is practically identical to that obtained by merely changing the value of k'. This is also illustrated in Fig. 7, as we see in Eq. (27)–(29) that changing k' at constant D_m and η is identical to changing D_m while keeping ηD_m and k' constant. Table VI, which contains parameter values used in Fig. 7, also lists those k' values at which the corresponding curve would be obtained if the diffusion coefficient is fixed at 3×10^{-5} cm^2/sec. It is seen that when k' increases it becomes more and more difficult to achieve a given plate number in a given time because the particles of the packing have to be smaller and smaller and as a consequence the pressure has to be larger and larger. Therefore it is difficult to achieve 5000 plates in 5 min for solutes having k' values larger than 10. It is important to note that the actual mobile phase flowrate increases with k' for a constant t_R according to our optimization scheme.

The minimum necessary pressure, however, is independent of the diffusion coefficient. A combination of Eqs. (28) and (30) to eliminate D_m gives the relationship

$$\Delta P = \frac{N^2}{t_R} \cdot \frac{\eta}{k_0} \cdot h^2 (1 + k) \tag{38}$$

It is seen that ΔP is independent of D_m, whereas the corresponding values of d_p and L increase in proportion to $\sqrt{D_m}$. Further calculations can easily be performed by using the equations which have been obtained by solving Eqs. (27)–(30) for d_p, L, and ΔP and are listed in Table VII.

Several additional conclusions can be drawn from examining the results obtained with Eqs. (27) to (30). *First,* excellent performance can be obtained with those columns which are packed uniformly to yield high efficiency. The parameter A used for quantifying the quality of the packing was assumed to have a good but not excellent value of 2 for calculating the data presented in Figs. 6 to 8. Better columns have been packed by many workers, at least with silica particles. Even with chemically bonded phases, reduced plate height lower than 5 can be achieved rather easily.

TABLE VII

**Calculation of the Column
Parameters and Pressure Drop**

$$d_{\mathrm{p}} = \left[\frac{t_{\mathrm{R}}}{N} \cdot \frac{\nu}{h} \cdot \frac{D_{\mathrm{m}}}{1 + k'} \right]^{1/2}$$

$$L = \left[N t_{\mathrm{R}} \cdot h\nu \cdot \frac{D_{\mathrm{m}}}{1 + k'} \right)^{1/2}$$

$$\Delta P = \frac{N^2}{t_{\mathrm{R}}} \cdot h^2 \cdot \frac{\eta(1 + k')}{k_{\mathrm{o}}}$$

Second, if we want to have a higher efficiency in a proportionally longer time by maintaining constant the number of plates per unit time, N/t_{R}, the same equations (28) and (34) show that the required d_{p} will remain the same. Hence both the column length and the inlet pressure increase linearly with the plate number. Therefore, only if the original pressure is low enough, can a large increase in plate number be practicable. The data shown in Figs. 5 to 8 demonstrate that for a solute having $k' = 3$ it is possible to achieve 50,000 plates in less than 1 h with present technology assuming we can pack a 1-m-long column in the same way as a short column. There are several indications in the literature that such a performance is indeed possible and can even be exceeded (*21, 52, 53*) by a factor of 2. Going to longer columns would be difficult and there are no reports in the literature on obtaining appreciably better performance with long columns. Recently, however, Scott and Kucera (*53*) have shown that 14 1-m-long columns can be combined to generate 650,000 plates with $t_0 \sim 9$ h. Although the specific performances are only fair ($d_{\mathrm{p}} = 5 \ \mu\mathrm{m}$, $h_{\mathrm{m}} = 4.3$, $\gamma_0 = 0.74$), the result is impressive and demonstrates a mastery of all the technological problems involved. These results are further discussed in Section XIII. An advanced packing technique which would yield a value of unity for the parameter A would be needed to pack long columns with relatively small particles yielding appreciably better performances.

The achievement of various kinds of performance goals is better discussed by assuming that the column is operated at the minimum pressure, a condition which is necessary in practice when the inlet pressure is in the hectobar range, that is, $\Delta P > 100$ atm. This problem is treated in the next section.

Finally we may ask the question how charts like those depicted in Figs. 5 to 8 can be used to determine whether a given column can meet some performance requirements. It is most probable that it will give either a too large or too small efficiency within a required analysis time. How can we find it out from the diagram?

Let us suppose we have a 10-cm-long column packed with 5-μm particles ($A = 2$, $C = 0.03$) and we want to study its possible performances for a given chromatographic separation ($\eta = 0.5$ cP, $k' = 3$, $D_m = 3 \times 10^{-5}$). From Eq. (25) we can calculate that if we want 5000 plates, h has to be less than 4 which is possible if ν is between 0.70 and 5.3 according to the h vs. ν plot in Fig. 3. This reduced velocity range corresponds to a range of flow velocity between 0.042 and 0.32 cm/sec or to an analysis time between 950 and 126 sec. The pressure drop will be between 8.5 and 64 atm. We can adjust both analysis time and pressure at will in this range and the efficiency is larger than 5000 plates. At the optimum ($\nu = 1.78$, $h = 3.31$) it yields 6040 plates and the analysis time and the pressure drop are 375 sec and 21.3 atm, respectively. It is of course impossible to achieve more than 6040 plates.

It should be emphasized also that the column is operated at a constant actual flow velocity for all compounds. Therefore the reduced velocity is different for solutes having different diffusivity because ν increases with

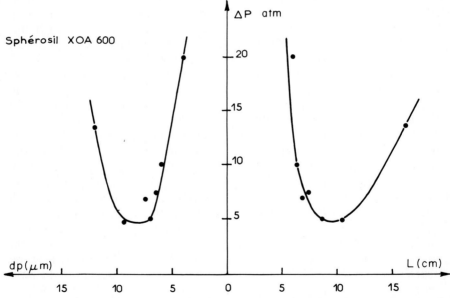

FIG. 9. Graph illustrating the experimentally observed relationship between the pressure drop across the column (vertical scale), the average particle diameter (left scale), and the column length (right scale). Mobile phase, n-hexane; solute, naphthalene ($k = 2.35$); stationary phase, Sphérosil XOA 600 (Rhône-Poulenc, Aubervilliers, France). Theoretical curve and experimental results after Guillemin *et al.* (*35*). Courtesy of Guillemin *et al.* and the *Journal of Chromatography*.

$1/D_m$. Thus if we work at $\nu = 1.78$ for a particular solute having D_m of 3×10^{-5} cm^2/sec, ν will be 5.3 for another solute for which $D_m = 1 \times 10^{-5}$ cm^2/sec, and the column efficiency for the second peak will still be 5000 plates. With solutes having lower diffusion coefficients, however, the value of ν will be larger than 5.3 and the efficiency drops below 5000 plates.

Figure 9 shows a comparison between experimental results and theoretical curves derived from the equations discussed above (35). The excellent agreement demonstrates the validity of our approach.

VIII. COLUMN PERFORMANCE AT MINIMUM INLET PRESSURE

Equations (29) and (30) allow us the calculation of column performance as a function of the particle diameter when the columns are operated at minimum inlet pressure. We find that the pressure gradient is inversely proportional to the cube of the particle diameter and the number of plates generated per unit time is inversely proportional to the square of the particle diameter.

Graphical illustration of the results is shown in Fig. 10. Such charts can easily be constructed provided the values of the constants in Eqs. (29) and (30) are available. The choice of parameters used to calculate the results represented by the three lines in Fig. 10 is somewhat arbitrary. They have been selected to be representative of a typical, a very favorable, and a very difficult case, and are shown by curves 1, 2, and 3, respectively.

The Wilke and Chang (36) correlation equation, which is somewhat approximative, is used most frequently to evaluate diffusion coefficients in liquids. This can be written as

$$\eta D_m = 7.4 \times 10^{-10} \cdot \frac{(\psi M_2)^{0.5} \, T}{V_1^{0.6}} \tag{39}$$

where M_2 and V_1 are the molecular weight of the solvent and the molar volume of the solute, respectively. The value of the association factor ψ is unity for nonassociated liquids and 2.6 for water. The extreme values of ψM_2 for solvents used in the present practice of HPLC are 41 for water ($M = 18$, $\psi = 2.3$) and 154 for CCl$_4$. Since for methanol ψ is larger than 1.3, the product ψM_2 is larger than 41 in this case also. The molecular volume of most sample components commonly separated by HPLC will be between 100 and 1000 cm^3. This does not include, however, polymeric substances, whose chromatographic separation is outside the scope of the present discussion. Therefore the choice of the numerical values used to

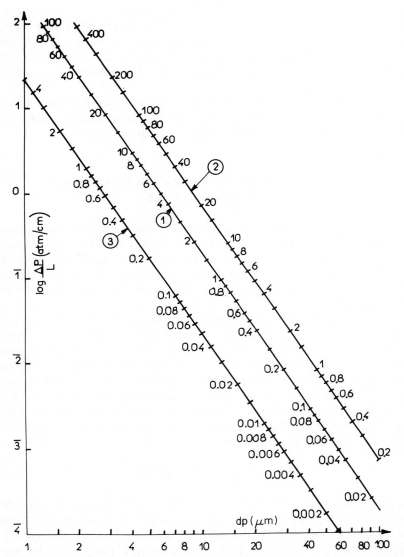

FIG. 10. Plot illustrating the dependence of the axial pressure gradient in a column operated at the optimum flow velocity on the average particle diameter of the packing. The marks and numbers represent the number of plates generated per second at $k = 3$. (1) $\nu_0 = 2$, $h_m = 3$, $\eta = 1$ cP, $D_m = 1 \times 10^{-5}$ cm^2/sec; (2) $\nu_0 = 4$, $h_m = 2$, $\eta = 0.44$ cP, $D_m = 4 \times 10^{-5}$ cm^2/sec; (3) $\nu_0 = 1$, $h_m = 5$, $\eta = 1$ cP, $D_m = 1 \times 10^{-6}$ cm^2/sec.

calculate the results shown in Fig. 10 does not apply well to GPC, although the theory itself is quite general and applies as well where lower diffusion coefficients have to be used. This is illustrated by the discussion of curve 3 on Fig. 10 at the end of this section. In our treatment we assume that ηD_m will be in the range from 1.7×10^{-7} to 2.2×10^{-8} cm g sec^{-2} in HPLC.

Curve 1 in Fig. 10 was obtained by using the following values: $\nu_0 = 2$, $h_m = 3$ (good column), $k' = 3$, $\eta = 1$ cP, $D_m = 1 \times 10^{-5}$ cm^2/sec, and $\eta D_m = 1 \times 10^{-7}$ cm g sec^{-2} which are typical of a moderately difficult separation problem. It is seen, for example, that in a column packed with 6-μm particles, we have a pressure gradient of about 1 atm/cm at the optimum flowrate which generates about 5 plates per second. To achieve 5000 plates we need about 1000 seconds (16 min and 40 sec); with $h_m = 3$ and $d_p = 6$ μm, the plate height hd_p is around 18 μm and the column length Nhd_p is 9 cm. The actual inlet pressure is 9 atm. This result is slightly different from the one obtained above because of the use of somewhat different values for the viscosity and diffusion coefficients: 1 cP instead of 0.5 and 1×10^{-5} cm^2/sec instead of 3, respectively. These conditions can easily be realized in practice, and the results are in agreement with those discussed in the previous section. We also see that using a column packed with 40-μm particles, 0.105 plates can be generated per second at a pressure gradient of 3.2×10^{-3} atm/cm. The separation, which requires 5000 plates, will be accomplished in 13 h and 14 min using a 60-cm-long column packed with 40-μm particles at an inlet pressure of 0.20 atm. This pressure is very low and by increasing the inlet pressure the analysis time can be greatly reduced. As seen from the h vs. ν curve (cf. curve 2 in Fig. 3), the reduced velocity can be increased to 25 before the plate height increases to twice of its optimum value with this type of column packing. If we double the length of the column, the analysis time is reduced by a factor of 6.3, whereas the pressure will be only 5 atm. In the late 1960s particles in the 30 to 50-μm range were frequently used in HPLC. Under such conditions the reduced velocities were very high: at $\nu = 250$, h becomes 20 which is 6 h_m on curve 2 in Fig. 1. The column length required to generate 5000 plates is 360 cm, the pressure 50 atm, and the analysis time 38 min under these conditions.

At the other end of curve 1 in Fig. 10, we see that we could generate 50 plates per second using 1.9-μm particles at a pressure gradient of 31.5 atm/cm. The analysis time is now 100 sec, the column length 2.85 cm, and the pressure 90 atm. Such column inlet pressure is fairly easy to use and the analysis time is quite attractive. On the other hand the column is somewhat short and it is difficult to make a column with a length of 2.85 ± 0.05 cm. The most difficult task, however, would be to prepare

1.9-μm-diameter particles since the size distribution attainable by using one of the available techniques is significantly wider than 0.1 μm. The assumption that particles have a very narrow size distribution does not hold yet for small particles. It is important to know, however, that we could in theory generate 10 to 40 plates per second by using available equipment and accomplish separations which require several tens of thousands of plates in a quite reasonable time.

In HPLC, however, we can encounter conditions different from those selected for the data shown by curve 1 in Fig. 10. Some are more favorable, others less. Curve 2 exemplifies a very favorable case of a very good column, $h = 2$ ($A = 1$, $C = 0.03$; cf. Fig. 4) at $\nu = 4$, a large diffusion coefficient, $D_m = 4 \times 10^{-5}$ cm^2/sec, and a low viscosity $\eta = 0.44$ cP typical for heptane, CH_2Cl_2, ethyl acetate, acetonitrile, or tetrahydrofuran. The value of D_m corresponds [cf. Eq. (39)] to a substance having a molar volume of 30 cm^3 and to a solvent much as acetonitrile, a very extreme case. Using the same particle size we can generate 11 times more plates per unit time at a 3.6 times larger inlet pressure than in the typical case discussed above. However, under such extremely favorable conditions, the performances have not been changed considerably.

Curve 3 in Fig. 10 is an example of an unfavorable case. We use a poor column ($A = 4$, $C = 0.03$ or $A = 3$, $C = 0.1$) that yields $h_m = 5$ at $\nu_0 = 1$, a moderately viscous solvent having $\eta = 1$ cP (cyclohexane, CCl_4, ethanol, water, pyridine) and a low diffusion coefficient, $D_m = 1 \times 10^{-6}$ cm^2/sec, which corresponds to a solute having molar volume of 6500 cm^3 and to a solvent like cyclohexane [cf. Eq. (39)]. Such a heavy molecule is more representative for the lower molecular weight solutes separated by gel permeation chromatography (GPC) at present. As expected the performance is quite low under such conditions. Using the same size particles we can generate 30 times less plates per unit time although the pressure gradient is 9 times smaller also. For example, a column packed with 6-μm particles yields only 0.15 plates per second instead of 5 plates per second in case 1. Consequently a separation that requires 5000 plates will take 9 h and 10 min. The column length is 15 cm and the inlet pressure is only 1.5 atm that corresponds to a pressure gradient of 0.010 atm/cm.

This is an example for a case where we have to increase the inlet pressure to reduce analysis time, but we also have to explore ways to improve the column packing technique. A comparison of the results shows that the very well-packed column generates ten times more plates per unit time than the poorly packed column. In Eq. (29) ν is multiplied by 4 and h is divided by 2.5. Thus, analysis time and column length are divided by 10 and

2.5, respectively. On the other hand, flow velocity and pressure gradient are multiplied by 4, so the actual inlet pressure is multiplied by only 1.6, cf. Eq. (30). The 5000 plates can now be generated with the same system in 55 min with a 6-cm-long column at an inlet pressure of 2.4 atm. Still better results with this system can be obtained by using smaller particles. A very well-packed column made with 1-μm particles will generate 5 plates per second with a pressure gradient of 22 atm/cm. A 20-cm-long column would generate 10^5 plates in 5 h and 33 min at a pressure drop of 440 atm. This shows that very high efficiencies can be obtained even for solutes having low diffusion coefficients.

Consequently the use of very fine particles, finer than those presently available with a reasonably good size distribution would permit significant improvement in the analysis of high molecular weight solutes which have low diffusion coefficients and for which the optimum reduced velocity, ν_0, is attained at a very low value of the actual flow velocity in columns packed with particles having the usual size.

IX. CRITICAL MINIMUM PRESSURE

The results in Figs. 3 and 4 shows that upon decreasing the flow velocity below its optimum value, ν_0, the plate height increases moderately at first. In difficult cases it it thus possible to achieve high column efficiencies with a given chromatographic system even if the upper pressure limit of the instrument is not sufficient to operate the column at its maximum efficiency. The price to pay for the high efficiency is an increase in analysis time.

Giddings (37–39) has shown long ago that there is a minimum pressure necessary to achieve a given separation. This critical pressure can be calculated as follows. Combination of Eqs. (27) and (30) and the use of Eq. (22) for the evaluation of h give for the inlet pressure the expression (40):

$$\Delta P = \frac{\eta D_m}{k_0} \cdot \frac{N}{d_p^2} (B + A\nu^{4/3} + C\nu^2) \tag{40}$$

The smallest possible inlet pressure, ΔP_c, is given by (41)

$$\Delta P_c = \frac{\eta D_m}{k_0} \cdot \frac{N}{d_p^2} \cdot B \tag{41}$$

where B is calculated from Eq. (21) as (42)

$$B = 2\gamma_m + 2k\gamma_s \cdot \frac{D_s}{D_m} \tag{42}$$

The value of B is usually about 1.5 but it may vary with k as discussed earlier. We have seen that the numerical value of the product ηD_m is between 1.7×10^{-7} and 2.2×10^{-8} in cgs units for most practical cases in HPLC. Since k_0 is always about 10^{-3} we can approximate the minimum pressure drop by (43):

$$\Delta P_c = (1.44 \pm 1.11) \cdot 10^{-4} \frac{N}{d_p^2} \tag{43}$$

In Eq. (43) both ΔP_c and d_p are given in cgs units. A more convenient relationship is given by Eq. (44):

$$P_c(\text{atm}) = (1.44 \pm 1.11) \cdot 10^{-2} \cdot \frac{N}{d_p^2 (\mu m)^2} \tag{44}$$

Both Eqs. (43) and (44) show that the minimum pressure is proportional to the plate number and inversely proportional to the square of the particle diameter.

Of course it is impossible to carry out a given separation at the critical pressure, ΔP_c, since the analysis time would be infinite. It is easy to show that if A is zero in Eq. (40), the pressure, ΔP_0, required to obtain the minimum plate height, h_m, is given by Eq. (45):

$$\Delta P_0 = 2\Delta P_c \tag{45}$$

If A is different from zero it is not possible to calculate exactly the numerical coefficient whose value is somewhat larger than 2, as we will show later in this section. As a rule of thumb, the pressure necessary to perform an analysis when working at minimum plate height can be approximated by Eq. (46) (ΔP in atm, d_p in μm)

$$\Delta P \simeq 6 \times 10^{-2} \frac{N}{d_p^2} \tag{46}$$

which gives a slightly larger value than would result from the combinations of Eqs. (44) and (45).

Equation (46) gives the correct value of ΔP within 50–70%. The pressure is higher for easy separation, characterized by low η and large D_m, and in such cases the analysis time will be shorter than predicted by Eq. (46).

Comparison of Eqs. (31), (41), and (45) shows that the pressure necessary to work at minimum plate height is given by Eq. (47):

$$\Delta P_0 = \frac{h\nu}{B} \cdot \Delta P_c \tag{47}$$

We have shown [cf. Eq. (31)] that the value of the product $h\nu$ is between 5 and 6 when C is small, hence $h\nu/B \simeq 4$. This relationship is in excellent agreement with the data shown by curve 1 in Fig. 9. The agreement is much less satisfactory with the results illustrated by curves 2 and 3 because the parameter values have been chosen to represent extremes and therefore they are not quite realistic.

Equation (44) can be used to determine the maximum plate number, which can be obtained at the highest pressure with a given equipment, and the minimum particle diameter, which may be used to carry out a given separation. For example, at $\Delta P = 300$ atm, the relationship between number of plate and the minimum particle diameter can be illustrated as follows: (*i*) $N = 10^4$, $d_p > 0.7$ μm; (*ii*) $N = 10^5$, $d_p > 2.2$ μm; (*iii*) $N = 10^6$ $d_p > 7$ μm. The values of d_p, at which the maximum available pressure of 300 atm suffices to maintain the optimum flow velocity, are twice as large as the values given above. The maximum pressure limit of commercial liquid chromatography permits the achievement of very large plate numbers, e.g., several hundred thousand plates, at the optimum flow velocity of the eluent using the now conventional 5-μm particles. Under such conditions the analysis time is very long and the exploitation of the high efficiency is seriously limited by the detection (*8–10*). If high efficiencies have been achieved only rarely, it is because we are missing difficult problems, apparently, and patience, certainly.

Let us emphasize that the value given by Eq. (47) is independent of the packing technique and only slightly affected by the choice of the mobile phase.

X. ACHIEVEMENT OF FASTEST ANALYSES

Most rapid separations are accomplished at the maximum column inlet pressure permitted by the design of the instrument. With most commercial equipment this pressure is on the order of several hundred atmospheres. Operating the chromatographic system at such pressures we find that the analysis time is shorter than practical in many instances and the width of the zones will be very small so that equipment problems related to extracolumn band spreading will arise, as discussed later in Section XII. In all cases the column will be longer than the one corresponding to the minimum plate height except if the maximum pressure happens to be equal to the necessary pressure. If it is larger, the analysis time will be shorter than that obtained at the minimum pressure. This result, of course, is valid only for a given column. If we wish to design the fastest possible column for a given equipment having fixed column inlet pressure, the problem has to be stated differently. Let us rewrite Eq. (40) as Eq. (48)

$$\Delta P = \frac{\eta D_{\rm m} N}{k_0} \cdot \frac{h\nu}{d_{\rm p}^2} \tag{48}$$

and Eq. (28) as Eq. (49)

$$t_{\rm R} = N \frac{1 + k}{D_{\rm m}} \cdot \frac{h}{\nu} \cdot d_{\rm p}^2 \tag{49}$$

The problem is to search for the minimum of $t_{\rm R}$ at constant ΔP, i.e., for the minimum of the term $h d_{\rm p}^2/\nu$ where the two variables ν and $d_{\rm p}$ are bound by $h\nu/d_{\rm p}^2 = $ const [cf. Eq. (48)] and h is given by Eq. (22).

The optimization of a function S of several parameters bound by one equation ($f = 0$) is solved by Lagrange multiplicator method. There must exist one numerical value of the parameter λ such that

$$dS + \lambda df = 0 \tag{50}$$

where dS and df are the total differentials of S and f, respectively. In our case Eq. (50) must be satisfied for any value of the differentials $d\nu$ and $d(d_{\rm p})$. We obtain therefore the following set of equations:

$$\frac{\partial S}{\partial \nu} + \lambda \frac{\partial f}{\partial \nu} = 0 \tag{51}$$

$$\frac{\partial S}{\partial (d_{\rm p})} + \lambda \frac{\partial f}{\partial (d_{\rm p})} = 0 \tag{52}$$

where λ is a numerical parameter. Equations (51) and (52) permit the calculation of the optimal values of ν and $d_{\rm p}$. The functions S and f are given by Eqs. (53) and (54)

$$S = N \frac{1 + k}{D_{\rm m}} \cdot \frac{h}{\nu} \cdot d_{\rm p}^2 = S' \cdot \frac{h}{\nu} \cdot d_{\rm p}^2 \tag{53}$$

$$f = \frac{\eta D_{\rm m} N}{k_0} \cdot \frac{h\nu}{d_{\rm p}^2} - \Delta P_{\rm M} = T \frac{h\nu}{d_{\rm p}^2} - \Delta P_{\rm M} \tag{54}$$

where $\Delta P_{\rm M}$ is the maximum pressure available and S' and T stand for $(1 + k)N/D_{\rm m}$ and $D_{\rm m}N/k_0$, respectively. Hence

$$\frac{\partial S}{\partial \nu} = -S' d_{\rm p}^2 \left[\frac{2B}{\nu^3} + \frac{2A}{3\nu^{5/3}} \right] \tag{55}$$

$$\frac{\partial S}{\partial (d_{\rm p})} = 2S' d_{\rm p} \left[\frac{B}{\nu^2} + A\nu^{-2/3} + C \right] = 2S' d_{\rm p} \cdot \frac{h}{\nu} \tag{56}$$

$$\frac{\partial f}{\partial \nu} = \frac{T}{d_p^2} \left(\frac{4}{3} A\nu^{1/3} + 2C\nu \right) \qquad (57)$$

$$\frac{\partial f}{\partial (d_p)} = \frac{2T}{d_p^3} \left[B + A\nu^{4/3} + C\nu^2 \right] = \frac{2T}{d_p^3} h\nu \qquad (58)$$

Equation (52) gives

$$2S'd_p \cdot \frac{h}{\nu} + 2\lambda \cdot \frac{T}{d_p^3} \cdot h\nu = 0 \qquad (59)$$

or

$$\lambda = d_p^4 \cdot \frac{S'}{T} \cdot \frac{1}{\nu^2} \qquad (60)$$

Combination of Eqs. (51) and (60) yields the relationship

$$-3B + A\nu^{4/3} + 3C\nu^2 = 0 \qquad (61)$$

which is identical to Eq. (33) obtained for the case when h is minimum ($\partial h/\partial \nu = 0$). Consequently the fastest possible separation is accomplished with a column in which the optimum flowrate, i.e., minimum plate height, is obtained at the maximum pressure available with the instrument. The particle diameter of the column is calculated from the following expression:

$$d_p = \left[\frac{\eta D_m N}{k_0 \Delta P} \cdot h\nu \right]^{1/2} \qquad (62)$$

According to Eq. (62), in a situation with $\eta D_m = 1 \times 10^{-7}$, $h\nu = 6$, and $\Delta P = 300$ atm, we find that for N equal to 5000 and 100,000 the value of d_p is 1 and 4.5 μm, respectively.

Under such conditions the analysis time is shorter than that obtained when using a longer column at a flowrate higher than the optimum. Combination of Eqs. (49) and (62) gives an expression for the minimum analysis time, that is, the same as that obtained by rearranging the last equation in Table VII. The above discussion suggests that if a given stationary phase is available only with a limited number of average particle diameters, as it usually happens in practice, we have to compromise. In this case we take particles having a diameter which is larger but closest to that given by Eq. (62) and pack a longer column. It is operated at a velocity larger than optimum and the analysis time will be longer than the theoretical minimum.

Figure 11 shows the variation of the minimum analysis time with the maximum pressure available for different values of the plate number. The

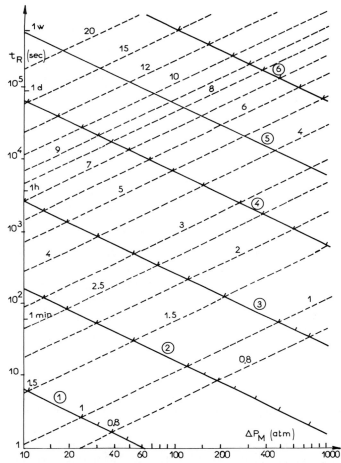

Fig. 11.   Minimum analysis time as a function of the maximum available pressure and the particle size. Solid lines represent constant plate numbers: (1) 1000; (2) 5000; (3) 20,000; (4) 100,000; (5) 300,000; (6) 1,000,000; dotted lines represent constant particle size stated in micrometers at each line. The columns are operated at minimum plate height of $h = 2$, and the other conditions are as follows: $\nu = 3$, $k = 3$, $\eta = 0.4$ cP, $k_0 = 1 \times 10^{-3}$, $D_m = 1 \times 10^{-5}$ cm^2/sec.

corresponding particle sizes are also depicted. Column length is given by Eq. (27). The experimental conditions chosen are favorable but realistic. As the previous discussion has shown that most routine separations can be conveniently carried out at low column inlet pressures ($\Delta P < 10$ atm) using 5- or 10-μm particles, in calculating the results shown in Fig. 11, emphasis is placed on difficult separation problems (cf. Section XIII).

XI. PRACTICAL CONSIDERATIONS

In many cases the analyst is not interested in the limiting performance of the chromatographic column, but prefers to work under less stringent conditions in order to facilitate sampling or detection, for example (19). In chromatographic practice most separations can be accomplished with 2 to 4×10^3 plates. In many cases, especially in reversed-phase chromatography, the column performance may deteriorate due to the slow kinetics, and the optimum plate height and flow velocity can be very poor, e.g., $h_m \simeq 15$, $v_0 \simeq 0.4$.

However, using more conventional techniques with $\eta = 0.5$ cP, $D_m = 2 \times 10^{-5}$ cm^2/sec, and $k = 3$, we can generate 2000 plates in 15 min, using a 7.5-cm-long column which is packed very well with 12-μm particles, at an inlet pressure of 0.85 atm. If the viscosity is increased to 2 cP, but the diffusion coefficient remains the same, i.e., $h = 10$ and $v = 0.6$, we still can generate 2000 plates in 15 min using an 8-cm-long column packed with 4.0-μm particles at an inlet pressure of 30 atm [cf. Eqs. (27), (29), and (30)]. Although this approach is more difficult because of the scarcity of 4-μm particles having uniform size distribution, it represents a viable alternative.

In many practical cases, a few plates per second, i.e., 5 to 20×10^3 plates/hour, suffice (21). It turns out that for $k = 3$ the corresponding value of d_p is about 10 μm for good and 5 μm for poor columns. A column is considered poor either because the packing is poor, as measured by a large A, or the kinetics of the retention process is slow, as measured by a large C. Nevertheless, under otherwise typical conditions, $A = 1$ and $C = 1$ still gives 1.1 and 3.5 for v_0 and h_m, respectively. In fact the peak is probably tailing, when the kinetics become slow enough to make C larger than unity, and as a result the previous equations are no longer valid. The columns characterized above can be operated at moderate pressures, even with aqueous buffers having high viscosity.

There is another way to look at the problem, however. Most analysts would like to have a simple column to solve separation problems in general. What column length and particle size would be optimum for such a purpose? A 30-cm-long column packed well with 10-μm particles, as now commercially available, can generate a maximum of 10,000 plates for a solute having $k = 3$ and $D_m = 2 \times 10^{-5}$ cm^2/sec at an optimum flow velocity of 0.04 cm/sec ($v = 2$). Under these conditions the analysis time is 50 min (3.3 plates/sec) and the pressure is 6 atm with an eluent having $\eta = 0.5$ cP. The column is suitable to attain difficult separations which require 10^4 plates, but it is slow. If we raise the pressure to 30 atm, which still is relatively low, the analysis time and efficiency are reduced to 10 min and 6000 plates, respectively, which are still good enough for most applications. The peak width is only 7.6 sec, however, and to avoid a deterioration due to extracolumn band spreading, we need an 0.4 sec response time data system that is faster than most commercially available equipment. Another way to reduce analysis time would be to cut the column in half and have two for the price of one. Now we can generate 5000 plates in 25 min with a 15-cm-long column at an inlet pressure of 3 atm. Raising the pressure to 5 atm reduces the analysis time to 15 min yet the efficiency is still higher than 4500 plates. The bandwidth is 13 sec and the time constant of the detector–recorder unit should be 0.65 sec, a specification that is somewhat easier to meet than that before.

These calculations suggest that commercial columns are too long. The reason for making columns longer than necessary is simple: they permit less tight specifications for the packing technique and the equipment. Quite a few commercial columns tested in the author's laboratory yielded $h_m \simeq 5$ and $\nu_0 \approx 1$. Under optimum conditions they generated 6000 plates when $P = 3$ atm, $t_R = 100$ min (1 h and 40 min), and $\tau < 4$ sec. On the other hand, at $P = 30$ atm, t_R is 10 min for $N = 3750$ plates and the time constant $\tau \simeq 0.5$ sec. The efficiency is still reasonable and the time constant manageable, although problems of band deformation will occur for less retained peaks as discussed in the following section.

One manufacturer, Dupont Instruments, Wilmington, Delaware, supplies certified columns with an individual performance report for several years. It is interesting to discuss their performances which shed light on the properties of some commercial columns. According to the specifications, a reduced plate height of at least 4.6 or 5.2 should be obtained, depending on the nature of the stationary phase which is silica or chemically bonded silica. This performance corresponds to 8 to 9×10^3 plates for a 25-cm-long column. In practice the reduced plate height of actual columns is often much better, however, and values between 2.4 and 4 are frequently encountered. It is striking that such reduced plate heights are obtained at flow velocities well above the optimum such as 0.2 cm/sec. For compounds like anthracene in $MeOH/H_2O$ (85:15), phenylacetonitrile in isopropyl alcohol/n-hexane (4:96), and chloronitrobenzene in $MeOH/H_2O$ (65:35), the diffusion coefficients vary between 0.5 and 1×10^{-5} cm^2/sec, so that, when $d_p = 6$ μm, the reduced velocity for such compounds at $u = 0.2$ cm/sec is between 12 and 25. Accordingly, these columns are not operated at the maximum efficiency, i.e., the minimum reduced plate height could have a value of approximately 2. Still, with $u = 0.2$ cm/sec, a compound with $k = 3$ is eluted in about 8 min at an inlet pressure of 140 atm ($\eta = 1$ cP) or less. These results demonstrate that the performance of the best commercial columns is close to the prac-

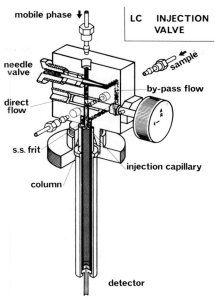

FIG. 12.   Schema of the Prolabo LC chromatograph showing the column squeezed by the column holder between the injector and the detector blocs and the details of the sampling valve.

tical maximum expected on the basis of the results of this study. In most cases the choice is between striving for an efficiency of 20,000 plates and an analysis time of about 1 h ($k = 3$) at low pressure or an efficiency of 12–16,000 plates and an analysis time of about 10 min ($k = 3$) at a higher pressure. Such a performance is more than enough to solve usual separation problems, but it is well above the manufacturer specifications, which only guarantee the column to be about half that good. Such specifications are not met by other commercial columns all the time—as of Spring, 1978.

Recently Prolabo (Paris, France) introduced a special assembly system which permits the use of short or very short columns by connecting them directly to the sampling device and the detector. The originality of the system lies partly in the design of the fittings which also allows minimal dead volumes at both ends (cf. Fig. 12). Columns of 15, 10, and 5 cm are standards. Using this system, Guillemin has been able to produce a good chromatogram, with a 2-cm-long column (59) and later with 1-cm and 0.5-cm-long columns! As an example of analytical results, Fig. 13 shows a chromatogram of polystyrene additives on a 5-cm-long columns, packed

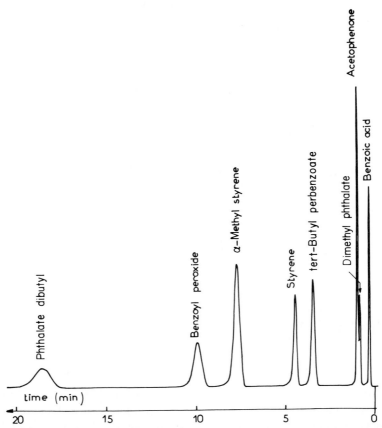

FIG. 13.   Chromatogram of polystyrene additives on a 5-cm-long column packed with Sphérosil/C_{18}, 5-μm particles. Mobile phase 98 (water 40/methanol 60)/acetonitrile 2 (v/v) 1 ml/min. Inlet pressure 56 atm. Valve injection 2-μl sample.

with Sphérosil C_{18}, 5-μm particles. The efficiency for α-methyl styrene ($k \simeq 20$) is 4300 plates ($h = 2.3$) and the linear velocity 0.2 cm/sec ($\nu \simeq 7$) which correspond approximately to the optimum conditions. The very long analysis time is due to the use of large surface area silica particles ($k = 40$ for dibutyl phthalate). For $k = 3$ the analysis time is about 100 sec. The inlet pressure, 55 atm, is rather large for this column length, because the viscosity of the water–methanol mixture is large ($\eta = 1.35$ cP). These performances exceed markedly those which have been assumed in drawing Figs. 5–11, which shows that their results are quite conservative: Eqs. (27), (28), and (31) (Table III, Section B) would give a column length of 4.3 cm, an analysis time of 16.7 min, and a pressure drop of 21 atm for $h = 2$, $\nu = 3$, $d_p = 5$ μm, $D_m = 1.5 \times 10^{-5}$ cm^2/sec, and $\eta = 1.35$ cP.

We can draw the following conclusions from the preceding treatment:

(*i*) For solving simple separation problems involving normal liquid phases and solutes, when the separation is not too difficult and the time not critical, a standard 30-cm-long column packed with 10-μm particles can give satisfactory results. Analysis time is adjusted depending on the required efficiency. The pressure will be in the 5 to 30 atm range and a gas pressurized pump can be used. Some improvement of the packing technique can be traded for additional efficiency or shorter column length or analysis time.

(*ii*) In the case of more difficult separation problems the particle diameter is reduced to 5 μm in order to obtain larger plate number or shorter analysis time. If systems with very low diffusion coefficients have to be used, the particle dimensions could be further reduced provided the finer particles are available in narrow enough size distribution.

The same conclusions were obtained by Halasz who illustrated them by similar plots, using the plate number instead of the HETP and column length (*54*). The paper being not published yet, it is not possible to comment on the degree of closeness of the conclusions and the possible difference in the results.

XII. EQUIPMENT SPECIFICATIONS

Most commercial liquid chromatographs have not been designed with the use of short, efficient columns in mind. While some changes are slowly taking place, significant improvement of most instrument modules such as sampling system, detector, connecting tubes, and data acquisition device, are necessary to bring them to the level of the column performances (*20, 27*). In fact, the main reason why large diameter columns are found to have the same or even higher efficiency than the conventional 4-mm-i.d. columns and why it is so difficult to prepare efficient, narrow bore columns is the liquid chromatograph itself. With the flow velocity being constant, the large flowrate through a large diameter column results in large band volumes that are not affected by the detector and the connecting tube, whereas the narrow bands obtained at the small flowrate through narrow columns have small volumes and can readily be disturbed by extracolumn effects.

In this section we discuss the contributions of the detector time con-

TABLE VIII

Equations Specifying Minimum Performances of Equipment and Sampling Conditions

Equation	Equation number
$\tau_M = \dfrac{\theta t_R}{\sqrt{N}}$	63
$V_{d,M} = \theta \dfrac{\pi d_c^2}{4} \cdot \epsilon(1 + k')\, h d_p \sqrt{N}$	67
$r^4 l < 6\theta^2 \cdot \epsilon(1 + k')^2 d_c^2 N d_p^3 \cdot \dfrac{h^2}{\nu}$	70
$V_{S,M} = \lambda V_{d,M} \left(\dfrac{d_S}{d_c}\right)^2$	
$t_{S,M} = \dfrac{\theta \lambda t_R}{\sqrt{N}}$	75

stant, detector cell volume, connecting tubes, and sampling device on band spreading as well as the effect of detection limits. The discussion of these phenomena in gas chromatography by Sternberg (45) is also applicable to extracolumn band spreading in liquid chromatography. A detailed experimental and theoretical study has been published recently (57). The equations used in the following discussion are given in Table VIII for easier reference.

A. Detector Time Constant

This is still the most serious problem at present, as the time constant of most detectors ranges from 0.5 to 3 sec including the time constant of the detector electronics to which the analyst has little access. Only a few commercial detectors have a time constant less than 0.5 sec whereas a few detectors, which are used mostly to build inexpensive homemade liquid chromatographs, have a time constant well in excess of 3 sec.

The effect of the detector time constant on the apparent efficiency depends only on the time width of the bands. It has been shown by Schmauch (41) and by McWilliam and Bolton (42) that the profile recorded with a detector having a time constant τ is wider than the actual profile by a factor $(1 + \tau/\sigma_t)$, where σ_t is the time standard deviation of the profile, provided this factor is less than about 1.2. Moreover, the peak height becomes smaller although the peak area remains unchanged. The first moment (retention time) of a peak increases by τ and the retention time of the peak maximum by a somewhat smaller amount. If the decrease in apparent efficiency is required to be smaller than θ^2, the detector time constant should be smaller than the maximum permissible time constant τ_M that is evaluated by

$$\tau_M = \theta \frac{t_R}{\sqrt{N}} \qquad (63)$$

Equation (63) (cf. Table VIII) shows that if this requirement is fulfilled, the relative increase in retention time is less than $\theta/\sqrt{N}$.

If we accept an efficiency loss of a few percent, we can tolerate a 1% contribution for each part of the equipment and let $\theta^2 = 0.01$. Then Eq. (63) shows that in order to obtain $N = 5000$ plates and $t_R = 5$ min, the time constant should be less than 0.4 sec, which is quite a demanding requirement in view of the performance of currently available commercial liquid chromatographs.

In most cases the time constant of the detector is due to the slowness of the electronics; this is especially true for optical detectors. There would be no technical problem to reduce the time constant to 20–50 msec, although the noise level is expected to increase somewhat. With such fast detectors computer data acquisition becomes necessary, as recorders with a time constant less than 0.5 sec are rare and expensive. Since they are difficult to maintain, they are impractical.

The time needed for the mobile phase to sweep the cell volume is not included into the time constant since the contribution of this volume will be discussed below.

If the detector is not linear, the recorded peak is smaller and broader than the actual peak. It has been shown that if the detector response can be expressed by

$$y = kC (1 - aC) \qquad (64)$$

the peak variance increases by a factor of 0.4 ah/k, where h is the actual peak height (43). This effect can be important especially with a UV detector.

The contribution of slow detector response can be neglected when the base peak width is at least 40 times larger than τ [cf. Eq. (63)]. In practice it is difficult to correct for such distortion because the time constant concept is only an approximation. It is not very reproducible and is sensitive to changes in the characteristics of the various elements of the electronics. Furthermore, detectors, amplifiers, and recorders are not first-order systems and their response is only approximated by an exponential function (44). The response time is therefore defined by the time necessary for the response to increase from 10 to 90% of the input signal. According to this definition it is 2.2 times greater than time constant characteristic for a first-order system.

Finally, the calculations should take into account the fact that in Eq. (63) the retention time of the first analytically significant peak should be used. It should be noted that for $k = 3$, as in the examples used throughout this review, the inert peak is eluted four times faster. For such an early peak the specification of the instrument is quite drastic because a response time smaller than 0.23 sec is required for $N = 5000$ plates and $t_m = 75$ sec.

B. Detector Cell Volume and Connecting Tubes

Depending on its design, the detector cell can be considered either as a tube with plug flow or as a mixing chamber. More probably, however, it is an intermediate system between these two limits (45). In the least favorable situation, the contribution to time variance, σ_t^2, is given by

$$\sigma_t^2 = V_d^2/F^2 \tag{65}$$

where V_d is the detector cell volume and F the flowrate. Since at a given linear flow velocity F is proportional to the squared column diameter, for a given detector, the contribution to variance is inversely proportional to the fourth power of the column diameter. This can explain at least in part why it is so difficult to obtain high efficiency with narrow columns.

It has been shown (20) that the maximum allowable detector volume $V_{d,M}$ can be expressed by

$$V_{d,M} = \theta \cdot \frac{\pi d_c^2}{4} \cdot \epsilon(1 + k) \cdot \sqrt{LH} \tag{66}$$

where d_c is the column diameter and ϵ is the packing porosity. Equation (66) can be rewritten as

$$V_{d,M} = \theta \cdot \frac{\pi d_c^2}{4} \cdot \epsilon(1 + k) \cdot h d_p \sqrt{N} \tag{67}$$

Equation (67) clearly shows that when d_p is decreased either the detector cell volume has to be decreased proportionally or the column diameter should be increased in order to avoid significant peak distortion. With typical values such as $\theta = 0.1$; $d_c = 0.4$ cm, $\epsilon = 0.8$, $k' = 0$ (inert peak), and $h = 2$, we obtain the following simple relationship:

$$V_{d,M} = 0.02 d_p \sqrt{N} \tag{68}$$

where V_d is given in milliliters and d_p in centimeters. Equation (67) shows that with $d_p = 5\ \mu$m and $N = 5000$ plates the maximum cell volume is 0.7 μl. With $k' = 3$ it will be four times larger as quoted in Ref. 20, but this is not a very realistic value as we are also interested in the efficiency of early peaks. When $d_p = 6\ \mu$m and $N = 20{,}000$ we still can afford a detector volume of 1.7 μl and a 5% loss of efficiency will occur by using a 2.2 times larger detector volume only. If we use a 1-cm-diameter column, the maximum allowable detector volume is calculated by

$$V_{d,M} = 0.125\ d_p\sqrt{N} \tag{69}$$

and the cell volumes become 4.4 and 10.5 μl for 5-μm particles, 5000 plates and for 6-μm particles, 20,000 plates, respectively. Whereas the first values calculated for the detector volume are impossible to meet with at the present time, the latter values are technically possible. Indeed the cell volumes of commercially available UV detectors range from 6 to 10 μl. As a consequence only peaks of retained compounds on standard columns or peaks eluted on larger diameter columns can be analyzed without significant extracolumn broadening using the commercial equipments presently available. In fact, in many cases detector cell volume and/or injection characteristics control the actual efficiency of chromatographic systems.

Scott and Kucera (46) have shown that the variance contribution of a connecting tube is proportional to its length and to the fourth power of its diameter. As long as the connecting tubes are straight, it can be shown (20) that the product r^4l should obey the following condition obtained by rewriting the corresponding condition in Ref. 20, using Eqs. (27) and (28) as follows:

$$r^4l < 6\theta^2 \cdot \epsilon(1 + k')^2 d_c^2 N \cdot d_p^3 \cdot \frac{h^2}{\nu} \tag{70}$$

where θ^2 is the relative loss of efficiency as defined above. Using the standard values of $\theta^2 = 0.01$, $\epsilon = 0.8$, $k' = 0$, $d_c = 4$ mm, $h = 2$, and $\nu = 3$, we find that

$$r^4l \le 0.01Nd_p^3 \tag{71}$$

This requirement is also very demanding because for $d_p = 5\ \mu$m and $N = 5000$ plates, we find that r^4l should be less than 6.2×10^{-9} cm^5 so that a 10-cm connecting tube should have a diameter smaller than 0.05 mm. The condition set forth in Eq. (71) is much easier to meet by moderately increasing one of the following parameters: d_p, k', d_c, h, or even N. For example with $d_p = 10\ \mu$m, $k' = 2$, $d_c = 1$ cm, $h = 3$, $\nu = 2$, and $N = 5000$, when the analysis time increases from 250 to 2250 sec for $k' = 2$ at $D_m = 1 \times 10^{-5}$ cm^2/sec [cf. Eq. (28)] we find that $r^4l < 1.94Nd_p^3$

so that a 60-cm-long and 0.2-mm-i.d. connecting tube can be used without appreciable contribution to extracolumn band spreading.

Efficient narrow bore columns are really difficult to handle in rapid analysis. For such columns the least detrimental connecting tube is no connecting tube at all. In order to accommodate different types of columns the equipment should be designed in such a way that the distance between sampling device and detector can be varied when the column length is changed.

Equation (70) is valid only for straight connecting tubes. When they are coiled, a radial secondary flow develops under the effect of centrifugal forces (47) and by enhancing radial mass-transfer the variance contribution of the connecting tube is attenuated. Thus a longer, but tightly coiled tube appears to yield less band spreading than a straight tube having the same inside diameter. The design of fittings which are used for the transition from the large diameter column to the narrow connecting tube is critical and can result in a severe efficiency loss.

C. Injection

The main contributions are the sample volume and the profile of the injected band characterized by both the injection volume and the injection time (16, 20, 27, 5). We want to calculate the maximum permissible sample size which causes a relative increase in the zone width by a factor θ, hence a decrease in the efficiency by θ^2. The sensitivity depends little on the column length at constant plate number if the maximum sample size is used in all cases.

The contribution of sample introduction to the band variance has been found experimentally to obey the following relationship:

$$\sigma_S^2 = V_S^2/\lambda^2 \tag{72}$$

where V_S is the sample volume and λ is a numerical factor. The latter depends on the mode of injection technique, and usually has a value of about 2. In the ideal case of plug injection the value of λ would be 3.5. Simple calculations (20) show that the sample volume should not exceed a maximum value, $V_{S,M}$, defined by the expression

$$V_{S,M} = \lambda\theta \cdot \frac{\pi d_c^2}{4} \cdot \epsilon(1 + k') \cdot \sqrt{LH} \tag{73}$$

which can be rewritten as

$$V_{S,M} = \lambda\theta \cdot \frac{\pi d_c^2}{4} \cdot \epsilon(1 + k') \cdot h d_p \sqrt{N} \tag{74}$$

Comparison with Eq. (28) shows that for a given chromatographic system (constant h, ν, k', D_m) $V_{S,M}$ is proportional to $\sqrt{t_R}$ (20). Combination of Eqs. (66) and (74) yields

$$V_{S,M} = \lambda V_{d,M}$$

Equations (67) and (68) show that the efficient columns available today tolerate much smaller sample sizes than the columns used a few years ago. Whereas for columns with $D_p = 25\ \mu m$, $h = 5$, and $N = 2000$ the maximum permissible sample volume $V_{S,M}$ is about 60 μl, the microparticulate columns can accept samples of only a few microliters without deterioration of the intrinsic column efficiency.

Other factors related to sample injection can also be important. It has been shown by Kirkland et al. (27) that the way of injecting the sample, especially by using a syringe, is critical. It seems that "isokinetic" sampling is necessary to establish a uniform profile of the sample band at the column inlet. Otherwise eddies can form at the needle tip and the resulting multimodal injection yields to band nonuniformities already at the column inlet. The time during which the sample is injected is also important and it should be less than a maximum value, $t_{S,M}$, given by the relationship

$$t_{S,M} = \theta \lambda t_R / \sqrt{N} \tag{75}$$

Comparison of Eqs. (74) and (75) shows that $V_{S,M}$ in Eq. (74) expresses a sample size that is injected across the entire column section during time $t_{S,M}$ at the same velocity as the mobile phase.

It has been recommended (27) to inject through a narrow tube at the center of the column in order to avoid that the sample band reaches the column wall by radial diffusion during elution. As at the wall the packing is prone to be heterogeneous and the resulting distortion of the solute band yields poor efficiency, this problem is avoided if we inject through a narrow coaxial tube and in this case if we want isokinetic injection, the maximum permissible sample volume obtained from Eq. (72) has to be multiplied by the ratio $(d_s/d_c)^2$ where d_s and d_c are the inner diameters of the sample tube and the column, respectively. The conditions of isokinetic injection of the sample and maximum sampling time are then both satisfied. Unless a rather wide syringe needle is used for injection, the permissible sample is very small. The sample size can be increased by injecting into the connecting tube at the inlet of the column. In this case, however, the column cannot be operated under conditions of "infinite diameter" (27) as discussed above.

It is not certain, however, that isokinetic sampling is the ideal solution.

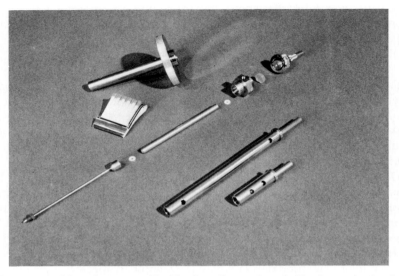

Fɪɢ. 14. Prolabo column assembly. Top line: 10-cm column holder fastened to handnut. Upper middle line: connecting tube to detector, Teflon O-ring, 10-cm column, Teflon O-ring, septum, nut with syringe needle guide. Lower middle line: 15-cm-long column in holder. Bottom line: 5-cm-long column in holder.

Experimental data (57) show that the efficiency is not markedly decreased if the sample is injected faster provided that the volume at the top of the column into which the sample is injected is filled with some kind of porous material to prevent convective mixing. This permits the injection of large sample volumes. In actual practice the injection time seems to be the critical parameter (57).

The connections of very short columns to injector and detector become especially difficult, as there is no room for the usual type of nuts and fittings. A special design permits the easy use of 1- to 15-cm-long columns (or possibly shorter) with the Prolabo chromatograph LC 50. A column holder fastened with a broad handnut squeezes the column between injector and detector block (cf. Fig. 14). Teflon rings maintain leak-proof seals. The column is conventionally closed by a stainless steel frit at the outlet and a Teflon frit at the inlet. The syringe needle guide is adjustable so the needle tip always reaches the middle of the Teflon frit. A specially designed microvalve can also be used (cf. Fig. 12). A bypass flow permits the injection of a 2-μl sample while limiting radial diffusion. The flowrate ratio is adjustable by a metering valve. Excellent analyses are obtained, as shown on Fig. 13, by the chromatogram of polystyrene additives on a

5-cm-long Sphérosil/C_{18} column.

This demonstrates that careful mechanical design permits the solution of the difficult equipment problems encountered in the use of short LC columns.

D. Detection Limits and Maximum Peak Number

The maximum number of peaks, n, which can be separated on a given column can be calculated from the following equation:

$$n = 1 + \frac{\sqrt{N}}{4} \cdot \ln(1 + k') \qquad (76)$$

This number, which is also called peak capacity, is infinite in theory, but limited in practice. Time considerations require that k' not exceed a certain limit, which in fact decreases when the column efficiency increases, because t_R should not exceed a few hours except in very special circumstances. The maximum peak number is also limited by the fact that with increasing retention the width and height of the peak increases and decreases, respectively. As a consequence of the limitations on sample size, peaks whose maximum concentration at the column outlet is lower than the detection limits remain unnoticed.

This problem has been discussed by Snyder (8, 10) and Scott (9, 11). A general solution is difficult to give since it would depend on the composition of the sample mixture, e.g., the concentration of the last eluting components, and the detection limit, which varies in liquid chromatography with the chemical nature of the sample component. Therefore some arbitrary assumptions have to be made. From Eq. (26), using the maximum permissible sample size given by Eq. (73), we can write for the retention volume, V_R, the following expression:

$$V_R = (1 + k')V_m = (1 + k') \frac{\pi d_c^2}{4} \epsilon L \qquad (77)$$

For the sample concentration we have

$$C_M = \frac{\lambda \theta}{\sqrt{2\pi}} \cdot C^* \qquad (78)$$

where C^* is the concentration of the solute in the sample. With $\theta = 0.1$ and $\lambda = 2$ we obtain from Eq. (78) that $C_M = 0.08 \, C^*$; that is, under optimum conditions the sample is diluted by a factor of 12.5 in the liquid chromatography. So the dilution is only slightly greater than an order of magnitude. We should emphasize first that the maximum sample concentration is also limited by stationary phase overloading which determines

C^*, and that C^* usually decreases with increasing retention of the sample component $(9-11)$. The maximum sample size is also a function of the retention, so that when the size of the sample is the maximum possible for the last compound of the mixture, then the column is overloaded by all the other sample components. If we use a sample size such that the column is not overloaded by any compound, we have to set k' equal to zero in Eq. (74). In this case we obtain for the maximum sample concentration, C'_M, the following expression

$$C'_M = \frac{\lambda \theta}{(1 + k')\sqrt{2\pi}} \cdot C_M \qquad (79)$$

It is seen from Eq. (79) that the dilution factor increases with $(1 + k')$ and that may limit the maximum peak number as shown by Scott (9) and Snyder (10).

With the very efficient columns presently available, large peak capacities can be achieved for the separation of complex mixtures, yet, this approach has not been fully explored.

The overloading of the stationary phase is related to the maximum solute concentration, C_M, at which the sorption isotherm associated with equilibrium distribution underlying chromatographic retention ceases to be linear. That deviation results in a broadening and deformation of the peak profile. Since this review deals with chromatographic phenomena and optimization we consider thermodynamics as beyond its scope.

XIII. CONCLUSIONS

We have seen that most separations in the practice of liquid chromatography can be performed by using 5- to 30-cm-long columns packed with 5- to 15-μm particles at inlet pressures in the $5-150$ atm range. Most important is the packing structure, which depends on the packing technique used to make the column, although the quality of the stationary phase, the mobile phase viscosity, and the diffusion coefficient of the solute also have some effect. The highest analytical performance possible, at least theoretically, due to recent advances in column technology, however, are difficult to attain in practice because of insufficient performance of the chromatographic equipment. Indeed, the instrument specifications are extremely stringent even for rather easy separations. Therefore, unless some technological breakthrough takes place, it will be practically impossible to exploit fully the potential of modern columns for fast analysis.

Advanced column technology could also be useful in the development of very efficient columns in terms of plate number. Figure 11 shows that it is indeed possible to prepare extremely efficient columns. For example a 10-m-long column packed with 5-μm particles could generate 10^6 plates at $k' = 6$ within 1 day and the pressure drop would be 10^3 atm. Under these conditions the peak capacity is 350 but if we accept waiting 5 days the peak capacity could be increased to 750. This is the ultimate performance possible in liquid chromatography. On all grounds we are getting dangerously close to the practical limits. In order to use higher column inlet pressures more sophisticated equipment is needed. Additional difficulties may arise from the compressibility of the mobile phase as well as the variation of eluent viscosity and the diffusion coefficient with pressure. Packing very long columns uniformly appears to be impossible, therefore, several shorter columns have to be packed separately and connected in series so that some efficiency will be lost in the fittings. We probably cannot coil the long column and do not benefit from the infinite diameter effect, and last but not least, we would find that the excessive analysis time is quite impractical.

The author does not see how the situation can be improved at present. A marked reduction of the particle size would not be possible without encountering extremely severe problems with the packing; 1-μm particles tend to stick together in suspension quite easily and are difficult to disperse. The inlet pressure of columns packed with such small particles would be enormous. At such large pressure gradients the heat generated by viscous friction (48) would create radial temperature gradients and have an adverse effect on the efficiency since the mobile phase viscosity would not be constant across the column anymore. In other words, a 100,000 plate column is relatively easy to prepare and operate; a 1,000,000 plate column is about the upper limit of what a chromatographer can dream of.

Scott and Kucera (49) have prepared a 500,000 plate column operated in GPC. High-efficiency GPC columns are easier to operate since all compounds are, in principle, eluted before the inert peak, i.e., by the definition used in this work, k' is negative.

More recently, the same authors (53) published experimental results on long, efficient, narrow bore packed columns. Columns that are 1 mm i.d. seem to pack well, have a slightly larger permeability than standard 4-mm-i.d. columns, and need a much smaller amount of mobile phase. The contributions of extracolumn sources of band broadening are most difficult to control, however, and the author feels that if Scott and Kucera are correct in pointing out the potential savings associated with the use of 1-mm-i.d. columns in routine analysis, they tend to underestimate the cost of equipment design and maintenance. Certainly when very high efficiency is at stake, wider columns could be a better compromise. Thus, in spite of all shrewd efforts to master the technologies involved and of their long-time proven ability to solve the most difficult experimental problems, Scott and Kucera could achieve only fair performances with 5-μm particles. The minimum reduced plate height of the 1-m-long column was 4.3, at a reduced velocity of 0.74. Fourteen of such columns combined gave a 650,000 plate efficiency for the last peak, eluted in about 9 h in one experiment of exclusion chromatography and 510,000 plates for a compound with $k = \sim 2$ in adsorption chromatography, with an analysis time of over 30 h ($t_0 \simeq 10$ h). The specific performances of the 20-μm-particle columns are much better, probably because the bands are wider, and the contributions of extracolumn sources of band broadening are relatively smaller and may even become negligible. A reduced plate height of 3.12 is achieved at a reduced velocity of 16, which is far larger than the optimum; the minimum reduced plate height is around 2. A 10-m-long column operated at $\nu = 16$ gives 160,000 plates, with a retention time for the inert peak of 205 min. Few data are given on the pressures used to operate these columns.

These results are in agreement with the results of the calculations made above if we take into account the values of the reduced plate height and velocity. From that we can safely predict that using wider columns it is possible to achieve about 1,000,000 plates with a 20-m-long column packed with 10-μm particles, operated under 250 atm, with an analysis time of 18 h.

It would be difficult to do much better. For a variety of reasons open tubular columns do not seem to offer a workable alternative (18, 54). This, however, leaves a large number of analytical problems to solve with the tools available now. In this field optimization is helpful as it permits large time savings.

REFERENCES

1. M. Tswett, "Chlorophylls in Vegetal and Animal Worlds." Warszaw, 1910.
2. E. Lederer, *J. Chromatogr.* **73**, 361 (1972).
3. J. J. Kirkland, "Modern Practice of Liquid Chromatography." Wiley, New York, 1971.
4. L. R. Snyder, *Anal. Chem.* **39**, 705 (1967).
5. I. Halasz and P. Walking, *J. Chromatogr. Sci.* **7**, 129 (1969).
6. L. R. Snyder, *J. Chromatogr. Sci.* **7**, 352 (1969).
7. J. H. Knox and M. Saleem, *J. Chromatogr. Sci.* **7**, 614 (1969).
8. L. R. Snyder, *J. Chromatogr. Sci.* **8**, 692 (1970).
9. R. P. W. Scott, *J. Chromatogr. Sci.* **9**, 449 (1971).
10. L. R. Snyder, *J. Chromatogr. Sci.* **10**, 187 (1972).

11. R. P. W. Scott, *J. Chromatogr. Sci.* **10**, 189 (1972).
12. L. R. Snyder, *J. Chromatogr. Sci.* **10**, 200 (1972).
13. L. R. Snyder, *J. Chromatogr. Sci.* **10**, 369 (1972).
14. J. F. K. Huber, F. F. M. Kolder, and J. M. Miller, *Anal. Chem.* **44**, 105 (1972).
15. I. Halasz and M. Naefe, *Anal. Chem.* **44**, 76 (1972).
16. B. L. Karger, M. Martin, and G. Guiochon, *Anal. Chem.* **46**, 1640 (1974).
17. R. P. W. Scott, and P. Kucera, *J. Chromatogr. Sci.* **12**, 473 (1974).
18. M. Martin, G. Blu, C. Eon, and G. Guiochon, *J. Chromatogr. Sci.* **12**, 438 (1974).
19. M. Martin, C. Eon, and G. Guiochon, *J. Chromatogr.* **99**, 357 (1974).
20. M. Martin, C. Eon, and G. Guiochon, *J. Chromatogr.* **108**, 229 (1975).
21. M. Martin, C. Eon, and G. Guiochon, *J. Chromatogr.* **110**, 213 (1975).
22. I. Halasz, H. Schmidt, and P. Vogtel, *J. Chromatogr.* **126**, 19 (1976).
23. M. Martin and G. Guiochon, *Chromatographia* **10**, 194 (1977).
24. B. Coq, C. Gonnet, and J. L. Rocca, *J. Chromatogr.* **106**, 249 (1975).
25. J. H. Knox, G. R. Laird, and P. A. Raven, *J. Chromatogr.* **122**, 129 (1976).
26. J. H. Knox, *In* "Practical High Performance Liquid Chromatography" (C. F. Simpson, ed.). Heyden, London, 1976.
27. J. J. Kirkland, N. W. Yau, H. J. Stoklosa, and C. H. Dilks, *J. Chromatogr. Sci.* **15**, 303 (1977).
28. J. H. Knox, *J. Chromatogr. Sci.* **15**, 352 (1977).
29. L. Karger, *Anal. Chem.* **39**, (8), 24A (1967).
30. M. Martin, G. Blu, and G. Guiochon, *J. Chromatogr. Sci.* **11**, 641 (1973).
31. R. B. Bird, W. E. Stewart, and E. N. Lightfoot, "Transport Phenomena." Wiley, New York, 1962.
32. J. C. Giddings, "Dynamics of Chromatography." Dekker, New York, 1965.
33. De Lignies, *J. Chromatogr.* **35**, 269 (1968).
34. J. N. Done, G. J. Kennedy, and J. H. Knox, *In* "Gas Chromatography 1972" (S. G. Perry, ed.). Applied Science, Barking, Essex, 1973.
35. C. L. Guillemin, J. P. Thomas, S. Thiault, and J. P. Bounine, *J. Chromatogr.* **142**, 321 (1977).
36. C. R. Wilke and P. Chang, *Am. Inst. Chem. Eng.* **1**, 264 (1955).
37. J. C. Giddings, *Anal. Chem.* **35**, 2215 (1963).
38. J. C. Giddings, *Anal. Chem.* **36**, 1890 (1964).
39. J. C. Giddings, *Anal. Chem.* **37**, 60 (1965).
40. M. Monroe, *In* "Some Like It Hot," conclusion.
41. L. J. Schmauch, *Anal. Chem.* **31**, 225 (1959).
42. G. McWilliam and H. C. Bolton, *Anal. Chem.* **41**, 1755 (1969).
43. T. Petitclerc and G. Guiochon, *Chromatographia* **7**, 10 (1974).
44. M. Goedert and G. Guiochon, *Chromatographia* **6**, 76 (1973).
45. J. C. Sternberg, *In* "Advances in Chromatography" (J. C. Giddings and R. A. Keller, eds.), Vol. 2, p. 205. Dekker, New York, 1966.
46. R. P. W. Scott and P. Kucera, *J. Chromatogr. Sci.* **9**, 641 (1971).
47. I. Halász, R. Endele, and J. Asshauer, *J. Chromatogr.* **112**, 37 (1975).
48. R. Endele, I. Halasz, and K. Unger, *J. Chromatogr.* **99**, 377 (1974).
49. R. P. W. Scott and P. Kucera, *J. Chromatogr.* **125**, 251 (1976).
50. P. A. Bristow, and J. H. Knox, *Chromatographia* **10**, 279 (1977).
51. R. E. Majors, *J. Chromatogr. Sci.* **11**, 88 (1973).
52. P. A. Bristow, *J. Chromatogr.* **149**, 13 (1978).
53. R. P. W. Scott and P. Kucera, *J. Chromatogr.* **169**, 51 (1979).

54. I. Halász, *Angew. Chem.*, *Int. Ed.* (1980, in press).
55. Cs. Horváth and H. J. Lin, *J. Chromatogr.* **126,** 401 (1976).
56. H. Colin, N. Ward, and G. Guiochon, *J. Chromatogr.* **158,** 183 (1978).
57. H. Colin, M. Martin, and G. Guiochon, *J. Chromatogr.* **185,** 79 (1979).
58. R. E. Majors, *J. Chromatogr. Sci.* **11,** 88 (1973).
59. C. L. Guillemin, *J. Chromatogr.* **158,** 21 (1978).

LIQUID CHROMATOGRAPHY ON SILICA AND ALUMINA AS STATIONARY PHASES

Heinz Engelhardt and Helmut Elgass*

Angewandte Physikalische Chemie
Universität des Saarlandes
Saarbrücken
Federal Republic of Germany

I.	Introduction	57
II.	Stationary Phases	60
	A. Silica	61
	B. Alumina	68
III.	Role of the Eluent	70
	A. Eluotropic Series	71
	B. Mixed Eluents	73
	C. The Role of Water and Other Modulators	77
	D. Partitioning	85
	E. Proper Choice of the Eluent	87
IV.	Effect of Sample Structure on Retention	90
V.	Anisocratic Analysis	92
	A. Pressure or Flow Programming	93
	B. Temperature Programming	95
	C. Programming of Stationary Phase	97
	D. Gradient Elution	98
VI.	Theory of Adsorption Chromatography	101
VII.	Selected Applications	104
VIII.	Prospects	106
	References	108

I. INTRODUCTION

Adsorption on polar stationary phases is the oldest separation principle applied in column chromatography. The chromatographic process is

---

* Present address: Hewlett-Packard GmbH., Waldbronn, Federal Republic of Germany.

57

governed by the energetics of the interaction between adsorbent, solute, and solvent molecules that determines the so-called "adsorption milieu" which can be represented as follows:

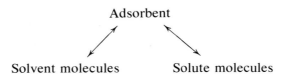

In principle all solids insoluble in common liquids can be used as stationary phases provided they have sufficiently high specific surface area. Only silica and alumina will be discussed here as examples of polar stationary phases inasmuch as the former is used almost exclusively, and the latter to a lesser extent, in current practices of HPLC.

Adsorption chromatography on polar stationary phases is especially suited for the separation of nonpolar to medium polar substances that have some solubility in solvents immiscible with water. More polar substances, which are soluble in polar solvents such as alcohols or water, of course, can be, and have been, separated on polar stationary phases, too. However, as nonpolar stationary phases are now easily available and widely used in reversed phase chromatography, such substances are preferentially separated on such phases because of better reproducibility and the convenience offered by this technique. (1)

For reproducible chromatographic work the retention parameters such as the retention time t_R, retention volume V_R, and the capacity ratio k, should be independent of sample size. This is only true as long as the sorption isotherm is linear (2), i.e., linear elution adsorption chromatography, LEAC, is practiced. Primarily Langmuir-type isotherms are encountered; thus, after a linear region, the relative amount adsorbed decreases with increasing sample concentration. Therefore the retention time decreases with increasing sample size in the nonlinear range. Furthermore, the region of highest sample concentration is shifted more and more to the front of the sample band and the concomitant tailing gives rise to asymmetrical elution peaks. For separations carried out in the nonlinear region of the isotherm, qualitative identification of peaks via retention volume is almost impossible, because the retention volume depends on sample size. The tailing reduces the resolution of two adjacent peaks and may hamper or prevent isolation of pure components in preparative work. Where an inhomogeneous surface with active sites differing widely in sorption energy is the reason for isotherm nonlinearity, the linear region of the isotherm will be increased if species like water or alcohols, which strongly interact with the stationary phase, are added to the mobile

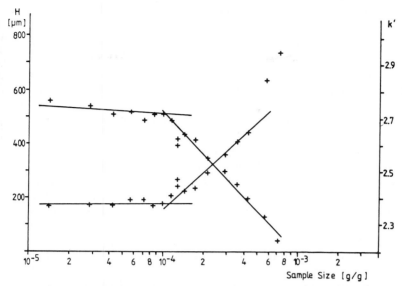

FIG. 1.   Effect of increasing sample loading on column efficiency and retention (*1*). Stationary phase, LiChrosorb SI 100, $d_p \sim 10~\mu$m; eluent, *n*-heptane containing 40 ppm H_2O; sample, nitrobenzene. (Courtesy of Springer-Verlag.)

phase. The components of the eluent are called "modulators" (*3, 4*) and it is assumed that their binding to active sites is responsible for decreasing surface inhomogeneity (*2*). Surface modification by chemical reaction as discussed by Engelhardt (*1*, Ch. 5) on bonded phases can also be used to extend the linear region.

For qualitative identification of substances reproducible retention data are obtained only in the linear region of the isotherm. To assure that one is working in this region it is recommended to check the column loading capacity of the particular chromatographic system given by the stationary phase, eluent, and sample. It can easily be determined experimentally by injecting increasing amounts of sample and determining both the retention time (or capacity ratio k), and the H values as shown in Fig. 1. Both parameters change drastically when sample size exceeds a certain value. In the case demonstrated here with silica as stationary phase, k and H values level off after sample size exceeds 10^{-4} g/g of adsorbent. As long as the sample size is below this maximum value, which is called the linear capacity of an adsorbent, chromatography is performed in the linear region of the isotherm (*2*).

In a first approach, it can be assumed that for silica or alumina as adsorbent, and for sample sizes of less than 10^{-4} g/g of adsorbent the retention time is independent of sample size. With adsorbents having highly in-homogeneous surface and/or with very dry eluents, however, the linear capacity of the system can be much smaller, i.e., less than 10^{-5} (2). On the other hand, in surface coated systems, physically or chemically bonded, the linear capacity of a system can be greater by up to an order of magni-tude. For preparative work, of course, the chromatographic system is usually overloaded, with concomitant loss of efficiency and resolution.

The physical sorption on polar, i.e., oxidic, sorbents is caused by spe-cific interactions between the polar surface groups and polar groups of the adsorbed molecule (5). These include dipole–dipole interactions between permanent and induced dipoles, hydrogen bonding as well as charge transfer or π-complex formation. Chemisorption of a sample component, which occurs occasionally with very active adsorbents and very dry eluents, is undesirable as it results in extremely long retention times, irre-versible sorption, and/or sample decomposition. Because of the com-plexity of the process involved, it is very difficult to describe adsorption chromatography completely by a simple theoretical model. It would be beyond the scope of this chapter to discuss details of different approaches used in the theoretical treatments of the subject. However an attempt will be made to present adsorption chromatography in a nonmathematical form and theoretical aspects will be included only to the extent deemed necessary.

Readers interested in a definitive treatment of linear adsorption chro-matography should consult the book *Principles of Adsorption Chromatog-raphy* by L. R. Snyder (2). Although written a decade ago, the book con-tains an extensive and still entirely valid presentation of the subject and forms one of the main sources used in writing this article.

II. STATIONARY PHASES

Among dozens of different solids used in classical column adsorption chromatography (6–9), silica and alumina are by and large the only polar adsorbents which have found employment in HPLC. As in classical col-umn chromatography, they are most frequently used as totally porous particles having large specific surface area and a high pore volume. In analytical columns the particle size is less than 50 μm, most commonly, 10 or 5 μm.

Additionally, they have been used as porous layer beads (PLB) (10,11), in which case a thin porous adsorbent layer is coated onto a fluid impervi-ous core, such as solid glass beads. The thickness of the porous layer is generally 1–3 μm, i.e., $\frac{1}{30}$ to $\frac{1}{40}$ of the particle diameter, and the particle size lies between 25 and 50 μm. Compared with totally porous materials in PLB columns, the amount of active stationary phase per unit column volume is small. The specific surface area of PLB coated with silica amounts to only 1–15 m²/g, compared to 200–400 m²/g for most of the totally porous materials. However, the actual surface area of the silica layer on PLBs is considerably greater and comparable to that of porous particles (*12*). Solute retention on columns packed with PLB is similar to that observed on columns packed with totally porous silica. The column efficiency with such particles is greater than with totally porous materials of the same diameter (*10, 13, 14*). Because of the low surface area of the stationary phase the linear capacity of PLB is very small, therefore, their importance is diminished since good and reproducible techniques became available for the packing of 5 and 10 μm totally porous adsorbent particles. In the following two sections the properties of the two most important polar chromatographic adsorbents, silica and alumina, will be described.

A. Silica

Silica is the most frequently used adsorbent and for use in HPLC it is sold under a variety of brand names such as LiChrosorb, LiChrospher, Nucleosil, Partisil, Porasil, Spherisorb, Spherosil, and Zorbax. For adsorption chromatography, silicas with a relatively large specific surface area (50–450 m²/g), a large pore volume (0.7–1.2 ml/g), and a moderate mean pore diameter (50–250 Å) are used.

The properties of some of the materials suitable for HPLC are summarized in Table I. It should be borne in mind that these materials, including the occasionally used porous glasses, are amorphous xerogels whose properties vary not only from brand to brand but may also vary from batch to batch, and may additionally change during storage and treatment used for purification.

It is generally accepted that the only important polar adsorption sites on the silica surface are the silanol functions, i.e., hydroxyl groups, that are attached to silicon atoms (*2*). They can interact with the sample molecules by hydrogen bonding and various physical observations can be used to prove this statement. Complete dehydration of silica by heating, i.e., removal of all surface hydroxyl groups, yields a hydrophobic silica which no longer shows adsorption for unsaturated and polar molecules and is no more wetted by water (*15*). Chemical modification of the surface hydroxyls such as used in the preparation of chemically bonded phases also eliminates the selective adsorption properties of the silica.

The data on the number of silanol groups per unit surface area vary widely, depending on the silica used and the method of their determina-

TABLE I

Polar Adsorbents as Stationary Phases for HPLC

Trade name or code		Chemical composition	Specific surface area (m²/g)	Mean pore diameter (Å)	Pore volume (ml/g)	Particle shape
A. *Totally porous materials*						
LiChrosorb[a]	SI 60	SiO_2	400	60	0.8	Irregular
	SI 100	SiO_2	300	100	1.0	Irregular
LiChrospher[a]	SI 100	SiO_2	250	100	1.2	Spherical
	SI 300	SiO_2	150	200	2.0	Spherical
	SI 500	SiO_2	50	500	0.8	Spherical
	SI 1000	SiO_2	20	1000	0.8	Spherical
	SI 4000	SiO_2	6	4000	0.8	Spherical
MikroPak[b]	Si	SiO_2	400	60	0.8	Irregular
μ-Porasil[c]		SiO_2	400	100	1.0	Irregular
Nucleosil[a]	50	SiO_2	500	50	0.8	Spherical
	100	SiO_2	300	100	1.0	Spherical
	100 V	SiO_2	430	100	1.5	Spherical
Partisil[e]		SiO_2	400	60		Irregular

Porasil[c]	A	SiO_2	400	150	1.05	Spherical, if $d_p > 40\ \mu m$ compare Spherosil[i]
	B	SiO_2	200	300	0.9	Irregular, compare Spherosil[i]
	C	SiO_2	75	600	0.7	
	D	SiO_2	30	1000	0.6	
	E	SiO_2	10	>2000	0.4	
	F	SiO_2	2		0.25	
SIL[d]	60–5	SiO_2	500	60	0.75	Irregular
Silica A		SiO_2	400	100		Irregular
SIL-X$_1$[f]		SiO_2	400	100		Irregular
Silica Woelm 18[g]		SiO_2	500	60	0.8	Irregular
Spherisorb[h]	S 5 W	SiO_2	200	80		Spherical
	S 10 W	SiO_2	200	80		Spherical
	S 20 W	SiO_2	200	80		Spherical
Spherosil[i]		SiO_2	600	60	0.8	Spherical
Zorbax SIL[j]		SiO_2	300	75		Spherical
Alox 60-D[d]		Al_2O_3	60	60	0.3	
Alox T[a]		Al_2O_3	80	100–200	0.3	
BioRad Aluminiumoxid[k]		Al_2O_3	200	100–200	0.3	
MikroPak Al[b]		Al_2O_3	80	100–200	0.3	
Spherisorb[h]	A 5 Y	Al_2O_3	95	150		Spherical
	A 10 Y	Al_2O_3	95	150		Spherical
	A 20 Y	Al_2O_3	95	150		Spherical
Woelm Aluminiumoxid N 18[g]		Al_2O_3	200	100–200	0.3	

(Continued)

TABLE I (*Continued*)

Trade name or code	Chemical composition	Specific surface area (m²/g)	Mean pore diameter (Å)	Pore volume (ml/g)	Particle shape
B. Porous layer beads (PLB)					
Actichrom[l]	SiO_2	25			Irregular
Corasil I[c]	SiO_2	7	50–70		Spherical
Corasil II[c]	SiO_2	14	50–70		Spherical
Liquachrom[l]	SiO_2	10			Irregular
Pellosil HS[e]	SiO_2	4			Spherical
Pellosil HC[e]	SiO_2	8			Spherical
Perisorb A[a]	SiO_2	~10	60		Spherical
Vydac	SiO_2	12	~60		Spherical
Zipax[j]	SiO_2	~1	400		Spherical
Pellumina HS[e]	Al_2O_3				Spherical
Pellumina HC[e]	Al_2O_3				Spherical
Pellidon[e]	Polyamide				Spherical

[a] Merck AG; [b] Varian; [c] Waters Associates; [d] Macherey & Nagel; [e] Whatman; [f] Perkin-Elmer; [g] Woelm Pharma; [h] Phase Separation; [i] Pecheney-St. Gobain; [j] Dupont; [k] BioRad; [l] Applied Science. Some of these materials are only available in packed columns.

tion (*15, 107*). For crystalline silica, a concentration of silanol groups can be calculated between 5 and 10 silanol functions per 100 Å², whether tridymite, cristobalite, or quartz is used for the calculation and which cut is brought through the crystal, An average value of 5 silanols per 100 Å², which corresponds to about 8 μmol/m² for amorphous silica, seems to be in good agreement with experimental measurements (*15, 16, 107*). For optimally only half of them are accessible to chemical substitution (*15*) or can react in the preparation of bonded phases. The silanol groups may be arranged so that they are either isolated from or are capable of forming hydrogen bonds with the adjacent silanol groups. The surface of wide pore silica is similar to that of crystalline tridymite which shows only free silanols. The narrower the pores, the more evident are the bonded silanols. (*2, 16*).

Thermal treatment of silica at temperatures below 150°C removes only physically adsorbed water and does not alter the silanol concentration at the surface. Dehydration due to surface reactions starts above 200°C, and neighboring hydroxyl groups condense to form siloxane groups. At temperatures above 600°C all silanols including the free silanol groups are split off to yield a hydrophobic silica covered by siloxane groups. Above 900°C the material starts to sinter. The chemical modification of heat treated silicas shows a relatively low concentration of reactive surface silanols, which are the only sites on the surface that can be esterified or otherwise modified. Consequently, such silicas allow the covalent binding of a lesser amount of organic function than the same weight of silica that has not been subjected to heat treatment (*17, 18*). The rehydration of thermally treated silica to reform silanol groups is a slow process and hardly occurs in contact with air at room temperature.

The silanol concentrations of different commercial silica preparations made from sodium silicate, silicon tetrachloride or an orthosilicic acid ester can vary distinctly despite the similarity of their other physical properties such as surface area, average pore diameter, and pore volume. As a result the chromatographic properties will be different. For example, the relative and absolute retentions of *m*- and *p*-nitroanilines were determined with three different brands of silica under otherwise identical chromatographic conditions. For the first two brands the relative retentions, α, were the same (*1, 36*) but the absolute retentions as measured by the k values differed by a factor of 2. With the second and third type of silica the absolute retentions were of the same magnitude but the relative retention obtained using the third silica was only 1.08 (*19*). Inversion of the elution order of some sample components upon changing brands of silica is not uncommon even when the eluent composition is invariant.

Actually the accessible silanol concentration per unit column volume governs the absolute retention of samples which is given by the surface concentration of the silanol groups times the surface area of sorbent per unit column volume. Generally the larger the surface area of the silica, the higher is the absolute value of the retention under otherwise fixed conditions. The relative retention should be independent of the surface area (20) as long as both the surface concentration of the silanol groups and the pore structure are unchanged.

The surface area is a function of the average pore diameter. In general, the smaller the pores the higher is the surface area. In Fig. 2 the influence of the variation of the specific surface area of the commercial spherical silica LiChrospher between 250 and 6 m²/g on the separation of oligophenyls is demonstrated. The eluent was n-heptane having the same water concentration in all cases. The average pore diameter of the silica seems to have no effect on the chromatographic behavior and column efficiency as long as it is much larger than the molecular dimensions of the solute. If the two are of the same order of magnitude, exclusion effects can be important. In adsorption chromatography, silica gels having average pore diameters in the range from 60 to 800 Å are used most commonly. When the

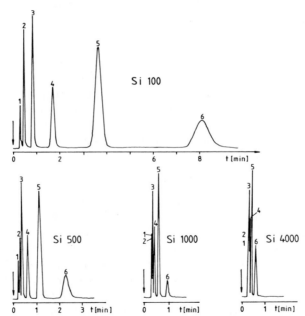

FIG. 2.   Effect of silica pore structure on separation. Stationary phases are 10 μm: LiChrospher SI 100, SI 500, SI 1000, and SI 4000; having 100, 500, 1000, and 4000 Å mean pore diameter, respectively. Flowrate and inlet pressure are 5 ml/min and 125 bar, respectively. Sample components 1, benzene; 2, diphenyl; 3, m-terphenyl; 4, m-quaterphenyl; 5, m-quinquephenyl; 6, m-sexiphenyl. (Courtesy of Merck AG.)

pore diameter is less than 60 Å, exclusion effects may be observed even with relatively small solute molecules (21, 22). The efficiency of such columns is lower than that of columns packed with silica having the same particle diameter but pore diameter larger than 60 Å.

In addition to the physicochemical parameters mentioned above the "history" of the silica is also an important factor which influences its chromatographic properties. The "history" includes the various factors to which the silica has been exposed during manufacturing, preparation, classification, and storage. The exposure of the silica to slightly different conditions can affect the silanol concentration and the detailed morphology of the surface and the difference cannot be quantified by the common physicochemical measurements used in practice. It is therefore not surprising that the absolute and relative retentions may change if another bottle of the same kind of silica is used.

Because the silica surface is weakly acidic, basic substances are more strongly retained than acidic or neutral solutes. Moreover, basic sample components may be adsorbed irreversibly or eluted as severely tailing peaks. In such cases the addition of a small amount of a basic substance, e.g., triethylamine, reduces tailing and prevents sample loss. By buffering the silica gel with a suitable buffer (e.g., boric acid/sodium borate buffer pH 8.0) either before packing it into the column or *in situ* by rinsing the pre-packed column with the buffer solution the tailing of the peak of a basic substance can be minimized (108). However, it should be kept in mind that at pH > 9 silica starts to dissolve. The efficiency of silica columns for the separation of acidic solutes can be improved in a similar way by adding either an acid (i.e., acetic acid) to the eluent or by coating the silica stationary phase with an acid buffer, e.g., sodium citrate/citric acid (108). For the efficient separation of acidic or basic solutes ion-pair chromatography is also recommended (109–111). In this case the variation of the counterion gives additional possibilities for optimizing selectivity.

Undesired chemical changes of sample components during separation on silica, however, are not a common problem in HPLC because of the usually short retention times. But as known from classical column chromatography (23) the acidic surface reaction can cause rearrangements in labile systems like terpenes, or may cause dehydration of hydroxyl compounds. Because the silanol groups are probable weakly acidic cation exchangers some heavy metal atoms with catalytic activity may also be present. Of course, these may also catalyze rearrangement of labile solutes (112); however, in such problematic cases the use of inert column material, e.g., glass, is highly recommended.

Chemical modification of the silica surface changes the properties and the selectivity of the adsorbent drastically. Nonpolar, hydrophobic derivatives of the reversed phase type are used by the majority of workers in the field, although polar derivatives having selectivity different from that of silica are widely used as well (24–27). Alteration of the surface by physical means such as silver ions (28) or trinitrofluorenone (29) can also yield a stationary phase having high selectivity. Argentation has been found particularly useful in the separation of saturated from unsaturated compounds.

By coating silica gel with a liquid phase either before packing into the column or *in situ* using thermodynamic control (30), a column suitable for liquid–liquid partition chromatography is obtained. The maximum loading with the liquid phase for partition chromatographic application is proportional to the specific pore volume of the solid. In practice silica with a pore volume of 1 ml/g can be coated with about 1 g of liquid phase per gram of silica. Such "heavily loaded" columns (113–115) exhibit several advantages: bleeding, i.e. the loss of stationary liquid phase, hardly influences the reproducibility of the retention of solutes if the eluent is presaturated with the liquid stationary phase. The peak capacity is high and the load capacity of such columns is at least an order of magnitude higher than that of "naked" silica. However, in preparative applications of such columns it should always be remembered that the isolated solute is contaminated with the stationary liquid phase. Therefore stationary liquid phases (like water, etc.) are recommended because they can be easily separated from the eluted samples.

The pore diameter of the silica support should lie between 100 and 500 Å. With pore diameters below 100 Å the H values of retarded peaks are much higher than for pore diameters of 100 Å or larger. At pore diameters above 500 Å, even at moderate eluent velocities, mechanical erosion of the stationary liquid phase diminishes column lifetime.

B. Alumina

Alumina suitable for chromatography is usually prepared from crystalline "bayerite" by dehydration and activation at 200–600°C. The alumina thus obtained is crystalline and consists mainly of γ-Al_2O_3. Upon heating at 900–1000°C, it is transformed into high-temperature forms. Finally at 1100°C α-alumina, i.e. corundum, which no longer possesses any chromatographic activity ("dead-burnt" Al_2O_3) is obtained. The specific surface area of γ-alumina ranges from 100 to 200 m²/g, and the pore volume is 0.2 to 0.3 ml/g. The average pore diameter is between 100 and 200 Å. Alumina prepared at higher temperatures has a smaller specific surface area (70–90 m²/g).

In its separation properties alumina resembles silica. However, the sorption mechanism is much more complex. In addition to the formation of hydrogen bonds to surface hydroxyl groups (~ 6 groups/100 Å^2 for alumina dried at 400°C), there is additional possibility of interactions of electron-rich molecules with Lewis acidic sites on partially coordinated aluminum ions. Thus, unsaturated molecules are generally more strongly retained on alumina than on silica. The relative retentions of condensed aromatic hydrocarbons are much higher on alumina. This may be illustrated for the separation of anthracene and chrysene on silica and on alumina (1). By varying the water content of the eluent n-heptane (cf. Section III,E), the k values of anthracene could be varied between 0.9 and 3.7 on both adsorbents. But whereas the relative retentions for this pair remained constant on silica ($\alpha = 1.9$), it increased strikingly from 2.3 to 8.6 on alumina on reducing the water concentration. This indicates that on changing the water concentration, the retention mechanism of these hydrocarbons remain the same on silica but certainly changes on alumina (2). Adsorption on these Lewis acid sites can be partially irreversible and may even cause sample decomposition or reaction (23). Deactivation with water, however, can eliminate such untoward phenomena (1, 2, 23).

Another kind of sample decomposition on alumina can be due to basic or acidic surface reactions. In the preparation of alumina the residual alkali content of bayerite (0.1–0.5% NaOH) is bound as sodium aluminate to the surface. As a result an aqueous suspension of such an alumina can have a pH value greater than 9. The sodium ions attached to the surface can be exchanged by other cations, consequently the so-called "basic" alumina is a cation-exchanger in water. Through careful neutralization, a "neutral" alumina, whose aqueous suspension has a pH of about 6.5, can be obtained. This is the material that should be used for chromatography because sample decomposition and irreversible adsorption are minimized. It should be mentioned that an "acidic" alumina, whose aqueous suspension has a pH of about 3.0, can be obtained by treating "basic" alumina with hydrochloric acid; "acidic" alumina acts as an anion-exchanger. These effects may also be noticeable in nonpolar eluents because water is almost always adsorbed on the surface. Retarded components may dissolve in the adsorbed water and thereby come in contact with an acidic or basic surface. This is illustrated by the fact that acidic compounds, like carboxylic acids and phenols, etc., are preferentially held on basic alumina compared to an acidic one or to silica (2), and are sometimes irreversibly retained. The basic or acidic surface reaction of alumina can catalyze the condensation reaction of carbonyl compounds like that of acetone to diacetone alcohol and to phorone, etc.

In contrast to silica, which loses its selectivity for the sorption of polar compounds when heated to temperatures above 200°C, the selectivity of

alumina increases with increasing the activation temperature up to 700°C
(*2, 62*). Such highly active alumina may be useful for gas chromatographic
separations but due to the omnipresence of water in the eluents, it can
hardly be used in liquid chromatography.

III. ROLE OF THE ELUENT

In liquid chromatography solute retention and resolution can be controlled in a wide range by selecting the eluent. In fact, the choice of the
proper solvent frequently affects a separation more than the selection of
the stationary phase. Generally speaking, the more strongly a given solvent is adsorbed by the stationary the higher is its eluent strength, i.e., its
capability to displace sample molecules adsorbed by the stationary phase.
Since the eluent is always present in great excess compared to the sample
and competes for the active surface sites, even relatively nonpolar solvents can be used to elute relatively polar substances.

There are several requirements an eluent must meet in order to be
useful in HPLC. First of all it should not hinder the detection of the
sample in the column effluent. The UV cut-off values of the most common
chromatographic eluents which can be used with UV detectors are shown
in Fig. 3. Most eluents listed in the conventional solvent series for chro-

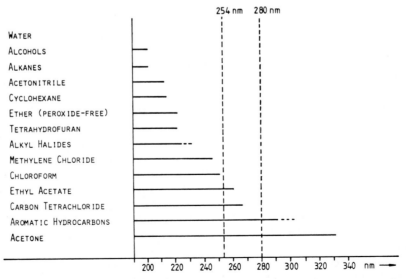

FIG. 3.   UV cut-off values for eluents most commonly used in HPLC (*1*). (Courtesy of
Springer-Verlag.)

matography are opaque in the UV region and cannot be used with UV detectors. Sometimes if for solving a specific separation problem such a solvent yields the best selectivity the use of another type of detector might be considered.

Of course, the sample components must be somewhat soluble in the eluent. This does not play as important a role for analytical as it does for preparative application. If the components are soluble only in solvents from which they are only weakly retained on a given stationary phase, and hence the separating efficiency of the system may not be sufficient, in such a case changing both the eluent *and* stationary phase is recommended. For instance either a reversed-phase chromatography or a partitioning system using ternary solvent mixture may be employed to attempt a separation. When two eluents or eluent mixtures have about the same "polarity" and selectivity, preference should be given to that having lower viscosity because it requires a lower pressure drop to achieve a given flow velocity. For instance in the case of aliphatic hydrocarbons cyclohexane or isooctane can easily be replaced by *n*-hexane or *n*-heptane.

It is essential that the eluent is inert and does not react with the sample components or the adsorbent. It must be borne in mind that aliphatic ketones like acetone or butanone easily undergo condensation reactions on active adsorbents and thus change the elution behavior of the chromatographic system.

A. Eluotropic Series

Empirical arrangements of solvents in order of increasing elution strength are called *eluotropic* series (*31*). The sequence is always arranged in order of decreasing sample retention; therefore, highest retention times are obtained by using the first members of the series. Conversely, the lower the position of the eluent in the series, the shorter are the retention times. For the last members of the series the "polarity" of the eluent becomes so high that the sample is no longer retained.

The most popular eluotropic series is that of Snyder (*2*) and shown in Table II. It encompasses most solvents used in liquid chromatography and differs only slightly from the first elutropic series of Trappe (*31*) that was established in 1940. The sequence is similar for silica and for alumina and the solvent strength increases roughly with increasing dielectric constant (*32*). The differences that appear in some cases may be attributed to selective or specific interactions between the individual eluents, adsorbents, and solutes used in establishing the series. In addition, traces of polar impurities may completely change the position of the solvent in the series (*9*). This is certainly the reason for the differences in the order with

TABLE II

Eluotropic Series of the Most Common Solvents Used in Liquid Chromatography[a]

Solvent	Eluent strength		Dielectric constant (Debye)	Viscosity at 20°C (cP)	Refractive index at 20°C	Cut-off wavelength (nm)
	Al_2O_3	SiO_2				
n-Pentane	0.00	0.00	1.84	0.235	1.358	200
n-Hexane	0.01	0.01	1.88	0.33	1.375	200
n-Heptane	0.01	0.01	1.92	0.42	1.388	200
Isooctane	0.01	0.01	1.94	0.50	1.391	200
Cyclohexane	0.04	—	2.02	0.98	1.426	210
Carbon tetrachloride	0.18	0.11	2.24	0.97	1.466	265
Diisopropyl ether	0.28	—	3.88	0.37	1.368	220
n-Propyl chloride	0.30	—	7.7	0.35	1.389	225
Benzene	0.32	0.25	2.28	0.65	1.501	290
Ethyl bromide	0.37	—	9.34	0.39	1.421	230
Diethyl ether	0.38	0.38	4.33	0.23	1.353	220
Chloroform	0.40	0.26	4.8	0.57	1.443	250
Methylene chloride	0.42	0.32	8.93	0.44	1.424	250
Tetrahydrofuran	0.45	—	7.58	0.46	1.407	220
Dichloroethane	0.49	—	10.7	0.79	1.445	230
Dioxane	0.56	0.49	2.21	1.54	1.422	220
Ethyl acetate	0.58	0.38	6.11	0.45	1.370	260
Nitromethane	0.64	—	35.09	0.65	1.382	380
Acetonitrile	0.65	0.5	37.5	0.37	1.344	210
Pyridine	0.71	—	12.4	0.94	1.510	310
n-Propanol	0.82	—	21.8	2.3	1.38	200
Ethanol	0.88	—	25.8	1.2	1.361	200
Methanol	0.95	—	33.6	0.6	1.329	200
Ethylene glycol	1.11	—	37.7	19.9	1.427	200
Water	Very large	—	80.4	1.00	1.333	180
Acetic acid	Very large	—	6.1	1.26	1.372	260

[a] The values of solvent strength, ϵ^0, were taken from Snyder (2). Some of the other solvent properties are listed.

some very important eluents such as dichloromethane, chloroform, and diethyl ether. Of course the moisture content also influences solvent strength.

The order of the eluents in Table II is in accordance with increasing ϵ^0 value, that is, the so-called "solvent strength" parameter of Snyder (2). The ϵ^0 values are determined by measuring the retention volumes of standard samples with different eluents and are related to those in n-pentane whose $\epsilon^0 = 0$ by definition. The difference in the ϵ^0 values of two eluents is proportional to the logarithm of the quotient of the two k values of the same sample in the two eluents, provided the adsorbent properties of the column packing are kept constant. It is essential that both eluents have an identical equilibrium water concentration (see Section III,C,3).

The numerical ϵ^0 values in Table II have been determined by using alumina as the stationary phase. Those for silica fall into the same sequence but are somewhat smaller, for instance, ϵ^0 is 0.32 for dichloromethane, 0.55 for acetonitrile, and 0.75 for methanol.

Perfluorinated hydrocarbons are much weaker eluents than aliphatic hydrocarbons and they can have negative ϵ^0 values. They are useful for separating nonpolar sample components such as aliphatic hydrocarbons, olefins, and benzene. However, their limited solubility for the higher alkanes ($> C_{15}$) and their high cost limits their use as eluents, particularly because the separation of such samples can also be performed by gas chromatography.

B. Mixed Eluents

Often the separation potential of adsorption chromatography can be fully exploited only by using solvent mixtures as mobile phases in chromatography. The elution strength ϵ^0 changes almost linearly with eluent composition when the difference in elution strength of the two solvent components is small, i.e., $\Delta\epsilon^0 < 0.2$. In these cases a nearly linear relationship of log k against eluent composition can also be expected. If the components of a mixture differ more widely in elution strength, the ϵ^0 value of the eluent increases markedly by adding small amounts of the polar component to the less polar solvent. For instance, addition of fractions of a percent of very strong eluting components like alcohols to n-alkanes will change the elution strength dramatically. As such eluent mixtures are difficult to prepare in a reproducible fashion and they may change upon storage (see discussion in Section II,C), adsorption chromatography has acquired the reputation of having poor reproducibility when such eluents are employed.

A further increase in the concentration of the stronger eluting component has less influence on elution strength. As was shown by Jandera and Churacek, the dependence of k on the volume percent of the stronger eluent yields a linear plot on double logarithmic graphs provided the $\Delta\epsilon^0$ values are not too small (33–35). For instance mixtures of n-heptane with dichloromethane give rise to a linear log k vs. log (vol %), plot especially at dichloromethane concentrations higher than 20% (v/v). Such a plot is shown in Fig. 4 for different samples with silica as stationary phase (36). Deviation from linearity is in these cases pronounced when the dichloromethane concentration is below 20%. The different slopes indicate that selective interactions of the samples with the silica surface also play a role.

Solvent mixtures of the same eluent strength can be prepared by adding a very small quantity of a strong eluent or a larger quantity of a less polar eluent to a weak eluent such as an aliphatic hydrocarbon. In Fig. 5, the compositions of eluents having equal eluotropic strength are depicted. The eluents are generated from the five most frequently used organic solvents

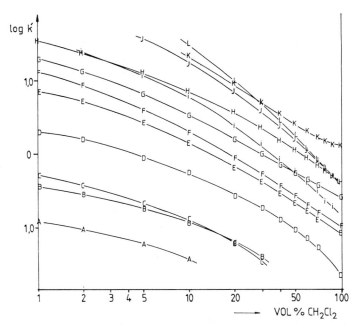

FIG. 4. Graph illustrating the dependence of capacity ratios on the composition of eluents containing n-heptane and dichloromethane. The stationary phase is LiChrosorb SI 100 silica gel. Sample: A, ethyl benzene; B, anthracene; C, m-terphenyl; D, nitrobenzene; E, benzonitrile; F, benzophenone; G, acetophenone; H, 1,4,5-xylenol; I, o-nitroaniline; J, m-nitroaniline; K, Δ^4-cholestenone; L, p-nitroaniline.

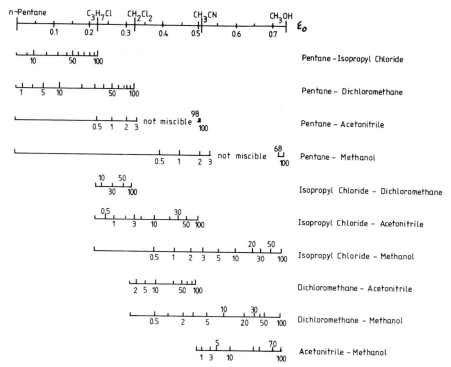

FIG. 5. Solvent strength of eluent mixtures for HPLC (*1*). The scales are similar to that given by Saunders (*38*). (Courtesy of Springer-Verlag.)

in HPLC and the composition scales are presented in a manner similar to that of Neher (*37*) and Saunders (*38*). Pentane may be substituted for by hexane or heptane without changing the eluotropic strength of such mixtures at all. Each scale corresponds to the full practical composition range of binary mixtures. As seen aliphatic hydrocarbons are not miscible with methanol and acetonitrile in every proportion. If eluents having eluotropic strengths intermediate between those of pentane and of propyl chloride are required, they can be produced by mixing propyl chloride and pentane in 5 or 10% increments. Of course the same eluotropic strength can be obtained by adding much smaller amounts of the more polar dichloromethane to pentane. It is seen that the eluotropic strength of a 1 : 1 mixture of pentane and propyl chloride is equal to that of an eluent obtained by adding about 10% dichloromethane to pentane. If acetonitrile or methanol is added to pentane to adjust the same solvent strength, amounts far less than 0.5% of these eluents are needed. Similar mixtures can be prepared from propyl chloride and dichloromethane, acetonitrile, or methanol, etc. This solvent strength scale gives only an approximation of the eluent strength. Sometimes the eluent has to be adjusted by slight changes in the composition to obtain the proper k values. Generally spoken an increase of 0.05 in elution strength causes the k values to decrease by a factor of 3–4. Ternary solvent mixtures are rarely required to control selectivity in LSC. Of course, other eluents such as ethyl acetate or diethyl ether (*38*) can also be included in these mixtures. However, eluents containing ethyl acetate at concentrations higher than 10% (v/v) cannot be used with UV detectors. One should, however, not be surprised if the absolute and relative retentions for a given sample mixture change when different eluent mixtures of the same eluotropic strength ϵ^0 are used. These phenomena have been attributed to "secondary solvent effects" (*2*). The different slopes and intersections of the curves obtained with different samples, shown in Fig. 4, are attributed to such secondary solvent effects. Another example was observed by Snyder in the separation of acetonaphthalene on alumina (*39*). With a 1 : 1 benzene–pentane mixture as the eluent, acetonaphthalene is retarded more strongly than dinitronaphthalene and the relative retention, *a*, is 2.0. On the other hand, with a

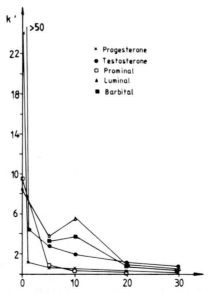

FIG. 6. Graph illustrating the effect of eluent composition on solute retention (*1*). The sample components are eluted with methylene chloride containing increasing concentration of ethyl acetate from LiChrosorb SI 100 with covalently bound dinitrophenyl groups at the surface. (Courtesy of Springer-Verlag.)

pentane–dichloromethane mixture containing 23% (v/v) CH_2Cl_2 as the eluent, the k value of acetonaphthalene (5.59) is almost unchanged, but in this case dinitronaphthalene is eluted later and the relative retention, α, is 1.05. When pentane containing 0.05% (v/v) dimethyl sulfoxide is used as the eluent, acetonaphthalene is only weakly retarded whereas dinitronaphthalene is eluted later and the selectivity is high as the value of α is 3.5.

Such drastic changes in selectivity may also occur with eluents prepared from two solvents of similar elution strength, such as dichloromethane and ethyl acetate. In Fig. 6 the effect of ethyl acetate concentration in dichloromethane on the retention behavior of some steroids and barbiturates is illustrated (1). The k values of the steroids are reduced drastically by the addition of 1% of ethyl acetate to methylene chloride. Some barbiturates behave similarly, whereas the behavior of others is dominated by other solvent effects that also manifest themselves in the widely different solubilities of the substances in methylene chloride and ethyl acetate.

The difficulties in reproducing adsorption chromatographic separations are illustrated by the following example of steroid chromatography in silica gel. With plain methylene chloride the k value for progesterone is 24, whereas that of testosterone exceeds 50. The addition of 1% (v/v) dry ethyl acetate to the methylene chloride reduces the k values to 1.2 and 4.4, respectively. A slight change in the ethyl acetate concentration, which may readily occur due to evaporation, produces a marked change in the k values and relative retentions may also change. In some cases the elution order may even be reversed. For sake of reproducibility, eluents should be prepared from solvents having only small differences in elution strength, so that no demixing occurs on the adsorbent and the k values are insensitive to minor changes in the eluent composition. On the other hand, the selectivity of a system is usually maximized by using eluent mixtures composed of eluents differing widely in eluotropic strength. As discussed above, very small concentrations [less than 3% (v/v)] of the very polar components are required in these cases to adjust selectivity. The price one pays for maximizing selectivity is a reduction in reproducibility, because such eluent mixtures are difficult to prepare, to preserve, and to maintain constant over a long period of time. Of course, the problems involved are similar to the problems of the so-called "modulated" systems, a discussion of which follows.

C. The Role of Water and Other Modulators

As shown above, additives to the eluent, which are much more polar than the solvent itself, influence the adsorption of substances strikingly even in the very low concentration range, as they are adsorbed preferentially. Such additives modulate the retentiveness of the chromatographic system and are therefore called modulators.

The most important and almost ubiquitous modulator is water. The modulating property of water was recognized early, and numerous attempts have been made to standardize the water content of adsorbents (9, 40) and to relate it to the equilibrium water content of eluents (41, 42). The procedures, which involve the addition of given amount of water to an adsorbent of a certain initial activity, have only limited applicability in HPLC. For the considerably greater amount of eluent flow per gram of adsorbent, as compared to that in classical column chromatography, and the longer use of the column, the water content of the actual eluent governs the adsorbent activity more strongly than the history of added water. In practice of HPLC equilibration is relatively fast between the adsorbed water and that present in the eluent. With columns packed by one of the customary suspension techniques, the activity of the stationary phase is adjusted by perfusion with an eluent having the proper water content.

The effect of the moisture level in the eluent on the retention behavior is most pronounced with the least polar solvents, such as the hydrocarbons, in which water has a very limited solubility, usually less than 100 ppm. Conversely, the greater the solubility of water in the organic solvent is, the smaller the effect. For example, the water content of methylene chloride may vary by several parts per million without appreciably changing the k values as the solubility of water in CH_2Cl_2 is about 0.2%.

In Fig. 7 the concentration ranges in which the different liquids have been used as modulators in aliphatic hydrocarbons and in dichloromethane are shown. The maximum concentration of the modulator water is limited by its solubility in the eluent. For instance water-saturated n-heptane contains a maximum of 0.01%, whereas the solubility of water in dichloromethane is about 0.2%. These are very low concentrations in comparison to those of the other organic modulators used. Mainly alcohols and other polar organic liquids have been used to modulate LSC systems and/or to suppress the effect of the omnipresent water on solute retentions. Glycol, glycerol (41), methanol and other aliphatic alcohols (43–45) up to n-butanol (61), tetrahydrofuran (46), acetonitrile (46, 47), and ethyl acetate (46) have been used as modulators in aliphatic hydrocarbons and dichloromethane. However, it is almost impossible to differentiate between the relative influence of water and the organic modulator purposely added to the eluent. Of course, if eluent and modulator are completely miscible, the concentration of the modulator can be increased to much higher concentrations; but such eluents are called mixed solvents or eluent mixtures rather than modulated systems. For instance, the range

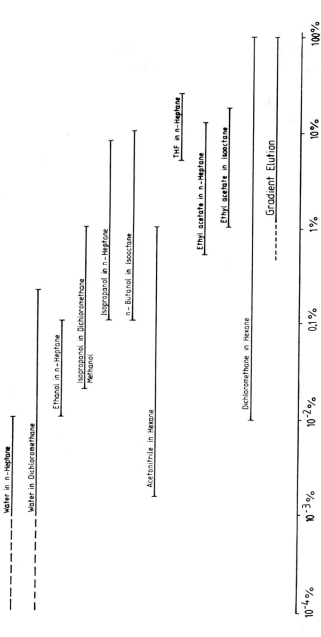

FIG. 7. Comparison of concentration ranges of modulators and eluent mixtures in liquid–solid chromatography.

in which dichloromethane has been used to modulate n-hexane as eluent corresponds to the usual range of eluent mixtures in isocratic or programmed analysis (gradient elution). Most of the problems involving "modulated eluents" stem from the difficulties inherent in determining and reproducing the very small concentrations of polar solvents that are usually added to the eluent as well as from the problem of controlling precisely the water concentration in the eluent.

Because the poor-reproducibility reputation of chromatography on silica and alumina can be, in part, traced to these problems with such modulated systems, broad space is given to their discussion.

1. Water-Modulated Systems

Unless subjected to a drastic drying procedure and stored under special conditions, each eluent contains water at a concentration level that can significantly affect retention behavior. Even commercial "pure solvents" have significant but variable amounts of moisture. To maintain an eluent with a definite water concentration is almost impossible because the dry or wet container surface alone may alter the concentration appreciably. Furthermore, changes in the water content on opening the flask or pouring the eluent into another container are virtually inevitable.

Ordinarily water content may be determined by a Karl Fischer titration, but for the low concentrations in aliphatic hydrocarbons this method is filled with difficulties. To obtain even modest accuracy such as ± 5 ppm at H_2O levels about or below 30 ppm as specified by the German Standard DIN 51777 relatively large eluent volumes (> 200 ml) are required. On the other hand, the absolute and relative retentions are sensitive to 1 to 2 ppm of water; consequently, two different batches of eluent with identical water content according to the above mentioned Karl Fischer titration can yield different retention values. Therefore, it is impossible to use titration as the sole method for eluent standardization. The reproducibility of adsorption chromatography, however, can be increased by continuously recycling a sufficiently large volume in a closed system to maintain equilibrium.

Snyder (44) recommends the use of eluents with the same percentage of water saturation, say 50%. These partially water-saturated eluents are called *isohydric* and can be prepared by blending corresponding amounts of "dry" and "water-saturated" mobile phases. In practice, however, the defined status "dry eluent" involves some of the problems with determination and control of water content already discussed. Surprisingly, it can be quite difficult to saturate solvents with water completely by simply shaking or stirring them with water (48) when the solubility of water is very low such as in hydrocarbons. According to Snyder, after conditioning a column with an isohydric eluent, an equilibrium is established between the water in the eluent and that on the adsorbent, thereby yielding a constant and reproducible covering of the surface by water. The equilibrium coverage remains unchanged when other eluents with the same degree of water saturation are substituted, and as a result the usually long reequilibration times associated with solvent changeover can be avoided. Similarly, the initial equilibration times with nonpolar eluents may be shortened by first conditioning the column to the desired water saturation with a mobile phase having an appreciable solubility for water, followed by the nonpolar eluent with the same relative saturation. For optimum surface coverage (2), assuming monolayer water adsorption, 25% water-saturated eluents are recommended for alumina and 50% saturated for silica (48).

A convenient way to improve the reproducibility of adsorption chromatography with regard to water content of the chromatographic system is to condition eluent and column together. This is accomplished by recycling the eluent through an appropriate apparatus so that the moisture becomes uniform in the entire chromatographic system. Such a moisture control system, MCS, can be readily assembled and installed upstream of the pump of the liquid chromatograph (49). The main component of MCS is a thermostated funnel mounted on the eluent reservoir filled with coarse silica or alumina which are impregnated with known amounts of water in the range from 3 to 20% (w/w). The eluent is introduced via the funnel and recycled through the MCS and the column. Depending on the initial water content, the eluent is either dried or moistened to an equilibrium value in this process. The original amount of water on the silica or alumina in the funnel and the temperature of MCS determine the absolute and relative water content of the eluent, and consequently, the absolute and relative sample retentions obtained with the column. The time required to attain equilibrium depends, of course, on the volumetric flowrate of the eluent and the solubility of water therein. However, it is independent of the nature of the adsorbents, i.e., whether silica or alumina is used in the column or in the MCS.

The time of the treatment required to "dry" a column, that is, to obtain increased k values is always longer than to "moisturize" the column. As a guideline for n-heptane, approximately 18 h is required to run "wet," whereas 48 h is needed to attain extreme dryness at a flowrate of about 4 ml/min. For methylene chloride the time required is substantially shorter, e.g. 6 and 14 h, respectively. If sample components are initially present in the column they are removed by the MCS so that a purging of the column is also accomplished.

As in the previous method, the equilibration times with hydrocarbon eluents can be shortened by first conditioning the column with an eluent having an appreciable solubility for water. After the column has attained equilibrium, a second MCS already preequilibrated with the desired eluent is connected in place of the first. Both MCS should hold adsorbent containing the same amount of water.

As a result of increasing the temperature of the MCS funnel the water concentration of the recycled eluent also increases. Consequently, the k values of the sample components can be varied by changing the temperature of the MCS funnel if the temperature of the chromatographic column is kept constant.

Figure 8 shows a separation of condensed aromatic hydrocarbons in which the water content of n-heptane used as the eluent was controlled by MCS. The water concentrations as determined by Karl–Fischer titration are also included for comparison. If all variables such as the water content of the silica in the MCS, the temperature of both MCS, and the chromatographic column are kept constant, the reproducibility of the k values is $\pm 2.5\%$.

Changing the water content of the chromatographic system permits the change of the k values of the samples. However, increasing the k values by decreasing the water concentration should be done carefully, because the linear region, i.e. the independence of k values on sample size, decreases also. The dryer the eluent the smaller the sample size has to be. Better and less dangerous, however, is to vary the k values by changing eluent composition.

The isohydric concept and the MCS are equivalent in adjusting and guarantee the constancy of the water concentration of the eluent.

2. Modulators Other than Water

Because of the problems encountered with the water system, the use of aliphatic alcohols, e.g., methanol, ethanol, and isopropanol, as modifiers of the adsorption strength has been recommended (44, 45, 50, 51). Usually, between 0.01 and 0.5% (v/v) alcohol is added to the eluent. As an example, the k values for the benzyl alcohols on a silica column are in the same range when eluted with dichloromethane containing either 0.1% water (50% water-saturated) or 0.15% methanol or 0.3% isopropanol (45). The preparation and preservation of these alcohol–eluent mixtures is accompanied by problems similar to those discussed with water-modified eluents. Also, column equilibration is slow (44). The efficiency of columns operated with alcohol-modified eluents is generally lower than that of water-modulated eluent system. At some alcohol concentrations, distorted peaks with tailing or frontal asymmetry have been observed (44), but other workers using another silica could not verify this observation (61).

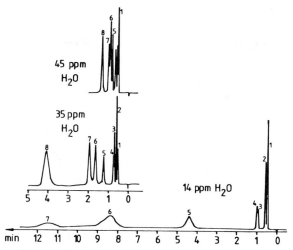

FIG. 8. Effect of the water concentrations in the eluent on the separation of benzologs with an alumina column. Water content was adjusted via a MCS containing neutral alumina (Woelm) coated with 4.5%, 6%, and 9% (w/w) water. The reservoir contained 500 ml of heptane and was maintained at 25°C. The analytical column is packed with neutral alumina (Woelm) $d_p \sim 5~\mu m$. Sample components: 1, tetrachloroethylene; 2, benzene; 3, naphthalene; 4, biphenyl; 5, anthracene; 6, pyrene; 7, fluoranthene; 8, 1,2-benzanthracene. [From Engelhardt and Böhme (49), courtesy of Elsevier.]

Less polar organic compounds such as acetonitrile and tetrahydrofuran have also been used as modulators with nonpolar aliphatic hydrocarbons as eluent. With decreasing polarity of the additive the concentration of "modulator" in the nonpolar eluent has to be increased in order to obtain similar retention values. Approximately the same k values for diphenoxybenzene are achieved with either 0.05% (v/v) of acetonitrile or 10% (v/v) of dichloromethane in n-hexane (47). As discussed above, the latter case corresponds to an eluent mixture rather than a "modulated" system.

Of course, in all these cases, the influence of the always-present "modulator" water cannot be neglected. Its influence increases with decreasing polarity of the added "modulator." Even with high modulator concentrations such as those used in the case of n-butanol or ethyl acetate, insufficient control of the water content leads to loss of reproducibility and especially to unpredictable selectivities (61).

3. Modulators and Selectivity

There is no doubt that even small modulator concentrations influence both absolute and relative retentions and that selectivity does change with the type of modifier used. With silica as stationary phase and by using a given type of modifier, e.g., aliphatic alcohols selectivity does not change appreciably by changing the modifier. In comparing water-modulated systems with those having acetonitrile as the modulator, significant variations in selectivity can be observed (47). Consequently, if the k values are of the same order of magnitude, the relative retentions of sequential peaks can differ significantly upon changing from water to an organic solvent as the modulator.

The influence of modulator concentration on selectivity also depends on the nature of stationary phase. As has already been discussed (cf. Section II,B), the relative retentions of polynuclear aromatic hydrocarbons are not affected by changing the water content in the n-heptane eluent with silica as the stationary phase, but vary widely with alumina as the stationary phase if the same moderator concentrations are used (1). Variations in the elution order are also likely to occur (2, 4, 39).

4. Problems with Modulated Systems

When gradient elution is employed with polar stationary phases, column regeneration is tedious and time consuming. Usually it is not difficult to remove the stronger eluting component of the mobile phase. However, as all eluents are contaminated with water and each solvent has a different equilibrium water content, which yields a given coverage of the stationary phase with water, the situation is particularly complex because the amount of adsorbed water in the column is constantly changing. Depending on the water content of the polar solvent, water is released or removed by the adsorbent. After returning to the starting eluent, some time is required to restore the initial conditions with respect to the adsorbed quantities of water. The time needed to achieve equilibrium is a function of the water content of the eluent, of the properties of the stationary phase, and, of course, of the flowrate. Reequilibration by running a reverse gradient is more rapid than an instantaneous return to the initial conditions (52, 53). Regeneration time may even exceed analysis time in the case of programmed analysis; therefore, it is recommended to use a standard procedure for column regeneration after each analysis instead of establishing column–eluent–modulator equilibrium prior to each analysis. Isohydric, i.e., half-saturated eluents, do not prevent these problems in gradient elution; on the contrary, they cause additional problems because the isohydric water concentration shows no linear dependence on eluent composition. If, for example, half-saturated dichloromethane is added to isohydric hexane or heptane, very soon the solubility of water in this mixture is attained, and water droplets will precipitate in the gradient mixer or on top of the column.

In adsorption chromatography with a plain eluent containing no modulator increasing column temperature results in a decrease of retention. Under practical conditions, however, solute retention can increase, decrease, or remain about the same with rising column temperature (50, 54, 55), and the effect depends on the adsorbent eluent and modulator combination employed. The observed behavior is due to the fact that temperature changes affect the equilibrium distribution of the modulator between the stationary phase and the eluent. Water and isopropanol modulators are desorbed from the silica stationary phase and removed from the column upon raising the temperature. Consequently, upon lowering the temperature to its initial value the retention values will be greater than those observed before the temperature increase.

If the mobile phase entering the column always has the equilibrium modulator concentration the polarity, i.e., the elution strength of the eluent, increases upon increasing the temperature. The modulator concentration can be maintained at its equilibrium value by pumping the mobile phase through a precolumn that is kept at the same temperature as the separating column. The precolumn should be packed with the stationary phase proper and have a sufficiently high capacity. With such a system "gradient elution" can be carried into effect simply by temperature programming. In a chromatographic system containing silica in equilibrium with n-heptane containing 0.1% (v/v) of isopropanol at 30°C, the modulator concentration increases to 0.5% upon raising the temperature to 80°C and having a sufficiently large precolumn (51). The disadvantage of this approach stems from the long time required for establishing equilibrium after returning to the initial conditions, a feature common to all modulated systems.

D. Partitioning

Distribution of a polar compound between the bulk eluent and the surface of the active adsorbent can be used to load the porous column packing with variable amounts of a stationary phase. Eventually, a column containing an active adsorbent can be transformed into a "liquid–liquid partition" column. In some cases, such as with prepacked columns, this is the only way to prepare a partition-chromatographic system. If ternary mixtures containing a hydrocarbon, e.g., heptane or isooctane, an alcohol such as ethanol or isopropanol, and water are used, the polar constituents of this mixture are preferentially adsorbed by the stationary phase, especially if its surface area is large. In this case the eluent mixture decomposes and forms a polar stationary liquid rich in water and alcohol in the pores of the stationary phase. The greater the polarity differences between the components of the eluent, and the greater the specific surface area of the support, the greater is the uptake by the support of the more polar component of the liquid phase. With water-saturated dichloromethane the total pore volume of silica can be filled with water (56). The coating procedure in preparing such *in situ* (30) or "naturally" (57, 58) coated columns can be followed by determining the consecutive decrease of the total porosity of the column. If the amount of silica in the column is known the coverage can be calculated (30, 56).

The sample capacity of such systems can be an order of magnitude greater than that of "pure" adsorption systems. Up to 10^{-3} g of sample/g of liquid stationary phase can be applied without appreciably increasing band broadening.

Such *in situ* coated columns with an equilibrium distribution of liquid phase have long operational lifetimes because the amount of liquid stationary phase remains constant in the column as long as its concentration in the mobile phase and the temperature remain constant. The column can be coated successively with different liquid phases after removing the stationary liquid phase with a suitable solvent and reactivating by treatment with dry eluent before recoating.

This approach is useful for routine studies only when equilibrium coverage of the support is maintained because only then do retention times and relative retentions remain constant. Obviously, one can operate above the "equilibrium coverage" if such coating can be achieved, for instance, by using precoated stationary phases. The k values of samples undergo a characteristic change during the coating procedure as shown in Fig. 9. The shape of the curves is typical for mixed adsorption and partition mechanisms (12, 60). Initially the k values decrease sharply with

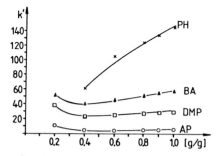

FIG. 9.   Graph illustrating the dependence of the capacity ratios on the amount of liquid loading in the column. Stationary phase, LiChrosorb SI 100; liquid phase, 3,3'-oxydipropionitrile; eluent, *n*-hexane. Samples: AP, acetophenone; DMP, dimethylphthalate; BA, benzyl alcohol; PH = phenol. [Reproduced with permission from Engelhardt and Weigand, *Analytical Chemistry* (21). Copyright by the American Chemical Society.]

increasing coverage of the support with the liquid stationary phase and after passing through a minimum they increase again. The decrease is caused by the coverage of the active surface sites by the stationary liquid while the rise corresponds to the increase in the amount of stationary phase in the column, i.e., to the increase in the so-called phase ratio. However, even in this region adsorption by the solid surface is frequently involved in the retention mechanism (21). Therefore, not only the absolute, but also the relative retentions, change by changing the liquid loading. This discussion is restricted to stationary phases with large surface areas (>20 m^2/g). The contribution of surface adsorption can be neglected when kieselguhr-based supports such as Chromosorb® are used since these packings have relatively small specific surface area and large pore volume. They also can be coated with large amounts of stationary liquid phases in the range from 30 to 50% (w/w); however, they have to be dry-packed as precoated particles, because with the commonly used suspension techniques such column materials would be stripped by the suspending medium and/or in the process of column conditioning.

E. Proper Choice of the Eluent

The selection of a solvent for a new separation problem, even today, is a matter of trial and error. The application of theory (2) with the additional application of the solubility parameters (63–65) makes it possible to estimate the composition of appropriate solvent mixtures for the separation of relatively simple compounds. In order to calculate the necessary solvent strength, however, a set of experimental data concerning the behavior of the sample components, the adsorbent, and the elution strength of the eluents with the specific adsorbent are necessary. Others (38) recommend a graphical method as a time-saving alternative to both calculation and the trial-and-error approach to obtain a first approximation of the eluent composition appropriate for the separation of a given sample.

The elution strength should be adjusted so that the sample components of interest are eluted within a reasonable time. Generally $k > 0.3$ are necessary to separate the sample components from unretained substances including the solvent. On the other hand, values of $k < 10$ are desired to reduce analysis time and minimize sample dilution which increases with increasing k value. In high-speed analysis optimum conditions are obtained if the k values of the sample components vary between 0.3 and 3.

In the case of mixtures of unknown composition, it is best to start with a moderately polar eluent such as methylene chloride. If the k values are too small the polarity of the eluent is too high and a less polar mobile phase, which is higher in the eluotropic series, must be selected. Another possibility is to reduce the water concentration in the eluent, thereby raising the activity of the adsorbent and increasing the k values. This alternative approach, however, should be employed with care because it may be accompanied by a pronounced nonlinearity isotherm of the adsorption and a concomitant tailing of the peaks. When the sample components are excessively retained, the elution strength, i.e., the polarity of the eluent, must be increased. This can be accomplished by using a solvent that is lower in the eluotropic series than the first. Excessive retention times can also be shortened by raising the water content of the eluent and thereby lowering the adsorbent activity. Usually pure eluents do not give optimum retention and resolution. Therefore eluent mixtures have to be prepared by blending eluents of lower eluting strength with those of higher eluting power. The eluting strength of the mixtures can be estimated by applying plots like shown in Fig. 5. The selectivity of the system can be optimized as discussed above. Adjusting the eluent in this fashion, however, is very tedious because of the long time required to reequilibrate the separation system after a solvent change.

Knowledge of the sample composition and the structure of its components simplifies the choice of the eluent and facilitates the prediction of the approximate elution order. Additionally, reference can be made to literature of classical column chromatography (7, 9, 66, 67). The exploitation of such results in HPLC represents no difficulties provided the eluent can be used with the detector of the liquid chromatograph. It should be kept in mind that the classical results are useful only to establish the chromatographic system for a particular separation but not to predict the exact retention data.

The transfer of thin-layer chromatographic (TLC) data to column methods is more difficult. The direct conversion of R_f values from TLC to the corresponding retention parameters such as k values or retention volumes, which are used in HPLC, is unreliable even though the R_f and k values are formally related by $R_f = 1/(1 + k)$ (68). This relationship holds only under identical equilibrium conditions. In a column the phase ratio remains constant, but in TLC the volume of the mobile phase decreases toward the solvent front. Similarly, the phase composition in a column is constant, whereas in TLC a demixing (frontal chromatography) of the developer solvent (eluent) is superimposed on the separation of the sample components. The migration rate of the solvent front, which is equivalent to the flowrate of the developer liquid, diminishes with increasing distance from the point of immersion into the solvent reservoir. Hence, it is impossible to operate TLC isocratically; therefore the relationship shown above for R_f and k cannot hold. For the same reason, the plate heights or plate numbers evaluated from the size of the spot in TLC, are not directly comparable to those measured on the chromatogram in column chromatography. The attainable efficiency in TLC is related to particle diameter in the same way as in column chromatography, but H values are defined only for constant eluent velocity, phase ratio, and isocratic conditions (69). In most cases the binders used in the fabrication of TLC plates may exert further effects to make the properties of the stationary phase different from that of the pure adsorbent.

For all these reasons it is not surprising that difficulties arise in attempting to transfer results obtained in TLC to the column chromatography. In addition, the R_f values depend on the developing technique used, e.g. linear or circular development, and on the devices used such as saturation tank, sandwich chamber, or streaming system. Nevertheless qualitative transfer of retention behavior from a TLC system with a nonpolar single-component solvent to a column system is relatively straightforward, although the water content of the layer, which is governed by the humidity, should be matched to that of the column preferably with MCS. The separation can then be optimized by minor adjustment of the eluent polarity. In agreement with the above discussion a transfer of TLC data obtained with multicomponent eluent, whose components have widely different polarities, is much more difficult inasmuch as solvent demixing alone prevents accurate estimation of the solvent and phase compositions that are required to characterize the spot migration. It was shown (70) that using a two-component developer to separate various azo dyes yielded different results with the saturated tank and sandwich chamber techniques, and that these also differed from those obtained in column chromatography with the same eluent. It is obvious that the transferability of TLC separations to columns diminishes with an increasing number of the components of the eluent used in TLC.

Specially developed equipment for so-called HPTLC (116) allows the complete control of humidity and the preequilibration of the thin layer with the mobile phase. This expensive device permits the simulation of column chromatographic conditions on thin-layer plates. Therefore, systems established with such equipment can be transferred to column chromatography with minor alterations.

A miniaturized circular chromatographic technique on a TLC plate can be conveniently used to select an appropriate mobile phase [117]. A few microliters of the sample solution are spotted on the plate. After the evaporation of the solvent in the center of each spot a few microliters of a pure eluent or an eluent mixture is transferred by using a glass capillary or a fine-tipped medicine dropper. On a 5 × 20 cm TLC plate about 10 different spots can be applied and, of course, the same number of different eluents can be checked. If the elution strength of the test mixtures is too great the sample components will sit unresolved on a concentric circle around the sample spot and the eluent starting point. If the elution strength of the eluent is too small, the sample components will remain on the initial spotting place. Only if the elution strength is properly selected will the sample components be separated from each other forming concentric rings around the initial sample spot and the application point of the eluent. Colorless samples can easily be visualized by inserting the plate in a chamber containing iodine vapor. Of course, all other reagents to develop the spots can be used alternatively.

To supplement the column chromatographic results, TLC equipment should be available in every HPLC laboratory where silica or alumina columns are routinely used. A rapid TLC run will readily establish whether all sample components can be eluted with the selected eluent; that is, no sample remains at the starting point, as seen after treatment by one of the many detection reagents available. Moreover, selective spray reagents can furnish additional information about the sample composition.

IV. EFFECT OF SAMPLE STRUCTURE ON RETENTION

As already discussed, chromatography on silica or alumina stationary phases nowadays should be restricted to the separations of organic soluble compounds. Of course, also ionic and other water-soluble substances have been and will be separated with such phases. But in order to obtain highly efficient separations (i.e., elution of solutes without tailing) the eluent composition has to be selected and adjusted carefully. Such polar components are better separated by applying bonded phases or ion exchange or ion pair chromatography.

Also separation of homologs is better achieved by applying bonded phases of the nonpolar or reversed phase type because the methylene groups are only weakly attracted by the polar stationary phase which is already covered by eluent molecules.

The selectivity of silica and alumina is optimum for the separation of compounds of different chemical type differing in the kind and numbers of functional groups.

Chromatographic retention is largely determined by the basic structure and number of functional groups in the solute molecule. In the simplest case every single atom and group of the molecule contributes to the strength of adsorption by specific interactions with the surface. The functional group must be able to interact with the adsorbent surface, i.e., there should be no steric hindrance. The type of the functional group determines the elution order. The retention of compounds (substituents) $R-X$, where R is the organic moiety and X a functional group, increases according to the following empirical sequence of the functional groups:

alkyls < halogens (F < Cl < Br < I) < ethers < nitro compounds

   < nitriles < tertiary amines < esters < ketones < aldehydes

   < alcohols < phenols < primary amines < amides < carboxylic acids

   < sulfonic acids (9).

Within this series transpositions may occur, depending on whether the functional group is attached to an aliphatic or aromatic residue. For example, if resonance with the benzene ring increases the electron density on the constituent, the interaction between the stronger "basic" group and the "acid" adsorption surface is enhanced considerably. The hyperconjugation of the alkyl side chain in toluene or ethyl benzene with the benzene ring is responsible for the stronger retention of these compounds as compared to that of benzene, for the alkyl groups themselves do not contribute appreciably to retention. The strength of retention is increased by the introduction of a second functional group. The magnitude of the change in retention, however, is difficult to predict. If the functional groups of a given molecule can interact with each other, e.g., via resonance or hydrogen bonding, the strength and type of their interaction with the adsorbent surface can also be altered. The retention behavior of o- and p-nitrophenols is often used to exemplify such phenomena. Because of intramolecular hydrogen bonding, the ortho isomer has a substantially lower retention value than the para isomer.

Chromatography on silica and alumina is unique among the liquid chromatographic methods in providing maximum selectivity for the separation of isomers. It is no problem to separated m- and p-dibromobenzene ($\alpha = 1.8$ in pentane) (2) or the three nitroanilines (19) on silica or alumina stationary phases with dichloromethane as eluent.

In aromatic compounds the effect of a functional group on retention may be enhanced or diminished by resonance. As illustrated in Fig. 4 the curves for monofunctional benzene derivatives exhibit a more or less parallel slope on the plot of log k against log eluent composition, whereas the multifunctional derivatives, e.g., nitroanilines, cholestenone, show distinctly different slopes. This demonstrates how difficult the prediction of retention behavior in adsorption chromatography is. The greater the deviation of the structure from the simple model compounds used for establishing the rules, the more difficult the prediction becomes.

The retention also depends on steric effects. For maximum interaction the adsorbed molecule must orient itself parallel to the adsorption surface. Bulky alkyl groups adjacent to the functional group diminish the retention. The cis isomer of a compound is always retained more strongly than the trans isomer and the separation of *cis-* and *trans-*azobenzene can serve as a classical example. Cyclohexane derivatives and steroids with substituents in equatorial position are stronger adsorbed than those having the same functional groups in axial position. Furthermore, the strength of adsorption increases with increasing molecular weight, i.e. molar volume, especially in nonpolar systems. Therefore a ketosteroid is, for instance, always more strongly retained than a comparable cyclohexanone derivative. The effect of the dispersive forces responsible for this effect, however, decreases with increasing polarity of the eluent.

To a first approximation samples containing a functional group like a nitro group or one preceding the nitro group in the empirical series presented above can be eluted with aliphatic hydrocarbons. By adjusting the water concentration to a medium level polynuclear aromatic hydrocarbons can also be eluted in reasonable time. With such an eluent, it is impossible to separate aliphatic and olefinic hydrocarbons for which fluorinated hydrocarbons should be the preferred eluent. Nitriles and carbonyl compounds such as aldehydes, ketones, and esters can be eluted with dichloromethane, regardless of whether or not they are substituents in an aromatic ring or attached to a large molecule, e.g., to a steroid skeleton. Eventually a few percent of ethyl acetate, methanol, or acetonitrile are required to reduce retention. If the functional groups increase the water solubility of the sample the use of nonpolar stationary phases is recommended in view of widespread employment of reversed phase chromatography.

V. ANISOCRATIC ANALYSIS

Complex mixtures containing components whose k values differ considerably cannot be separated and eluted within reasonable time by using a single eluent under isocratic conditions. To resolve both weakly *and* strongly retained components and to elute them as easily recognizable peaks, the separation conditions have to be changed during the run by one of the various programming techniques. In this way, each sample peak may be eluted under optimal conditions. A detailed discussion and comparison of the different methods has been given by Snyder (*71*); therefore only some of practical aspects will be treated here.

In HPLC the following variables may be programmed, that is, changed in the course of a chromatographic run: (*a*) mobile phase velocity (flow or pressure programming); (*b*) column temperature (temperature programming); (c) stationary phase (i) by changing adsorbent activity and (ii) by means of coupled columns; and (d) eluent composition (gradient elution).

Both the fundamental and the technical difficulties, which are associated with the use of these programming techniques, increase in the above order. However, the enhancement of separation capability and the range of applications also increase in the same order.

It should be stressed again, that the *resolution* of a given pair of sample zones is always lower in programmed than in nonprogrammed analysis under otherwise identical conditions. The analysis time, however, can considerably be shortened for the separation of mixtures whose components are strongly retarded at isocratic elution and the resolution under such conditions would be much greater than necessary. Moreover, when the elution bands are sharpened in the process, e.g., temperature programming and gradient elution, the samples are eluted as more concentrated zones with a concomitant increase in the apparent detection sensitivity. In the general chromatographic practice gradient elution is used most widely and the other techniques have found a very limited acceptance.

A. Pressure or Flow Programming

To a good approximation, the retention time is inversely proportional to the pressure drop along a column under otherwise constant conditions. A linear increase in the inlet pressure and, hence, in the flow velocity results in a linear decrease in the retention time (*1, 72*). Since the retention times increase exponentially within a homologous series, it is most expedient to employ exponential pressure programming to maintain a constant distance between peaks.

Flow programming may easily be achieved with any equipment capable of gradient generation at high pressure, applying two different pumps from which only one is used. No special demands are placed on the stationary phase and the column. Differential refractometers and UV detectors can be used for flow-programmed analysis. Differential refractometers are likely to exhibit baseline drift at comparatively low programming rates due to changes in the flow velocity. The functions of UV detectors are not expected to show strong dependence on the flowrate.

The advantages of pressure programming are demonstrated in Fig. 10 by comparing isobaric separations at 30 and 100 bar with a separation obtained by pressure-programming at column inlet pressures increasing from 30 to 225 bar (*73*). At low inlet pressure the early peaks are separated and eluted within 5 min, whereas the last one is eluted with tailing after an additional 5 min. The linear flow velocity is 1.1 cm/sec. At 100 bar the

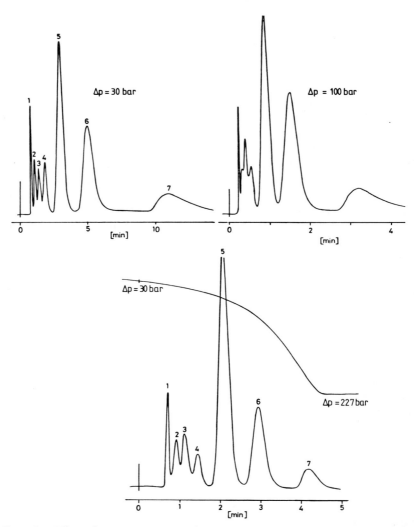

FIG. 10. Effect of pressure programming on chromatographic separation. Stationary phase, nitrobenzyl-silica, SI 200; $d_p \sim 35$ μm; column, 50 × 2 mm; eluent, n-heptane, temp., 23°C. Sample components: 1, unretained; 2, bromobenzene; 3, toluene; 4, naphthalene; 5, anthracene; 6, brasan; 7, a,h-dibenzanthracene.

analysis is completed after 5 min but the resolution of the early peaks is inadequate. In the pressure-programmed run the resolution of these peaks is good and analysis is completed in reasonable time.

Whereas an exponential program is optimal for the separation of members of a homologous series, any type of program may be employed and it may be interrupted as often as desired in the course of the run. The primary advantage of pressure programming lies in the ability to return in a few seconds to the initial conditions upon completion of an analysis by depressurizing to the original pressure. The main disadvantages are the relatively small range in which pressure can be varied and in particular, the large peak volumes of later eluting peaks. The apparent detection sensitivity is not increased by this programming technique; in fact, the sample concentration in late peaks is lowered by the increase of the flowrate.

B. Temperature Programming

As already discussed in Section III,C, the influence of temperature changes on solute retention is governed by the eluent and the "modulator" dissolved in it. An increase in temperature should always *shorten* the retention times as predicted by thermodynamics. With many eluents this is the case indeed. Yet, at elevated temperatures the modulator is stripped off the stationary phase by the eluent, and after a temperature programmed run cooling to the original temperature the adsorbent will be more active than before (4). To restore column equilibrium, i.e., to obtain identical k values before and after temperature programming requires a long time, as discussed already. However, with polar chemically bonded phases, e.g., dinitrophenyl-substituted silica, temperature programmed analyses can be carried out without these complications (1, 73).

Because column packings are poor heat conductors, it is insufficient merely to thermostat and program the column; the eluent must be raised to the proper temperature before entering the column. For this purpose an approximately 1-m-long capillary (i.d. 0.5–1 mm) should be installed as a heat exchanger and maintained at the same temperature as the column. As liquids possess a considerably higher heat capacity than air, and heat transfer from the air to the eluent is not sufficiently rapid, liquid thermostats are recommended for temperature-programmed analyses. Programming with a preheated eluent has the additional advantage of avoiding a radial temperature gradient, inside the column, as significant temperature nonuniformities contribute to band spreading. Since the heat capacities of eluents and column packings are of the same order of magnitude, temperature equilibration between them occurs rapidly. Unlike refractive index

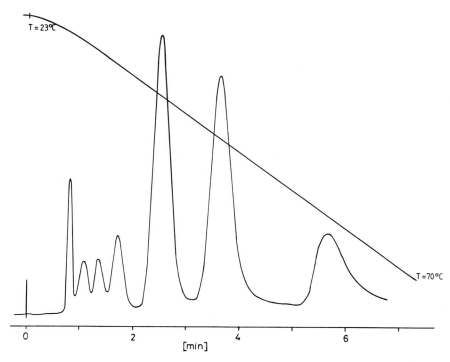

FiG. 11.   Effect of temperature programming on the separation. The same conditions as stated in Fig. 10 but an initial temperature of 23°C and a linear temperature increase at a rate of 6.3°C/min were used.

monitors UV detectors are suitable for temperature-programmed analyses. To avoid bubble formation in the detector it should be kept pressurized by using a restriction at the outlet. The permissible upper temperature limit is about 20°C below the boiling point of the eluent.

Temperature-programmed analysis of the same mixture shown in Fig. 10 is illustrated in Fig. 11. The program rate was 6.3°C/min starting from 23°C and a pressure drop of 30 bar. A temperature increase of 50 to 70°C leads to a reduction of the k values by a factor of around 100, i.e., samples that have k values in the range 100–200 at room temperature elute at the higher temperature with $k < 5$.

With respect to the analytical results, in this case there is no major difference between flow and temperature programming. Both programming modes are appropriate when the resolution of the late peaks is excessive and wastes analysis time. Which approach is preferable depends on the particular conditions.

C. Programming of Stationary Phase

An increase in the water concentration of the eluent causes the adsorbent activity to decrease and as a result the elution of the more polar components is accelerated. When a dry eluent is replaced by a wet one demixing occurs on the adsorbent, i.e., more water is taken up by the adsorbent at the *front* of the column than by that at the end. The transition zone from the less active to the more active adsorbent is more or less continuous and moves gradually down the column as more and more water is added. Because of this, the sample components are compressed into

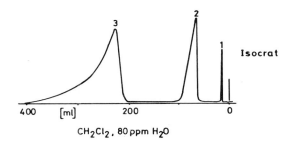

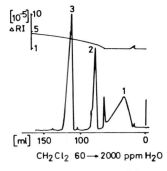

FIG. 12. Programming of the activity of the stationary phase by programmed modulator concentration. Stationary phase, LiChrosorb SI 200, $d_p \sim 27$–32 μm; column, 500×2.1 mm; flowrate, 1.6 ml/min; 22°C; inlet pressure, 30 bar. Top chromatogram: isocratic elution with dichloromethane containing 80 ppm H_2O. Bottom chromatogram: elution with CH_2Cl_2 whose water content increased from 60 to 2000 ppm during the chromatographic run. [From Berry and Engelhardt (56), courtesy of Elsevier.]

sharper zones. In extreme cases when the breakthrough of the water front occurs the zones are displaced by the water front (56). That may cause several components to elute as one peak or one component to elute as two peaks, when the tail end of a peak is detached and compressed into a separate peak. This problem of *band splitting* has already been observed in classical column chromatography (74), and is demonstrated in Fig. 12. The upper chromatogram shows the unsatisfactory separation of steroids with constant, low moisture content eluent at isocratic elution. The lower chromatogram shows the same separation with a continuously increasing moisture content of the dichloromethane eluent. The small sharp peak on the tailing end of the first eluted component coincides with the water breakthrough and illustrates band splitting, which can be readily misinterpreted. Therefore this programming technique should be applied with sufficient care, despite the fact that the k values on silica can be decreased by a factor of up to 1000 when the water concentration in the dichloromethane is increased from 60 to about 2000 ppm (56, 19).

Coupled columns packed with different stationary phases can be used to optimize the analysis time (71, 75). In this approach the different columns are connected in a series or in parallel. The sample mixture is first fractioned on a relatively short column. Subsequently the fractions of the partially separated mixture are separated on other columns containing the same or other stationary phases in order to obtain the individual components. Columns differing in length (number of theoretical plates), adsorptive strength or phase ratio (magnitude of specific surface area), and selectivity (nature of the stationary phase) can be employed, whereas, the eluent composition remains unchanged. Identification of the individual sample components via coupled column technique requires a careful optimization of each column and precise control of each switching step.

D. Gradient Elution

The term gradient elution refers to a programmed increase in the elution strength of the mobile phase during the chromatographic run. It has the greatest potential among all programming techniques to facilitate separation of very complex mixtures. As the use of a large variety of eluents is possible with this technique, sample mixtures having components with wide-ranging polarities can be separated. In practice, however, most instruments permit gradients to be prepared from two eluents only, but the two components, which themselves can be solvent mixtures, can be mixed together according to various programs. The increase in concentration of the second component as a function of time can be described by concave, linear, or convex curves. In practice the use of a linear gradient

should be attempted first. The actual gradient given by the time dependence of the eluent strength is only partially dependent on the programming mode for the solvent composition and it is a function of the polarity
or elution strength differences of the two solvent components. When two
solvents with a large difference in their elution strengths are used, even a
small increase in the concentration of the polar components produces a
sharp rise in elution strength. Such rapid initial increase in eluent strength
is rarely desirable because the components are almost always eluted close
together at the beginning of a separation. Furthermore, displacement effects may appear as a result of demixing of eluent mixtures that consist,
for instance, of pure *n*-hexane and *n*-hexane containing a few percent of
an alcohol such as isopropanol.

For unknown and complex sample mixtures, a linear increase in the solvent strength was shown to be optimum for gradient elution (*2*, *76*) because such a gradient produces peaks having identical peak widths, and
the overall resolution of the sample components is optimum. Only if the
eluent components used in gradient elution have relatively small differences in polarity will the elution strength increase in an approximately
linear fashion in linear solvent programming. With a polar stationary
phase this can be accomplished with solvent gradients from *n*-hexane to
chloroalkanes such as propyl chloride. With a gradient from *n*-hexane to

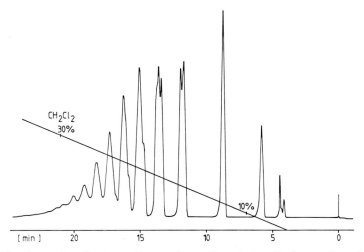

FIG. 13.   Separation of styrene oligomers by gradient elution. Stationary phase, LiChrosorb SI 100, $d_p \sim 10\ \mu m$; column, 300 × 4.1 mm; Linear solvent gradient from *n*-heptane
with increasing concentration of dichloromethane; gradient-volume, 70 ml; flowrate, 1
ml/min. Sample: polystyrene, $\overline{M}_w = 600$.

dichloromethane, a linear increase in solvent strength can be obtained by using a slightly concave "gradient." This can be inferred from the data in Fig. 4 which shows a quasi-linear dependence of log k on the logarithm of the eluent composition over a wide range. Figure 13 gives an example for the separation potential of gradient elution. A polystyrene standard with an average molecular weight of 600 is separated on silica and the peaks show the individual homologous oligomers. A gradient from pure n-heptane to 30% (v/v) dichloromethane in n-heptane was used and the composition change with time is also shown on the chromatogram.

The components can be identified from a knowledge of the eluent composition at which they elute, if the gradient volume, i.e., product of volume flowrate and program time is kept constant. Because of dead volumes in the system represented by the volume of the mixing chamber and connecting tubing, as well as by the volume of the mobile phase in the column, the actual eluent composition in the column is not the same as that indicated by the instrument. By taking this effect into account to evaluate at which eluent composition certain peaks are eluted we can use gradient elution as a tool for quickly scouting the optimum eluent composition for corresponding isocratic analysis (36).

Gradient elution places special demands on *solvent purity*. Only carefully purified solvents should be used, and, it is recommended that prior to use they be passed over activated alumina or silica (1). The column acts as a collector of impurities which may elute as sharp peaks at a certain eluent composition and can be mistaken for sample components. It is therefore advisable to run the gradient first without injecting the sample in order to recognize the impurity peaks.

Column regeneration, i.e., the return to the original conditions by removing the polar eluent component, requires a long time. Difficulties stem less from flushing out the residual polar eluent than from reestablishing equilibrium between the water on the adsorbent and in the eluent (4). Since this frequently requires more time than the gradient–elution separation itself, it is often more practical not to return to the original conditions, that is, reestablish equilibrium, but to set a standard regeneration time. The latter is a necessary precaution anyway because of the eluent impurities. The regeneration time can also be shortened considerably by reversing the gradient (52, 53).

The equipment for gradient elution in HPLC is expensive because the solvents are mixed on the high pressure side. When such equipment is unavailable, gradient mixing may be performed before the pump on the low pressure side (53). Alternatively, the solvent composition may be changed stepwise by simply changing the reservoir. When the steps are sufficiently small, the result may be equivalent to that obtained in gradient

elution. Care has to be taken, however, not to use large polarity differences between the individual steps. Too large differences in polarity
can be recognized on the chromatogram by the very large peaks appearing
shortly after eluent change or the breakthrough of the second component
of the eluent that acts as a displacing agent. Therefore, such peaks may
arise from simultaneous displacement of several unresolved compounds.

Gradient elution with polar stationary phases can be readily used to
separate mixtures whose components would encompass a k range of 10^4
or greater under isocratic elution conditions (2, 71). A definite treatment
of gradient elution is given by Snyder in Volume 1 of this series.

VI. THEORY OF ADSORPTION CHROMATOGRAPHY

Several different physicochemical models have been proposed to predict and explain the retention behavior in liquid–solid chromatography.
The models can be divided into two groups depending on the assumptions
made concerning the fundamental mechanism of the chromatographic
process. The two assumptions are as follows:

(i) The solute competes with eluent molecules for the active adsorption
sites on the surface of the stationary phase. Interactions between solute
and solvent molecules in the liquid phase are cancelled by similar interactions in the adsorbed phase. This model has been introduced by Snyder
(2) and by Soczewinski (77, 78) and is called the "competition model."

(ii) Solute–solvent interactions in the mobile phase are solely responsible for solute retention as in practice the surface of the adsorbent is assumed to be completely covered by a modulator. This "solvent interaction" model was introduced by Scott and Kucera (79–81).

The competition model of Snyder assumes that the adsorption surface
is completely covered by adsorbed eluent molecules forming a monolayer. When solute molecules are adsorbed they displace solvent molecules. Due to the size differences one or more eluent molecules are displaced by the solute molecules. The adsorbent surface is assumed to be
homogeneous and each molecule tends to interact totally with the surface,
i.e., it is adsorbed flatwise. Thus, the adsorbent surface area that the molecules require can be calculated from their molecular dimensions. Ne-
glecting the interactions between solute and eluent molecules in the liquid
and the adsorbed phase, the retention of an adsorbed molecule (expressed
as net retention volume per unit weight of adsorbent: K) can be related to
the properties of the stationary phase, the eluent, and the sample by

$$\log K = \log V_a + \alpha'(S^0 - A_s\epsilon^0) \tag{1}$$

where K is the net retention volume per unit weight of adsorbent. V_a corresponds to the volume of an adsorbed eluent monolayer per unit weight of adsorbent, and is, with common eluents, mainly a function of the adsorbent surface area. The surface activity coefficient, α', depends on the chemical nature of the adsorbent and its water coverage. For active alumina $\alpha' \equiv 1$ by definition, and for silicas its value varies between $0.7 < \alpha' < 1$ (2). S^0 is the sample adsorption energy on a surface having standard activity, i.e., $\alpha' = 1$. A_s is the surface area a sample molecule requires on the adsorbent surface and ϵ^0 is the adsorption energy of the solvent per unit surface area of standard activity adsorbent.

The influence of solvent change on sample retention is governed by

$$\log(K_1/K_2) = \log(k_1/k_2) = \alpha' A_s(\epsilon_2^0 - \epsilon_1^0) \qquad (2)$$

where the subscripts 1 and 2 represent two different eluents. It is seen that the effect of solvent change on K or k is related to the differences in solvent strength, provided that adsorbent activity and surface requirement for the solute molecule are unaffected. The ϵ^0 values of most commonly used eluents has already been presented in Table II. By using n-pentane as reference eluent ($\epsilon^0 \equiv 0.00$) the solvent strength of other eluents can be evaluated from the retention data by applying Eq. (2).

The model assumes an effectively homogeneous adsorbent surface. Hence all sites of the surface are equivalent and the entire surface can be covered completely by solute or solvent molecules. It can be readily imagined that with increasing surface inhomogeneity the solute molecules tend to prefer the energetically more favorable sites and the number of effective binding sites available decreases. It results in a lower surface coverage and hence in a greater effective area occupied by an adsorbed molecule. This was taken into account by Snyder (2) by using A_s values somewhat larger than the actual molecular areas of the species involved.

If one develops this concept further it becomes evident that the general adsorption of the molecules at the surface proceeds continuously to a localized adsorption on distinct and specific surface sites. This appears entirely reasonable as an extreme case, i.e., for ion exchange interaction with basic or acidic surface sites. The consequence, of course, is that the effective area required by an adsorbed sample molecule increases.

The adsorption of the solutes by discrete one-to-one complexes was discussed by Soczewinski (77, 78). In this simple model equations for k value are derived by the application of the law of mass action to the competitive adsorption equilibria between solute and solvent molecules for the active sites of the adsorbent. It follows then that with an eluent mixture containing a polar solvent in an "inert" nonpolar diluent, a linear relationship holds so that

$$\log k = \text{const} - n \log X_B \tag{3}$$

where X_B is the mole fraction of the stronger (more polar) component of the eluent. Equation (3) was experimentally verified with different solvents and solutes on silica (78, 82) and even on magnesium silicate (83) in both TLC and HPLC (33, 35). The results represented by Eq. (3) are not in disagreement with Snyder's approach because his model also predicts that the k values show a similar logarithmic dependence on the concentration of the stronger eluent if some boundary conditions are valid (33). In fact Snyder gives a more detailed treatment for dependence of elution strength on eluent composition (2).

The predicted linear $\log k$ vs. $\log X_B$ plot is only achievable with solutes having relatively simple chemical structure as seen in Fig. 4. Especially in the region of less than 10% of the more polar component in the eluent, deviations from the behavior predicted by Eq. (3) are observable.

Both approaches represent limiting cases for the same model to describe adsorption on polar stationary phases. Snyder's treatment applies to weakly adsorbing eluents and solutes and to the use of water-deactivated adsorbents. In such cases the ideal situation of a continuous homogeneous surface is closely approached. For the opposite case involving strongly adsorbing solutes and solvents and discrete, separated active surface sites, Soczewinski's model gives a good description of the adsorption process. In order to treat adsorption systems whose behavior is between these limits it is necessary to modify the limiting models. Taking into account a two-point attachment of certain sample molecules to the adsorbent surface the validity of Soczewinski's approach can be improved. In the case of strong interactions between adsorbate molecules and adsorbent surface, when strong eluents are used in chromatographic practice, Snyder fits his equation by taking into account "secondary solvent effects," that is, by introducing another term into Eq. 1 (2). Furthermore, steric and electronic effects of the sample molecule on adsorption are considered additionally (2) in order to provide a better quantitative description of chromatographic retention.

The model of Scott and Kucera (79–81) deals with the chromatographic process assuming that the surface of the polar stationary phase is completely covered by the strongly polar eluent component. When more than 2% (w/v) of either isopropanol or tetrahydrofuran is dissolved in n-heptane, for instance, the surface of silica is completely covered by the polar modulator and the activity of the stationary phase is assumed to be constant. By adding more of the polar component to the eluent a linear relationship between the reciprocal net retention volume V_R' and the concentration of the polar components in the eluent is obtained. According to

Scott this observation demonstrates that only solute–solvent interactions due to the polarizability of solute or solvent and/or dispersion forces influence the retentions in these systems. However, the variations in the retention volumes in this concentration range of stronger eluting components are relatively insensitive to changes in eluent composition, and therefore, changes in selectivity are difficult to discern.

When the concentration of the polar solvent in the eluent is so low that it cannot deactivate the silica, the linear relationship between $1/V'_R$ and polar solvent concentration is no longer observed. In this region, as discussed in Section III, the retention is strongly dependent on *small* changes of the concentration of the strongly eluting component, e.g. isopropanol in n-heptane.

The retention behavior of solutes in "adsorption" chromatography can be described either by the "competition" model or by the "solute–solvent interaction" model depending on the eluent composition. It appears that both mechanisms are operative but their importance depends on the composition of the eluent mixture (*84*).

Considering the complexity of adsorption on polar stationary phases it is not surprising that no theory, which would be generally applicable, is available. Nevertheless the time and effort expended on exploring the mechanism of the chromatographic process has produced the limiting models which contribute to an understanding of the processes involved. The words of F. Schiller, *"wir gelangen nur selten anders als durch Extreme zur Wahrheit,"* appear to be particularly fitting to express the situation.

VII. SELECTED APPLICATIONS

The primary area of application of adsorption chromatography on polar stationary phases is in the separation of nonpolar to moderately polar organic compounds. A preliminary decision on whether or not this system is adequate can be based on sample solubility in such "nonpolar" solvents as aliphatic or aromatic hydrocarbons, haloalkanes, perhaps with the addition of a few percent of esters, acetonitrile, or even alcohols. When the sample is soluble or miscible with these eluents, the use of a polar stationary phase may be the best approach to chromatographic separation.

Polycyclic aromatic hydrocarbons represent a group of compounds that is most frequently separated by using silica or alumina as the stationary phase (cf. Figs. 2 and 8). The separation of 25 of the most important polynuclear aromatic hydrocarbons (*49*) is illustrated in Fig. 14. To en-

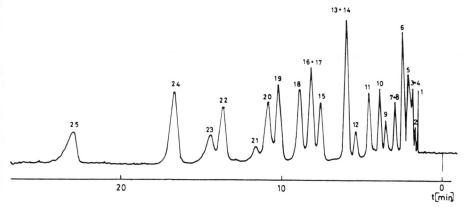

FIG. 14. Separation of polynuclear aromatic hydrocarbons. MCS: alumina with 10% (w/w) of water at 25°C. Column: Spherisorb alumina, 5 μm; 20° × 4.1 mm; temp. 25°C; eluent, *n*-heptane with 37 ppm of water; inlet pressure 30 atm; flowrate, 1.6 ml/min; linear flow velocity, 2.4 mm/sec. Sample components: 1, tetrachloroethylene; 2, benzene; 3, naphthalene; 4, biphenyl; 5, dichloromethane (solvent of the sample mixture); 6, acenaphthene; 7, anthracene; 8, phenanthrene; 9, pyrene; 10, fluoranthene; 11, 1,10-benzfluoranthene; 12, triphenylene; 13, 1,2-benzanthracene; 14, chrysene; 15, 1,2-benzpyrene; 16, 3,4-benzpyrene; 17, perylene; 18, 3,4-benzfluoranthene; 19, 11,12-benzfluoranthene; 20, 1,12-benzperylene; 21, anthanthrene; 22, 2,3-phenylenepyrene; 23, coronene; 24, 1,2,7,8-dibenzanthracene; 25, 1,2,5,6-dibenzanthracene. [From Böhme and Engelhardt (*49*), courtesy of Elsevier.]

hance selectivity, the adsorbents can be coated with appropriate complexing agents such as silver nitrate or trinitrofluorenone (*28, 29*).

Pesticides (*14, 85*), polychlorinated biphenyls (*86, 87*), and herbicides (*88*) are usually separated by this technique also. In analytical work, however, the detection sensitivity of the selective detectors used in gas chromatography could not be achieved (*89*). Nevertheless, such substances can be separated by liquid chromatography with no attendant decomposition problems and no derivatization, making the procedure significantly simpler.

From the beginning, the principal application of HPLC has been in the analysis of pharmaceuticals and pharmaceutical preparations (*90–92*), steroids (*93, 94*), cardiac glycosides (*95, 96*), alkaloids (*97*), and aflatoxins (*98*). Diastereomers have also been separated (*99, 100*) and a recycling system can be employed to increase the separation efficiency (*101*). Silica with a relatively large particle size (40 μm) was used and the mixture was pumped several times through the same column. With microparticulate (4–8 μm) silica a relatively short 30-cm-long column was found to suffice for the resolution of diastereomers (*100, 102*).

It would be beyond the scope of this article to present all separations described in HPLC in the last decade. Reference is made to monographs and bibliographies that discuss the separation systems described for individual classes of substances (*8, 66, 67, 103*). As discussed, the older monographs on column liquid chromatography (*7*) as well as those on thin layer chromatography (*104*) may provide valuable information for the design of an appropriate HPLC system.

Of course, many separation problems can be solved equally well by using more than one separation system. Thus, for example, the separation of steroids is effected by adsorption on polar as well as on nonpolar stationary phases or by partitioning in a ternary system. The choice of the chromatographic system depends not only on the nature of the compounds to be separated, but also on the familiarity of the analyst with a particular separation system.

VIII. PROSPECTS

Many problems that arise in the application of adsorption chromatography can be related to the slow attainment of equilibrium distribution of the omnipresent water and other modulators between eluent and stationary phase. With suitable precautions, such as moisture control, reproducible work is possible with both silica and alumina as the stationary phase.

By covalently bonding organic functions via the hydroxyl groups at the surface of silica or alumina, the chromatographic behavior of the solid stationary phase can be drastically modified and the problems mentioned above seem to be less important with such bonded phases.

Besides the fact that column equilibrium is not so readily disturbed by external effects like temperature and the moisture content of the eluent, bonded phases possess some additional advantages over conventional phases or those coated with a stationary liquid. These are:

(*i*) the eluent need not be saturated with the liquid phase;

(*ii*) no precolumn is necessary;

(*iii*) there is no contamination of the effluent by the stationary phase, a condition that usually hampers preparative application of partition systems;

(*iv*) the stationary phase cannot be strippped off because it is covalently bonded to the solid, thereby, problem of mechanical erosion at higher flowrates is eliminated;

(*v*) various programming techniques, particularly gradient elution, can be employed without the difficulties encountered when using conventionally coated supports and bare silica or alumina.

Silica is used almost exclusively as the substrate for the preparation of chemically bonded phases. If the organic moiety exerts no specific selective effect, i.e., an alkyl group bound to silica, the k values on such stationary phases are always smaller than on bare silica, provided the same nonpolar eluent is used. By varying the chemical nature of the bonded organic moiety, for example by introducing functional groups in the organic residue, preferably in the ω-position, the selectivity of the stationary phase phase can be modified. Since it is impossible to react *all* silanol groups on the silica surface the selectivity of the phase is determined by both the functional groups in the bonded moiety *and* the unreacted silanol groups at the surface. It is, therefore, very difficult to attribute unequivocally the selectivity of the stationary phase to the bonded organic residue. With polar eluents such as methanol–water mixtures the nonpolar (reversed phase) properties of the stationary phase seem to pre-

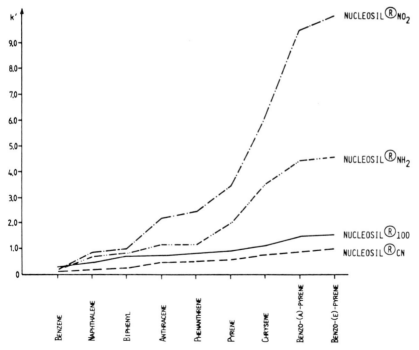

FIG. 15.   Selectivity of certain polar bonded phases for benzologous solutes using *n*-heptane as the eluent. (Courtesy of Dr. Sebestian of Macherey & Nagel A.G.)

dominate, whereas with nonpolar eluents the bonded polar functional groups *and* the unshielded silanol groups of the support are both responsible for selectivity.

Figure 15 compares the selectivity of chemically bonded phases containing various functional groups with that of the original silica gel (*105*). On the stationary phase containing covalently bound nitrophenyl groups the k values of the polynuclear aromatic hydrocarbons are about five times greater than those observed on the silica gel, proper when water-saturated n-heptane is used as eluent. A bonded amino phase also showed higher selectivity for the aromatics than did the bare silica. It is interesting to note that with the aqueous eluents this phase is eminently suitable for the separation of sugars. In Fig. 15 we can see that on the nitrile phase the k values of the benzologs are below 1 when water-saturated n-heptane is used as the eluent.

The influence of the bonded organic moiety on solute retention has not yet been elucidated and only a very small number of papers discuss the properties and use of such phases so far. The numerous advantages of chemically bonded phases make the application of polar chemically bonded phases with nonpolar eluents quite attractive even if the standardization of these phases may pose problems (*106*) similar to those encountered in the standardization of adsorbents as well as of polymeric liquid phases in gas chromatography. A detailed discussion of the properties and chromatographic use of bonded stationary phases is given by Melander and Horváth (this volume).

Acknowledgments

This paper contains unpublished results from doctoral theses by W. Böhme, D. Mathes, U. Neue, and H. Wiedemann submitted to the University of Saarland in Saarbrücken. Financial assistance by the Deutsche Forschungsgemeinschaft, the Verband der Chemischen Industrie, and the Graduiertenförderung is gratefully acknowledged.

REFERENCES

1. H. Engelhardt, "High Performance Liquid Chromatography." Springer-Verlag, Berlin and New York, 1978.
2. L. R. Snyder, "Principles of Adsorption Chromatography." Dekker, New York, 1968.
3. R. P. W. Scott, and J. G. Lawrence, *J. Chromatogr. Sci.* **7**, 65 (1969).
4. H. Engelhardt, *J. Chromatogr. Sci.* **15**, 380 (1977).
5. A. V. Kiselev, *J. Chromatogr.* **49**, 84 (1970).
6. M. S. Tswett, *Ber. Dtsch. Bot. Ges.* **24**, 316, 384 (1906). Compare: G. Hesse, and H. Weil, *In* "Woelm-Mitteilungen Al 8." Eschwege, 1954.

7. E. Lederer, (ed.), "Chromatographie en chimie organique et biologique," Vols. I and II. Masson, Paris, 1959, 1960.
8. E. Heftman, "Chromatography." Van Nostrand—Reinhold, Princeton, New Jersey, 1975.
9. G. Hesse, "Chromatographisches Praktikum." Akad. Verlagsges. Frankfurt/Main, 1968.
10. I. Halász, and P. Walkling, *J. Chromatogr. Sci.* **7**, 129 (1969).
11. C. Horváth, B. A. Preis, and S. R. Lipsky, *Anal. Chem.* **39**, 1422 (1967).
12. B. L. Karger, H. Engelhardt, K. Conroe, and I. Halász, *In* "Gas Chromatography 1970" (R. Stock, ed.). Institute of Petroleum, London, 1971.
13. J. J. Kirkland, *Anal. Chem.* **41**, 218 (1969).
14. J. N. Little, D. F. Horgan, and K. J. Bombaugh, *J. Chromatogr. Sci.* **8**, 625 (1970).
15. L. Boksányi, O. Liardon, and E. sz. Kováts, *Adv. Coll. Inter. Sci.* **6**, 95 (1976).
16. K. Unger, *Angew. Chem.* **84**, 331 (1972).
17. M. Uihlein, and I. Halász, *J. Chromatogr.* **80**, 1 (1973).
18. R. P. W. Scott, "Contemporary Liquid Chromatography." Wiley, New York, 1976.
19. W. Böhme, Ph.D. Thesis, Saarbrücken, 1976.
20. J. F. K. Huber, and F. Eisenbeiss, *J. Chromatogr.* **149**, 127 (1978).
21. H. Engelhardt, and N. Weigand, *Anal. Chem.* **45**, 1149 (1973).
22. A. V. Kiselev, Yu. S. Nikitin, I. I. Frolov, and Ya. I. Yashin, *J. Chromatogr.* **91**, 187 (1974).
23. G. Hesse, *Z. Anal. Chem.* **211**, 5 (1965).
24. I. Halász, and I. Sebestian, *Angew. Chem.* **81**, 464 (1969).
25. E. Grushka (ed.), "Bonded Stationary Phases in Chromatography." Ann Arbor Science, Ann Arbor, Mich., 1974.
26. V. Řehák, and E. Smolkova, *Chromatographia* **9**, 219 (1976).
27. H. Engelhardt, and D. Mathes, *J. Chromatogr.* **142**, 311 (1977).
28. R. Vivilecchia, M., Thiebaud, and R. W. Frei, *J. Chromatogr. Sci.* **10**, 411 (1972).
29. B. L. Karger, G. Guiochon, M. Martin, and J. Lohéac, *Anal. Chem.* **45**, 496 (1973).
30. H. Engelhardt, J. Asshauer, U. Neue, and N. Weigand, *Anal. Chem.* **46**, 336 (1974).
31. W. Trappe, *Biochem. Z.* **305**, 150 (1940).
32. J. Jacques and J. P. Mathieu, *Bull. Soc. Chim.*, p. 94 (1946).
33. P. Jandera and J. Churáček, *J. Chromatogr.* **91**, 207 (1974).
34. P. Jandera and J. Churáček, *J. Chromatogr.* **93**, 17 (1974).
35. P. Jandera, M. Janderová, and J. Churáček, *J. Chromatogr.* **115**, 9 (1975).
36. H. Elgass, Ph.D. Thesis, Saarbrücken, 1978.
37. R. Neher, *In:* "Thin Layer Chromatography" (G. B. Marini-Bettolo, eds.), p. 75. Elsevier, Amsterdam, 1964,
38. D. L. Saunders, *Anal. Chem.* **46**, 470 (1974).
39. L. R. Snyder, *Anal. Chem.* **46**, 1384 (1974).
40. H. Brockmann and H. Schodder, *Ber. Dtsch. Chem. Ges.* **74**, 73 (1941).
41. G. Hesse and G. Roscher, *Z. Anal. Chem.* **200**, 3 (1965).
42. H. Engelhardt and H. Wiedemann, *Anal. Chem.* **45**, 1649 (1973).
43. R. E. Majors, *J. Chromatogr. Sci.* **11**, 88 (1973).
44. L. R. Snyder, *J. Chromatogr. Sci.* **7**, 595 (1969).
45. J. J. Kirkland, *J. Chromatogr.* **83**, 149 (1973).
46. R. P. W. Scott, *J. Chromatogr.* **122**, 35 (1976).
47. D. L. Saunders, *J. Chromatogr.* **125**, 163 (1976).
48. L. R. Snyder and J. J. Kirkland, "Modern Liquid Chromatography." Wiley, New York, 1974.
49. W. Böhme and H. Engelhardt, *J. Chromatogr.* **133**, 67 (1977).

50. R. J. Maggs, *J. Chromatogr. Sci.* **7**, 145 (1969).
51. R. P. W. Scott and J. G. Lawrence, *J. Chromatogr. Sci.* **8**, 619 (1970).
52. R. E. Mayors, *Anal. Chem.* **45**, 755 (1973).
53. H. Engelhardt and H. Elgass, *J. Chromatogr.* **112**, 415 (1975).
54. L. T. Chang, *Anal. Chem.* **25**, 1235 (1953).
55. R. J. Maggs, and T. E. Young, *In:* "Gas Chromatography 1968" (C. L. A. Harbourn, ed.). Institute of Petroleum, London, 1969.
56. L. V. Berry and H. Engelhardt, *J. Chromatogr.* **95**, 27 (1974).
57. J. F. K. Huber, *J. Chromatogr. Sci.* **9**, 72 (1971).
58. G. Frank and W. Strubert, *Chromatographia* **6**, 522 (1973).
59. G. Rössler and I. Halász, *J. Chromatogra.* **92**, 33 (1974).
60. I. Halász and E. E. Wegner, *Nature (London)*, **189**, 570 (1961).
61. J. E. Paanakker, J. C. Kraak, and H. Poppe, *J. Chromatogr.* **149**, 111 (1978).
62. J. M. Bather and R. A. C. Gray, *J. Chromatogr.* **156**, 21 (1978).
63. R. A. Keller, B. L. Karger, and L. R. Snyder, *In:* "Gas Chromatography 1970" (R. Stock and S. G. Perry, eds.), p. 125. Elsevier, Amsterdam, 1971.
64. B. L. Karger, L. R. Snyder, and C. Eon, *J. Chromatogr.* **125**, 71 (1976).
65. L. R. Snyder, *J. Chromatogr. Sci.* **16**, 223 (1978).
66. Z. Deyl, K. Macek, and J. Janak, (eds.), "Liquid Column Chromatography." Elsevier, Amsterdam, 1975.
67. Z. Deyl and J. Kopecky, Bibliography of Liquid Column Chromatography, *J. Chromatogr. Suppl.* **6** (1976).
68. F. Geiss, "Paramter der Dünnschichtchromatographie." Vieweg, Braunschweig, 1972.
69. G. Guiochon, A. Siouffi, H. Engelhardt, and I. Halász. *J. Chromatogr. Sci.* **16**, 152 (1978).
70. E. Soczewinski and J. Kuczmierczyk, *J. Chromatogr.* **150**, 53 (1978).
71. L. R. Snyder, *J. Chromatogr. Sci.,* **8**, 692 (1970).
72. H. Wiedemann, H. Engelhardt, and I. Halász, *J. Chromatogr.* **91**, 141 (1974).
73. H. Wiedemann, Ph.D. Thesis, Saarbrücken, 1973.
74. L. R. Snyder, *J. Chromatogr.* **13**, 415 (1964).
75. J. F. K. Huber, R. van der Linden, E. Ecker, and M. Oreans, *J. Chromatogr.* **93**, 267 (1973).
76. L. R. Snyder, and D. L. Saunders, *J. Chromatogr. Sci.* **7**, 195 (1969).
77. E. Soczewinski, *Anal. Chem.* **41**, 179 (1969).
78. E. Soczewinski and W. Golkiewicz, *Chromatographia* **4**, 501 (1971).
79. R. P. W. Scott and P. Kucera, *J. Chromatogr.* **112**, 425 (1975).
80. R. P. W. Scott, *J. Chromatogr.* **122**, 35 (1976).
81. R. P. W. Scott, and P. Kucera, *J. Chromatogr.* **149**, 93 (1978).
82. E. Soczewinski, *J. Chromatogr.* **130**, 23 (1977).
83. E. Soczewinski, W. Golkiewicz, and T. Dzido, *Chromatographia* **10**, 221 (1977).
84. E. H. Slaats, J. C. Kraak, J. T. Brugmann, and H. Poppe, *J. Chromatogr.* **149**, 255 (1978).
85. R. J. Dolphin, F. W. Willmot, A. D. Mills, and L. P. J. Hoogveen, *J. Chromatogr.* **122**, 259 (1976).
86. K. A. Th. Brinkman, J. W. F. L. Seetz, and H. G. M. Reymer, *J. Chromatogr.* **116**, 353 (1976).
87. K. A. Th. Brinkman, A. De Kok, G. De Vries, and H. G. M. Reymer, *J. Chromatogr.* **128**, 101 (1976).
88. F. Eisenbeis and H. Sieper, *J. Chromatogr.* **83**, 439 (1973).

89. H. Jork, and B. Roth, *J. Chromatogr.* **144**, 39 (1977).
90. O. N. Hinsvark, W. Zazulak, and A. I. Cohen, *J. Chromatogr. Sci.* **10**, 379 (1972).
91. G. J. Krol, C. A. Mannan, F. Q. Gemmill, Jr., G. E. Hicks, and B. T. Uko, *J. Chromatogr.* **74**, 43 (1972).
92. M. H. Stutz, and S. Sass, *Anal. Chem.* **45**, 2134 (1973).
93. F. A. Fitzpatrick, S. Siggia, and J. Dingman, *Anal. Chem.* **44**, 2211 (1972).
94. J. C. Touchstone and W. Wortmann, *J. Chromatogr.* **76**, 244 (1973).
95. M. C. Castle, *J. Chromatogr.* **115**, 437 (1975).
96. F. Nachtmann, R. W. Spitzy, and R. W. Frei, *J. Chromatogr.* **122**, 293 (1976).
97. F. Erni, R. W. Frei, and W. Lindner, *J. Chromatogr.* **125**, 285 (1976).
98. L. M. Seitz, *J. Chromatogr.* **104**, 81 (1975).
99. G. Helmchen, G. Haas, and V. Prelog, *Helv. Chim. Acta* **56**, 2255 (1973).
100. G. Helmchen and H. Strubert, *Chromatographia* **7**, 713 (1974).
101. M. Koreeda, G. Weiss, and K. Nakanishi, *J. Am. Chem. Soc.* **95**, 239 (1973).
102. C. G. Scott, M. J. Petrin, and T. McCorkle, *J. Chromatogr.* **125**, 157 (1976).
103. E. Johnson, "Liquid Chromatography Bibliography." Varian, New York, 1977.
104. E. Stahl, "Dünnschicht-Chromatographie." Springer-Verlag, Berlin and New York, 1967.
105. I. Sebestian, personal communication.
106. C. W. Qualls, Jun., and H. J. Segall, *J. Chromatogr.* **150**, 202 (1978).
107. H.-P. Boehm, *Angew. Chem.* **78**, 617 (1966).
108. R. Schwarzenbach, *J. Liquid Chromatogr.* **2**, 205 (1979).
109. S. Eksberg and G. Schill, *Anal. Chem.* **45**, 2092 (1973).
110. B. L. Karger and B. A. Pearson, *J. Chromatogr. Sci.* **12**, 521, 678 (1974).
111. J. C. Kraak and J. F. K. Huber, *J. Chromatogr.* **102**, 333 (1974).
112. J. J. C. Schoeffer, A. Koedem, and A. Baerheim, *Chromatographia* **9**, 425 (1976).
113. I. Halász, H. Engelhardt, J. Asshauer, and B. L. Karger, *Anal. Chem.* **42**, 1460 (1970).
114. H. Engelhardt and N. Weigand, *Anal. Chem.* **45**, 1149 (1973).
115. G. Rössler and I. Halász, *J. Chromatogr.* **92**, 33 (1974).
116. R. E. Kaiser, *In* "High Performance Thin-Layer Chromatography" (A. Zlatkis and R. E. Kaiser, eds.). Elsevier, Amsterdam, 1977.
117. E. Stahl, *Chem. Ztg.* **82**, 323 (1958).

REVERSED-PHASE CHROMATOGRAPHY

Wayne R. Melander and Csaba Horváth

Department of Engineering and Applied Science
Yale University
New Haven, Connecticut

I.	Introduction	114
	A. Significance of Reversed-Phase Chromatography	114
	B. Evolution of Reversed-Phase Chromatography	117
	C. Description of the Technique	120
II.	The Stationary Phase	123
	A. General Properties	123
	B. Preparation and Characterization of Bonded Hydrocarbonaceous Stationary Phases	131
	C. Column Packing	145
	D. Stationary Phase Properties and Chromatographic Behavior	148
	E. Other Nonpolar Stationary Phases	162
III.	The Mobile Phase	165
	A. Physicochemical Properties of the Solvents	165
	B. Eluent Strength	170
	C. Effect of the Composition of Mixed Solvents	176
	D. Gradient Elution	184
	E. Special Considerations	188
	F. Selection of the Appropriate Mobile Phase	189
IV.	Effect of Temperature	192
	A. Retention Time and Resolution	192
	B. Enthalpy of Binding	195
	C. Secondary Equilibria in the Mobile Phase	199
	D. Multiple Retention Mechanisms	200
V.	Theory of Reversed-Phase Chromatography	201
	A. Hydrophobic and/or Solvophobic Effect	201
	B. Energetics of Solvent Effect	204
	C. Free Energy Change in the Chromatographic Process	211
	D. Solvophobic Interpretation of Experimental Observations	215
	E. Mechanism and Kinetics of Retention	224
	F. Conclusions	227
VI.	Modulation of Selectivity by Secondary Equilibria	229
	A. General Considerations	229
	B. Phenomenological Model for the Process of Retention	231

HIGH-PERFORMANCE LIQUID
CHROMATOGRAPHY, Vol. 2

        C. Protonic Equilibria: pH Control 238
        D. Ion-Pair Chromatography 240
        E. Analytical Applications 260
   VII. Physicochemical Measurements 266
        A. Effects of Pressure on the Equilibria Involved in
           Chromatography 266
        B. Evaluation of Enthalpy and Entropy Changes 268
        C. Correlation between Retention Factor and Partition
           Coefficients Used in Quantitative Structure–Activity
           Relationships 273
        D. Measurement of Equilibrium Constants for Association
           Processes in Solution 276
  VIII. Selected Analytical Applications 279
        A. General Remarks 279
        B. Analysis of Physiological Samples 280
        C. Biochemical Applications 286
        D. Analysis of Environmental Samples 292
        E. Analysis of Natural Products and Foodstuffs 293
        F. Analysis of Industrial Products 296
        G. Miscellaneous Separations 298
        Notation 299
        References 303

I. INTRODUCTION

A. Significance of Reversed-Phase Chromatography

Someone who wants to know chromatography today is likely to be advised to learn about HPLC. It is also probable that this individual will have difficulty in understanding why so many other chromatographic systems were used before the advent of reversed-phase chromatography (RPC) in the form we know it today. In fact almost anyone who is starting to work in HPLC begins with RPC and will probably stay with it. Nonpolar sorbents, which mimic stationary phases employed in HPLC columns, are increasingly employed even in thin-layer chromatography where their use is restricted as far as the composition of hydroorganic eluents is concerned.

The name reversed-phase chromatography was a rational choice at a time when partition chromatography was practiced almost exclusively by using a polar stationary phase and a nonpolar eluent. However, today an estimated 80–90% of chromatographic systems used in HPLC work con-

sist of nonpolar stationary phases and polar eluents. Yet, the name reversed phase is still in use. The predominance of RPC thus necessitates
the somewhat curious description of the former technique as "normal,"
"straight," or perhaps "forward" phase chromatography. Paying tribute
to the relative importance of RPC we may facetiously term the other technique as "reversed reversed-phase" chromatography.

What is responsible for the phenomenal success of RPC? The following
comparison of RPC with gas chromatography might be helpful in answer-
ing this question. Just like gas chromatography HPLC represents a fully
instrumented chromatographic method. Both are subject to the same fundamental theory of linear elution chromatography even if the disparity in
the respective apparatus and columns reflects the well-established differences between the physicochemical properties and handling of the gas
and condensed phases. Besides those, however, there is one significant
factor that distinguishes liquid chromatography. In the present practice of
gas chromatography the mobile phase (carrier gas) is practically invariant;
consequently, the magnitude of retention and selectivity is usually adjusted by manipulating the nature of stationary phase and/or the column
temperature. In contradistinction, the variation of eluent composition
alone extends both retention and selectivity in HPLC over an extremely
broad range. It is not surprising then that in gas chromatography a plural-
ity of stationary phases has found practical application whereas HPLC
tends toward the use of a very limited number of columns and the optimization of the separation by manipulating the composition of the mobile
phase.

In this regard RPC has unique advantages and it is not surprising that
applications of reversed phase chromatography showed a dramatic increase after 1974 when columns packed with microparticulate octadecyl
silica became widely available from various sources for use in HPLC. The
popularity of the technique rests with the convenience, versatility, and reproducibility offered by the use of hydrocarbonaceous bonded phases.
The convenience is mainly due to the fact that the employment of a single
column suffices and relatively short time periods are required for the equilibration of the chromatographic system upon changing eluent composition. Furthermore aqueous eluents having high optical transparency as
well as low flammability and toxicity can be used to accomplish most separation goals. The reproducibility of the method stems not only from the
relatively rapid equilibration of the system but also the stability of the
column in comparison with other microparticulate bonded phases that
yield high efficiency and high speed of separation. In comparison to silica
based bonded phases with covalently bound polar functions, hydrocar-

bonaceous bonded phases such as octyl and octadecyl silica are relatively stable in contact with most aqueous eluents having a pH value less than 8. Nevertheless, further improvement in the stability of such stationary phases would be highly desirable in order to obtain columns of extended operational life.

Probably the most important feature of reversed-phase chromatography is its versatility, i.e. its broad scope in terms of the classes of chemical compounds which can be separated efficiently by using this technique. Whereas in terms of speed, efficiency, and loading capacity columns of other types can compete successfully, the scope of RPC is probably the widest among all chromatographic techniques developed so far as far as the chemical nature of the substances to be separated and the means of manipulating the selectivity of the chromatographic system are concerned. As a result, nonpolar bonded phases have largely replaced silica gel and ion exchangers that were previously the most widely used stationary phases in HPLC. The interest in nonpolar sorbents for chromatographic applications, of course, is not new. In fact, the competition between silica gel and charcoal, the major nonpolar solid stationary phase of the past, had been very vigorous before the introduction of ion-exchange chromatography. In the comprehensive treatment of chromatography by Lederer and Lederer (1) which was published in 1956, charcoal was the most frequent entry in the subject index after silica gel.

In this chapter we attempt to give a definitive treatment of most aspects of RPC. The use of HPLC, which allows very precise chromatographic data to be collected, is relatively new and a great deal more effort will be needed to find satisfactory interpretations for the chromatographic retention of complex molecules. Nevertheless, the experience accumulated in the brief history of RPC can guide in finding the appropriate conditions for solving most separation problems.

It is hoped that a closely related family of columns will evolve as a universal column set to satisfy the need not only for separations but also for the measurement of physicochemical properties of diverse solutes over a wide molecular weight range by RPC. The wide use of RPC has intrigued the analysts to find out more about the factors governing selectivity and the theoreticians to establish a firm theoretical framework for the interpretation and prediction of retention data. This endeavor, albeit still in its infancy, may expand the scope of HPLC beyond that of an analytical or separative tool and turn it into an instrument for physicochemical measurement of molecular properties, particularly those of biological substances, in solution. Thus, RPC is responsible for a renewed focus on chemical phenomena in chromatographic science where research in instrumentation and column efficiency were in vogue for the last 20 years.

B. Evolution of Reversed-Phase Chromatography

The birth of the term reversed-phase chromatography can be traced back to the treatment of kieselguhr with dimethyldichlorosilane vapor by Howard and Martin (2). After obtaining a nonpolar porous support this way, these authors could use for the first time a nonpolar liquid such as paraffin oil or octane as the liquid stationary phase in conjunction with a polar eluent such as methanol–water or methanol–acetone mixtures. The chromatographic system thus obtained was the reverse of conventional systems in terms of phase polarity and the respective retention orders were also found to be reversed; consequently, the technique was christened reversed phase partition chromatography. According to a personal account by Martin, he gained the idea of employing silanization for the preparation of a nonpolar support by reading an article in an electrical engineering journal about the advantages of silanizing the surface of insulators. In retrospect we may say that his discovery led not only to reversed-phase chromatography proper but, also, to the genesis of bonded stationary phases that have been largely responsible for the success of HPLC in this decade.

The idea of using a nonpolar stationary phase and a polar mobile phase was first put forward in the literature by Boscott (3), who advocated the use of cellulose acetate as the stationary phase. Shortly thereafter Boldingh (4) separated C_8–C_{18} fatty acids by using moderately vulcanized rubber saturated with benzene as the stationary phase and water–methanol as the mobile phase. This was rapidly followed by other work (5–7) and the method was widely practiced by lipid chemists until the advent of HPLC. The main problem with the use of rubber for the stationary phase is that the degree of swelling is extremely critical in determining the magnitude of retention and memory effects are substantial. Reproducible results can be obtained only with precise temperature control and batch to batch variations of the column material have significantly impeded the wide acceptance of the technique that served well lipid chemists, who mastered the method, for many years all these difficulties notwithstanding.

In many respects the partition method pioneered by Howard and Martin (2) was easier to control and it is not surprising that it enjoyed popularity and that the technique was extended to the separation of higher molecular weight fatty acids (8). Reversed-phase paper chromatography using silicone oil or vaseline-treated filter paper was also introduced (9). The first precolumn derivatization in conjunction with RPC was the hydroxylation of unsaturated fatty acids by permanganate (10) in order to facilitate their separation by the method of Howard and Martin

(2) and the conversion of volatile carbonyl compounds into 2,4-dinitrophenylhydrazones (11). The first study of structure and retention relationships (12) in RPC focused on the effect of the number and positions of double bonds in fatty acids and the substituents on the retention behavior of fatty acids in a partition system. The results are, of course, similar to those observed later with hydrocarbonaceous bonded phases in HPLC.

From another perspective chromatographic methods employing charcoal or other similar carbonaceous sorbents can also be considered as precursors of present day RPC. Following the early studies by Abderhalden and Fodor (13) on charcoal adsorption of amino acids and peptides, and that of fatty acids by Cassidy (14, 15), the technique was used for the separation of amino acids (16) and unsaturated acids (17) in aqueous ethanol. The potential of the method has been most thoroughly explored by Tiselius and co-workers (18–20). They recognized that the heterogeneous surface of charcoal is responsible for irreversible retention of solutes and poor resolution. Therefore, a "saturator," usually n-decanol, was employed to obtain a "soft" surface and facilitate chromatographic separation of complex biological substances on charcoal. Thus, adrenocorticotropic peptides could be chromatographed on charcoal with decanol as saturator (21) or on charcoal diluted with kieselguhr by using 0.4% of trimethyl lauryl ammonium salt in the eluent (22). This sorbent, which preferentially adsorbs aromatic compounds (23) could have gained broad use in chromatography save for its property of catalyzing the oxidation of organic substances. Whereas treatment with KCN or H_2S has been successfully used to abate the catalytic activity (24, 25) the subsequent development of ion-exchange chromatography greatly reduced the need for other solid adsorbents. Nevertheless attempts to use polymeric sorbents of nonpolar character has continued and, among the early works, the employment of polyethylene powder (26) merits mentioning.

The term reversed-phase chromatography has also been used for the technique which employs a water immiscible strong electrolyte such as di(2-ethylhexyl)orthophosphoric acid (27) or a methyl tricapryl ammonium salt (Aliquat 336) as the liquid stationary phase and an aqueous eluent. This technique has mainly been used in the field of inorganic chemistry and the interested reader is advised to consult the review by Cerrai and Ghersini (28). The above-mentioned stationary phases have also been called "liquid ion-exchangers" and are predominantly employed in the extraction of metals. Therefore, this type of chromatography has also been termed extraction chromatography.

Horváth and Lipsky (29) used first this kind of reversed phase chromatography for the separation of biological compounds by HPLC and the technique acquired prominence as the most powerful method for nucleic acid separations (*30, 31*). In these methods Aliquat 336 or the structurally similar Adogen 464 "liquid ion-exchangers" were used as the stationary phase. As reversed phase extraction chromatography in the inorganic field evolved from metal separations, extraction methods developed in the pharmaceutical field played a formative role in exploiting this technique for drug separations (*32, 33*) and gave rise to the use of the term "ion-pair chromatography." We shall discuss later ion-pair chromatography with nonpolar bonded stationary phases that is fundamentally different from chromatography using liquid ion exchangers as the stationary phase (*34*).

Perusal of the chromatographic literature shows that a wide variety of methods in partition chromatography, including the fractionation of ribonuclease on cellosolve as the liquid stationary phase (*35*), were called reversed-phase chromatography, yet, the term was generally not employed for the use of charcoal as the stationary phase.

Efficient and relatively stable columns, which are responsible for the present popularity of RPC, have evolved from the concept of bonded phases. Treatment of siliceous surfaces with reactive silanes gained wide technological importance by the middle of the 1960s when nonpolar bonded phases, which were obtained by reacting silica gel with hexadecyltrichlorosilane, were introduced in gas chromatography (*36*). Soon thereafter Stewart and Perry (*37*) suggested that such a material would yield a particularly useful stationary phase for liquid chromatography. They envisioned "that a family of anchored phases offers the best (if not the only) present possibility for liquid–liquid chromatography of lipophilic mixtures without the necessity for precision thermostating and careful preequilibration of conventional (immiscible) phase pair."

The practical use of hydrocarbonaceous bonded phases in HPLC began when Kirkland and DeStefano (*38*) developed a pellicular column material which was commercialized under the trade name Permaphase ODS. It was prepared by forming polyoctadecylsiloxane in the cavities of pellicular silica support and by simultaneously binding the finely dispersed polymer via siloxane bonds to the porous siliceous layer on the surface of glass beads. The approach represented a combination of the pellicular configuration introduced by Horváth *et al.* (*39*) with the siloxane chemistry suggested by Abel *et al.* (*36*). Since the actual stationary phase was confined to the periphery of the fluid impervious pellicular support, it possessed favorable mass transport properties and yielded relatively high column efficiency. Furthermore, the stationary phase was stable toward aqueous eluents even at relatively high temperatures due to the resistance of siloxane bonds to hydrolysis.

Subsequently, Majors (*40*) popularized microparticulate bonded phases made with reactive silanes. Such column materials are used almost exclusively in RPC at present. It should be noted that in these bonded phases, which are prepared by siloxane chemistry, the organic functions are attached by $\equiv$Si—C$\equiv$ bridges to the surface. Consequently, they are fundamentally different from the "estersil"-type (41, 42) "brushes" which gained wide popularity in gas chromatography after the work by Halász and Sebastian (43). In the latter type of stationary phases organic functions are bound to the surface via $\equiv$Si—O—C$\equiv$ bridges that are not stable in the presence of nucleophilic agents such as water or alcohols; therefore, they have not found application in HPLC.

Although to most chromatographers today RPC appears to be a relatively new branch of HPLC, we have seen that, on the time scale of chromatographic history, it has a long and a venerable past. After the beginning of HPLC with pellicular ion exchangers (39, 44) and microparticulate as well as pellicular packings for liquid–liquid chromatography (45, 46) the spectacular rise of RPC came to most experts in the field as a total surprise. Nonetheless, as we showed above, the exploitation of hydrophobic interactions for chromatographic separations is not so new. Even ion-exchange chromatography, particularly with resinous stationary phases, features the hydrophobic retention mechanism prevalent in RPC in addition to specific ionic interactions at the fixed charges on the stationary phase. In fact when it is employed for the separation of nonionic substances, such as in "salting-out" (47) and "solubilization" (48) chromatography, "ion-exchange" chromatography properly can be considered a kind of RPC.

On the other hand electrostatic interactions in the mobile phase largely contribute to the selectivity of RPC when ionized eluites are involved. As a result RPC has the potential to replace ion-exchange chromatography as far as biochemical separations are concerned. Hydrophobic interactions have been almost always involved in the chromatographic process used for separation and purification of complex biological substances and biopolymers. Examples are numerous and scattered throughout the biochemical literature. It should be noted, however, that the recent evolution of hydrophobic affinity chromatography (49, 50) parallels that of RPC although the technique, which employs agarose with covalently bound hydrocarbonaceous functions as the stationary phase for the separation of biopolymers, is not expected to have the same wide analytical significance as the method used in HPLC.

C. Description of the Technique

At present the term RPC is mostly associated with the use of a nonpolar solid stationary phase and a polar eluent. Consequently the name RPC encompasses a wide range of chromatographic systems and lacks a precise operational definition as do the adjectives polar or nonpolar. Nevertheless the present usage of the method in HPLC is closely associated with the employment of hydrocarbonaceous bonded phases such as octadecyl silica. In HPLC the column material is preferentially microparticulate and has a narrow particle size distribution with a mean diameter less than 10 μm in analytical work. For columns used in preparative scale RPC, of course, larger particles are used.

Both spherical and irregularly shaped silica particles are used as supports and n-octadecyl as well as n-octyl functions are the more popular organic ligates attached covalently to the support surface. The inner diameter of analytical columns is usually about 5 mm and the length varies between 50 and 500 mm. They are operated at eluent flowrates ranging from 0.5 to 3 ml/min. At present a variety of nonpolar bonded stationary phases are commercially available. They differ in particle shape, particle and pore size distribution, the nature of the ligate, the morphology of the bonded surface layer, and the magnitude of surface coverage. Fundamentally all hydrocarbonaceous bonded phases having an alkyl chain length of eight or more carbons are very similar. Yet, differences between the various commercial products can occasionally be quite significant in practice. The technology used for the preparation of these column materials is quite new and we can expect further improvements in the chromatographic properties of the products. Moreover advances in specification and characterization of these stationary phases should facilitate their classification according to the practical needs of the analyst. Most analytical work can be carried out with a 150-mm-long column packed with 5 μm octadecyl or octyl silica of good quality and the retention behavior of various types of solutes is manipulated by varying the composition of the eluent.

In fact, it is the very broad useful range of eluent strength together with the employment of specific solvent effects and complexation in the mobile phase that is responsible for the wide scope of RPC. In general, the lower the polarity is the higher the strength of the eluent is at a given set of conditions. The changes in eluent strength upon going from a plain aqueous mobile phase to a nonpolar solvent such as hexane may be equivalent to a difference of several hundred degrees in column temperature. For convenience we may distinguish between the techniques according to the mobile phase used: PARP, MARP, and NARP (51) where the acronyms stand for plain-aqueous, mixed-aqueous, and nonaqueous reversed-phase chromatography, respectively. For the separation of polar substances the weakest eluent is used. Since neat aqueous solutions without organic solvent have the lowest eluent strength polar mixtures are usually separated in the PARP mode. In most cases, however, hydroorganic eluents are used. Water–methanol and water–acetonitrile mixtures are most popular and the strength of such eluents drastically increases with the concentration of organic solvent. Therefore MARP finds the widest practical application. In both PARP and MARP it is advisable to use a buffer in the eluent in order to maintain an ionic strength greater than 0.01 and a constant pH. In RPC of ionogenic substances the use of buffer which serves as a background electrolyte and maintains the pH constant is essential to obtain reproducible results and quasi-symmetrical peaks. For the separation of nonpolar substances on hydrocarbonaceous bonded phases nonaqueous eluents are used; i.e. the elution mode is NARP. This approach has not yet been fully explored. It is believed that in contact with a nonaqueous eluent the residual silanol groups at the surface of the stationary phase play a dominant role in determining retention behavior. In the limit separations of certain substances can be obtained by using a neat hydrocarbon such as hexane in the eluent and the question may be legitimately raised whether such a chromatographic system still can be considered "reversed phase."

A great deal of flexibility in adjusting chromatographic selectivity can come from exploiting special solvent or complexation effects. A change from water–methanol to water–acetonitrile can already give a change in elution order. The employment of a ternary eluent containing three different solvents can be a particularly powerful tool to alter selectivity and bring about the resolution of solute pairs having low relative retention values, α, when eluted with a binary eluent in RPC. As discussed by Karger *et al.* (52) in great detail and shown later in this chapter, secondary equilibria in the mobile phase can specifically alter the retention behavior of certain eluites. In the simplest case protonic equilibria can be exploited to change the degree of ionization and thereby the retardation factor of ionogenic solutes upon controlling the pH of the eluent. The greater is the degree of ionization of a solute the faster it is eluted under otherwise fixed conditions. Thus, weak acids and bases are retarded longer at low and high eluent pH, respectively.

The formation of a complex can have dramatic effect on the retention of an eluite. So far the use of sulfonic acids or quaternary ammonium compounds, bearing substantial hydrophobic moieties, is exploited most frequently to enhance selectively the retention of basic and acidic eluites due to "ion-pair" formation. Since these complexing agents which are present in the eluent are detergents the technique is also referred to as "soap" chromatography (53), although the most commonly used name is ion-pair reversed-phase chromatography, IP-RPC. The formation of other types of complexes, for instance, those of unsaturated compounds with silver ions present in the eluent has also been exploited to augment chromatographic selectivity.

In order to enhance peak capacity, gradient elution is frequently used in RPC. Hydrocarbonaceous bonded phases facilitate rapid equilibration and reequilibration of the column with the eluent. In the most convenient gradient elution system a predominantly aqueous hydroorganic solvent such as a water–acetonitrile mixture is used first and the concentration of the organic component, e.g., acetonitrile, is increased during the gradient run. The peculiar aspects of gradient elution are treated in detail by Snyder (54) in the first volume of this series.

The above brief description of RPC should serve the need of those who want to get a general idea of the technique before exploring some special problems that are discussed in the following parts of this chapter. The scope of the technique, however, is expected to broaden in the near future. Introduction of novel column materials as well as chromatographic systems with novel properties, for instance, may facilitate the rapid separation of biopolymers in analytical practice. Although it is difficult to predict further advances, RPC, perhaps under another name, seems sure to remain one of the most important separation systems for general chromatographic use in HPLC.

II. THE STATIONARY PHASE

A. General Properties

1. The Role of the Stationary Phase

It has been customary to classify the various techniques of liquid chromatography according to the stationary phase employed. Thus, we traditionally distinguish between liquid–liquid and liquid–solid chromatography denoting by the second adjective the physical state of the stationary phase. As we have seen previously, the name reversed phase originates from a liquid–liquid chromatographic technique employing paraffin oil as the stationary phase on a nonpolar porous support and a hydroorganic mixture such as water–methanol or water–acetone as the mobile phase.

The present popularity of RPC, however, is due to the development of hydrocarbonaceous bonded phases in microparticulate form which yield columns particularly suitable for use in sophisticated HPLC equipment. Although the column constitutes only a small part of an analytical liquid chromatograph, both in dimension and price, it occupies a supremely important position. First of all, the chromatographic process takes place in the column proper; therefore, we cannot obtain a better separation of the sample components than that yielded by the column. Besides efficiency, the stability, rapid response to changing conditions, and reproducible behavior are also very important features of the column. Dependable operation of the liquid chromatograph to a large extent rests upon the reliability and durability of the column. In view of the intrinsic instability of liquid–liquid chromatographic systems, it is surprising that much time was spent on attempts to develop such systems for HPLC in the early 1970s, even after a variety of bonded phases was developed to replace the actual liquid by a molecular fur covalently bound to the surface of a porous support. Stationary phases made by following the same concept but a different approach have been eminently successful in gas chromatography (43). After the introduction of bonded phases a natural selection process has taken place and finally those column materials that were best suited for use in sophisticated HPLC equipment have replaced the others.

In retrospect it is perhaps easy to say that silica-based hydrocarbonaceous bonded phases have had the greatest potential in HPLC, although this was not readily recognizable at the outset. The use of mesoporous silica gel for the preparation of the column materials has numerous advantages. The reader is referred to the recent book by Unger (55) for detailed information on the preparation, properties and chromatographic applications of silica. It suffices to say that silica gel is probably the best known aerogel and the technology of controlling pore size and pore size distribution in a very wide range is available. Silica can be produced in bulk and subsequently crushed and size graded to obtain uniformly sized, irregularly shaped particles. On the other hand microspheres of silica gel having controlled pore size can be grown to obtain a quasi-monodisperse product which may not require size grading. Advances in silicon chemistry and polymer technology after World War II led to the development of a large variety of reactive organosilanes with various functional groups. Some of them have been used to treat the surface of glass and other siliceous materials used as reinforcement or filler in plastics in order to promote adhesion of the polymers to the surface of the inorganic material (56).

Before a family of bonded phases found its way into HPLC, methods for covalently binding organic moieties to siliceous surfaces via bridging groups were successfully employed for the immobilization of biological molecules such as enzymes at solid surfaces. Recently heterogeneous catalysts have also been prepared by a similar approach. Nevertheless it has been the field of HPLC where silanization methods found a salient application for modifying silica surfaces and thereby obtaining a family of novel stationary phases of widely varying properties that are in many respects superior to that of silica. The properties and use of some of these materials are also discussed by Majors (57) in the first volume of this series.

In spite of the great variety of bonded phases, many of which have been developed to replace silica gel and resinous ion-exchange resins, prominence has been attained mainly by those with nonpolar ligates. The reason for this is, of course, the broad scope of RPC, which could be explored and exploited because efficient and stable stationary phases had been available.

To a great extent, column efficiency depends on the particle size and size distribution as well as on the pore structure and surface properties of the stationary phase. Major advances in HPLC had been made possible by the development of technology for making uniform microparticulate silica gel, either spherical or irregularly shaped, so that suitable siliceous support materials were available for the preparation of bonded phase yielding high efficiency. Covalent binding of hydrocarbon chains, such as octadecyl and octyl functions, to the surface results in a soft stationary phase surface having favorable kinetic properties for mobile phase equilibration and elute adsorption. Such column materials are relatively stable toward hydrolytic decomposition. On the other hand, siliceous bonded phases with polar functional groups often have poor stability in contact with aqueous eluents, and as a result column life is short. Hydrolytic cleavage of surface siloxane bonds is much faster when they serve as bridges to the silica surface for polar groups rather than for nonpolar groups. Moreover, the polar organic moiety, being water soluble, is rapidly removed by the eluent and the exposed silica is available for further attack by the aqueous medium and further erosion of the support material occurs.

It is instructive to estimate the equivalent sphere diameter which corresponds to the silica gel used for the preparation of bonded phases. Assuming specific surface area of 360 m^2/g for silica having a bulk density of 2.5 g/cm^3 we find that the surface to volume ratio is about the same as for hypothetical 6 nm diameter particles. At such a high degree of dispersity it is not at all surprising that the column material rapidly degrades even if the rate of dissolution per unit surface area is relatively low.

The column packing structure is a delicate arrangement of the particles and some experience is required to produce efficient columns with small, e.g. 5 μm, particles in a reproducible fashion. The particle size distribution is usually not very uniform so that some small particles, called fines, are also present. If such small particles occupy an important position in the structure and thereafter degrade upon exposure to eluent flow, perturbances in the packing may occur with a concomitant deterioration of column efficacy and the whole packing structure may even collapse. Usually the packing is formed in a high-speed, high-pressure filtration process and, as a result, some of the particles in the packing can be under high stress. As we know from tribochemistry, chemical reactions in solids accelerate under stress so that the smallest particles are subject to a relatively rapid degradation with the consequences mentioned above.

Nonpolar phases have exhibited the highest stability among bonded phases toward aqueous eluents in a relatively wide pH range and this finding has greatly contributed to the popularity of reversed phase chromatography. For sake of simplicity we can picture the surface of these sorbents as a thin, greasy film that is covalently attached to the silica backbone. The unctious layer would not be soluble in aqueous or mixed aqueous eluents even in the absence of covalent bonds to the surface. Consequently it shields the silica from water and protects it from hydrolytic degradation to a much greater extent than do other functional groups. Recent improvements in silanizing technology facilitate the preparation of products with a minimum amount of accessible free silanol groups at the surface so that such bonded phases can be used in contact with aqueous eluents up to pH 9.0 for an extended period of time. Hydrocarbonaceous bonded organosilica type stationary phases represent composite materials which combine the meritorious morphology, as well as mechanical and chemical (toward organic solvents) stability of silica gel with the inertness and uniformity of the hydrocarbonaceous molecular fur.

2. The Ideal Stationary Phase

Earlier in this chapter we contrasted RPC with gas chromatography on the basis of the pluralities which characterize the two techniques with respect to the mobile and stationary phases, respectively. Ideally, all HPLC work could be performed by using RPC and a column packed with one type of nonpolar stationary phase. It is relatively easy to establish the specifications for this material as follows.

(i) It has high efficiency and facilitates rapid analysis. This is accomplished by relatively short, i.e. 5 to 30 cm, microparticulate columns packed with 5 μm porous particles. Such columns yield a sufficiently high number of plates and short separation times to carry out most chromatographic analyses in practice. Uniformly sized silica gel particles, both spherical and irregularly shaped, are available and the technology for packing them uniformly has been developed.

(ii) It has adequate stability and yields long column life even under widely varying operating conditions. Column packings made of organic polymers are usually not stable in contact with organic solvents as they may swell or shrink. Silica gel is not stable in contact with aqueous eluents, particularly at high pH. However, a porous siliceous matrix, the surface of which is coated with a covalently bound long-chain aliphatic hydrocarbon offers a composite material with nearly ideal behavior.

(*iii*) The column should permit the modulation of retention behavior over a very wide range of conditions. This requirement in fact means that the stationary phase is inert, that is it does not facilitate specific interactions with certain molecular functions of solute molecules with the concomitant advantage of a relatively clean and rapid adsorption–desorption kinetics. Preferably then the stationary phase has no functional groups such as fixed charges that would have strong affinity to counterionic solutes and exclude solutes of co-ionic nature. In this regard the properties of well-prepared hydrocarbonaceous bonded phases indeed approach those that we would expect from an ideal phase.

(*iv*) In order to facilitate efficient separation of samples falling in different molecular weight ranges, it is necessary that the column material be available with different mean pore diameters and pore size distributions. The technology for controlling these properties of silica gel for the preparation of silica particles having uniform particle size is available. For chromatography of small molecular weight (MW < 3000) substances, silica support with 10 nm mean pore diameter is sufficient. Silica support with controlled 20, 30, and 40 nm average pore size is available for use in higher molecular weight ranges and, in fact, is used today in exclusion chromatography.

(*v*) Chromatographic efficiency also depends on the uniformity of the stationary phase surface in that the free energy change and concomitantly the adsorption–desorption rate constants associated with the chromatographic process on the molecular level fall within a rather narrow range. Experience shows that the molecular fur of alkyl chains (C_8–C_{18}) covalently bound to the surface yields a rather uniform nonpolar surface which in this regard is superior to that of other nonpolar sorbents such as macroreticular polymeric, carbonaceous, and fluorocarbonaceous sorbents.

In view of the foregoing discussion of the properties of hydrocarbonaceous bonded phases on silica support it is readily appreciated that well-prepared stationary phases presently used in RPC approach the ideal with the exception of their relatively poor stability in contact with aqueous eluents, particularly at high pH, and the fact that surface silanol groups cannot be completely eliminated. The latter may interact with polar solutes, particularly when the dielectric constant of the eluent is relatively low. Nevertheless, residual surface silanols can be masked by alkylamines in the eluent with the result that peak tailing, when it is due to silanophilic interactions, is substantially reduced. By and large, the presence of silanols may not be as undesirable as thought originally as long as they do not decrease the stability of the column.

3. Hydrocarbonaceous Bonded Phases

Silanization of siliceous materials originally served to make the surface hydrophobic by eliminating silanol groups. With the advent of HPLC, covalent binding of organic functions, which are conveniently referred to as ligates, to the surface of silica gel supports has given rise to a family of bonded phases that are discussed by Majors (57) in detail.

Bonded phases with hydrocarbonaceous ligates are employed in RPC almost exclusively. A particular feature of such stationary phases is that the hydrocarbon ligates are covalently attached to the surface of silica gel by siloxane bridges and could be considered to behave as a liquid layer bonded to the surface. As mentioned earlier, such a surface is expected to be more homogeneous than the hydrocarbonaceous surface of porous polymer such as macroreticular polystyrene or polyethylene, or the surface of porous carbon. It may be said that the hydrocarbonaceous surface fully covered with ligates is a much "softer" surface than that of the other sorbents mentioned above that have a "hard" surface. The idealized structure of octadecyl and octyl silica bonded planes is illustrated in Fig. 1, where an attempt was made to depict the various structural elements by

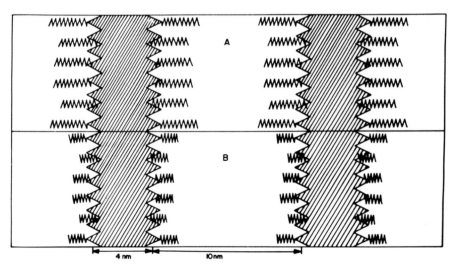

Fig. 1.   Schematic illustration of the structure of (A) octadecyl and (B) octyl silica. The cross-hatched and the open areas represent the solid and open areas of the porous siliceous support.

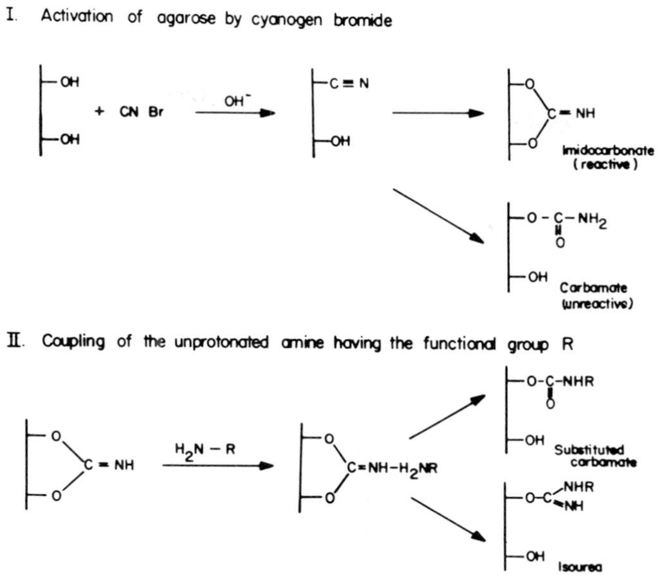

I. Activation of agarose by cyanogen bromide

II. Coupling of the unprotonated amine having the functional group R

FIG. 2.   Chemistry employed in the attachment of hydrophobic functions, R, to agarose for use in hydrophobic chromatography of biopolymers. Redrawn from Egly and Porath (59), with permission from Pergamon Press.

using the same scale. In the next section the preparation and properties of such phases will be discussed in detail.

It should be mentioned here that another family of hydrocarbonaceous phases has been developed for the hydrophobic chromatography of biopolymers (49, 50). For the preparation of such stationary phases agarose beads are first activated with cyanogen bromide and subsequently reacted with a primary amine having the appropriate hydrocarbonaceous moiety. The sequence of reactions that is used for preparation of alkyl agaroses such as ethyl, butyl, and hexyl agaroses is illustrated in Fig. 2. This type of hydrocarbonaceous stationary phase has pores sufficiently large to be penetrated by biopolymer molecules having a molecular weight of 10^6 daltons, due to the highly porous nature of swollen agarose gels. A main advantage of agarose (58), besides penetrability by large molecules, is inertness as acidic groups are removed in the course of purifying this naturally occurring carbohydrate. However, some carboxylic functions still may be present. The swollen spherical gel particles contain only 1 to 4% of solid material and consequently have very poor mechanical stability. Agarose and its derivatives, therefore, cannot be used in columns exposed to significant pressure gradients.

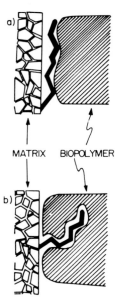

MATRIX   BIOPOLYMER

FIG. 3. Schematic illustration of two possible ways for protein binding to the alkyl agarose stationary phase in hydrophobic chromatography. (a) The alkyl chain intrudes into a hydrophobic pocket of the biopolymer; (b) contact between the alkyl chain and the nonpolar regions (greasy patches) at the protein surface.

Chromatography with hydrocarbonaceous agarose conjugates as the stationary phase is traditionally called hydrophobic chromatography and it should be considered as a kind of reversed phase chromatography. As schematically shown in Fig. 3 the interaction between the hydrocarbon chain of alkyl agarose and the nonpolar regions in the biopolymer can involve either some external hydrophobic (''greasy'') patches on the surface or some internal pockets. The magnitude of chromatographic retention is determined by the magnitude of such hydrophobic interactions and hence the name of the technique (60).

Steric effects are believed to play a more significant role in determining the magnitude of solute–stationary phase interactions with molecules having complex three-dimensional structure such as biopolymers than with small molecules having relatively simple structure. Nevertheless steric effects always play a certain role; therefore the morphology of the hydrocarbonaceous stationary phases can have a significant effect on the hydrophobic or solvophobic interactions underlying solute retention in reversed-phase chromatography.

Besides hydrocarbonaceous bonded phases, other nonpolar stationary phases can also be employed in RPC. Although some of them will be briefly surveyed at the end of this section, main attention is paid to hydro-carbonaceous bonded phases on siliceous support as these materials have become the workhorses of HPLC as it is practiced today.

B. Preparation and Characterization of Bonded Hydrocarbonaceous Stationary Phases

1. Silica Support

The properties of the silica aerogel used for the preparation of bonded stationary phases have a far-reaching effect on their chromatographic behavior. The recent book by Unger (55) gives a definitive treatment of the properties of silica with particular regard to its use in modern liquid chromatography. On the other hand, valuable information on silica gel can be found in chapters written by Majors (57) and Engelhardt and Elgass (57a) in this series. Nevertheless it appears to be worthwhile to recapitulate some properties of the silica support that have a significant influence on the chromatographic behavior of hydrocarbonaceous bonded phases. Certain specifications of a popular silica support and an octyl silica bonded phase are listed in Table I.

First the particle size and shape should be mentioned. As columns packed with nominally 5 or 10 μm particles are used most commonly in HPLC, the particle size distribution has become an important character-istic of the packing because with such small particles it is difficult to ob-tain a monodisperse product. Size distribution of the particles can be rep-resented in many different ways. Most commonly the cumulative par-ticle size distribution curve obtained with an instrument such as the

TABLE I

Specifications of a Commercial Silica and an Octyl Silica[a]

Specifications	LiChrosorb Si-100 (silica)	LiChrosorb RP-8 (octyl silica)
BET surface area (m²/g)	275	245
Average pore diameter[b] (nm)	10	9.5
Nominal particle diameter (μm)	10	10
$d_{10\%}(\mu m)$	5.5	7.6
$d_{50\%}(\mu m)$	9.0	9.6
$d_{90\%}(\mu m)$	11.8	11.8

[a] $d_{10\%}$, $d_{50\%}$, and $d_{90\%}$ are the upper limits of particle diameter for 10%, 50%, and 90% of the particle population. Compiled from data reported by Brownlee and Higgins (62).

[b] Measured by mercury porosimetry.

Coulter counter is used. One leading manufacturer of microparticulate silicas prefers to characterize the size distribution by the three ogives $d_{10\%}$,
$d_{50\%}$, and $d_{90\%}$ which denote those particle diameters that represent the
upper bound for 10, 50, and 90% of the total population, respectively. For
instance, a powder having nominally 10 μm particle size may be characterized by the ogives for the three percentiles as $d_{10\%} = 5.5$ μm, $d_{50\%} =$
9.0 μm, and $d_{90\%} = 11.8$ μm. Thus 10% of the particles are smaller and
10% are greater than 5.5 and 11.8 μm, respectively. On the other hand,
40% of the particles are as great or greater than 9.0 μm and less than
11.8 μm. The particle size distribution affects the packing structure and,
as a result, the stability of the column. The effect of size distribution on
efficiency is not entirely clear, whereas column permeability is adversely
affected by the presence of particles having diameters significantly
smaller than the average. Consequently the removal of fines is advisable
to augment column permeability and stability as well. In general the difficulties in obtaining narrow size distribution by the usual classification
techniques increase as the particle diameter decreases, so that the uniformity of size distribution of particles having an average diameter of
5 μm is usually inferior to that of those with 10 μm mean particle diameter. One of the main reasons why HPLC started with pellicular column
materials instead of microparticulate packings was the lack of adequate
techniques for classifying (size grading) subsieve particles in the early
1960s. The relatively large pellicular stationary phase particles could be
obtained with a very narrow size distribution by a regular sieving process.

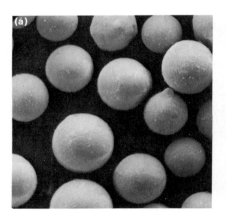

Fig. 4. Electron micrograph of (a) spherical (Spherisorb) and (b) irregularly shaped
(Partisil) octadecyl silica particles. The scale, given on the micrograph of the irregularly
shaped particles, is the same in each figure.

Still and all, the diffusion distances in the stationary phase proper were quite short, a few microns, due to the pellicular configuration.

It has been assumed in the above discussion that silica gel is prepared in bulk and crushed thereafter. The irregularly shaped particles are then size-graded by using a suitable air classifier. The fines that adhere to the particles are removed by washing and sedimentation. However, one can also obtain uniform particles by letting colloidal silica coalesce into microspheres. Proper adjustment of conditions such as the nature and concentration of the starting material, pH, temperature, additives, and stirring rate can result in uniform spherical silica microparticles which do not require size grading. Nevertheless in most instances the size distribution is not sufficiently narrow so that the spherical particles are also air classified. The uniformity of the spherical product is generally higher than that of irregular particles. Figure 4 shows electron micrographs of two kinds of commercial C_{18} reversed-phase packing and illustrates the morphological differences between irregularly shaped and spherical silicas. Conversion of the siliceous packing material into a hydrocarbonaceous bonded phase does not have an apparent effect on the structure at that magnification.

Pore size distribution of the support can significantly affect the chromatographic behavior of bonded phases. Size exclusion may also play a role in interactive liquid chromatography (63). Consequently, size distribution of the pores which can be explored by the eluite can significantly affect selectivity due to size exclusion and may have far-reaching effects on overall chromatographic behavior of the stationary phase. On the other hand, pores commensurable to the molecular dimensions of the eluite constitute relatively large mass transfer resistances as well as "traps" and therefore may be responsible for low column efficiency and tailing.

Figure 5 shows the pore size distribution of two siliceous column materials that are commercially available. At present no satisfactory account of the effect of the pore size distribution on the properties of hydrocarbonaceous bonded phases can be found in the literature. Nonetheless, the differences observed in reversed phase chromatographic responses of various commercial bonded phases having the same ligate but different siliceous substrates may be attributed—at least in part—to differences in the pore structure of the support material.

The surface of the silica is covered with weakly acidic silanol groups and it is assumed that the pK_a value is in the range of 5 to 7. The maximum surface concentration of silanol functions is about 8 μmol/m^2. As the binding of the organic functions takes place via the accessible silanol groups it is desirable to maximize their surface concentration and to pre-

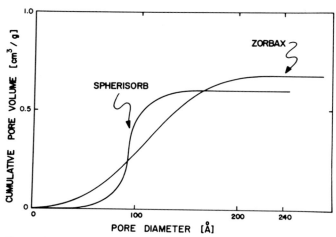

FIG. 5.   Pore volume as a function of the pore diameter for the commercial microparticulate silicas (a) Spherisorb S5W silica and (b) Zorbax silica used for the preparation of C_{18} reversed phase packing. It is seen that the pore size distribution is much narrower for Spherisorb S5W than for Zorbax Silica.

treat the silica prior to silanization. Hydrothermal treatment or soaking the silica in 0.1 N hydrochloric or nitric acid are most commonly employed (62, 64, 65). Both procedures may result in an opening up of the pore structure and the latter treatment also removes metallic contaminants frequently present in silica. For the removal of adsorbed water the product is dried *in vacuo* at temperatures not higher than 180°C so that the surface concentration of silanols does not decrease due to loss of water and formation of siloxane bonds.

2. Silanizing Agents and Their Reactivity

The chemical structure of silanizing agents employed in the preparation of hydrocarbonaceous bonded phases for RPC can be represented by the general formula

$$X_3-\underset{\underset{X_2}{|}}{\overset{\overset{X_1}{|}}{Si}}-R$$

where R is the hydrocarbonaceous function of interest. At least one of the substituents X_1, X_2, X_3 has to be a reactive group such as chloro or methoxy although all three may be reactive. For the preparation of these compounds the reader is referred to the literature of organosilicon chemistry (56).

TABLE II

Chlorosilanes for the Preparation of Hydrocarbonaceous Bonded Phases

Trialkylchlorosilanes	MW	BP[a]	Dialkylchlorosilanes	MW	BP[a]	Alkyltrichlorosilanes	MW	BP[a]
Trimethylchlorosilane	108.7	57	Dimethyldichlorosilane	129.1	70	Methyltrichlorosilane	149.0	66
t-Butyldimethylchlorosilane	150.7	124–126	Hexylmethyldichlorosilane	199.2	204–206	n-Butyltrichlorosilane	191.55	142–143
n-Octyldimethylchlorosilane	206.8	222–225	Octylmethyldichlorosilane	227.3	94/6	n-Hexyltrichlorosilane	219.6	191–192
Octadecyldimethylchloro-silane	347.1	185/0.2	Octadecylmethyldichloro-silane	367.5	185/2.5	n-Octyltrichlorosilane	247.7	224–226/730
Tri-n-butylchlorosilane	234.9	93–94/4	Dibutyldichlorosilane	213.2	212	n-Dodecyltrichlorosilane	303.8	155/10
Benzyldimethylchlorosilane	184.8	75–76/15	2-Adamantylethylmethyl dichlorosilane	277.4	125–126/1.5	Octadecyltrichlorosilane	388.0	160–162/3
Phenyldimethylchlorosilane	170.7	193–194	Dihexyldichlorosilane	269.3	111–113/6	Phenyltrichlorosilane	211.6	201
Diphenylmethylchlorosilane	232.8	295	Phenylmethyldichlorosilane	191.1	205	Cyclohexyltrichlorosilane	231.6	90–91/10
Triphenylchlorosilane	294.9	207–210/12	Diphenyldichlorosilane	253.2	309–310	β-Phenethyltrichlorosilane	239.6	93–96/3
Tribenzylchlorosilane	336.9	140–141[b]	Methyl(2-bicycloheptyl)-dichlorosilane	209.2	115–118/3	Bicycloheptyl-2-trichloro-silane	229.6	63–64/9.5
2-(4-Cyclohexenyl)ethyl-dimethylchlorosilane	202.8	105–110/10	2-(4-Cyclohexenyl)ethyl-methyldichlorosilane	223.2	110/10	2-(4-Cyclohexenyl)ethyltri-chlorosilane	243.6	74–75/0.7
						Benzyltrichlorosilane	225.6	140–142/100
						Adamantylethyltrichloro-silane	297.7	135–136/3
						Eicosyltrichlorosilane	416.0	225–227/3

[a] Boiling point (°C/mm Hg); 760 mm if unstated.
[b] Melting point.

The reagents can be conveniently classified according to (*i*) the chemical nature of the organic moiety of interest, (*ii*) the number of reactive substituents in the molecule, and (*iii*) the nature of the reactive functions. Table II shows chlorosilane reagents that can be used for the preparation of hydrocarbonaceous bonded phases. The first commercial stationary phases were prepared with trichlorooctadecylsilane. In recent years, however, the employment of octadecyl- and octyldimethylchlorosilanes has been on the increase.

Silane derivatives having only one reactive group react with surface silanol groups in a relatively straightforward fashion. Reagents with substituents $X_1 = X_2 =$ methyl and $X_3 =$ chloro, e.g., octyldimethylchlorosilane are the most popular among this type of silane for the preparation of hydrocarbonaceous stationary phases. The surface reaction can be represented as follows:

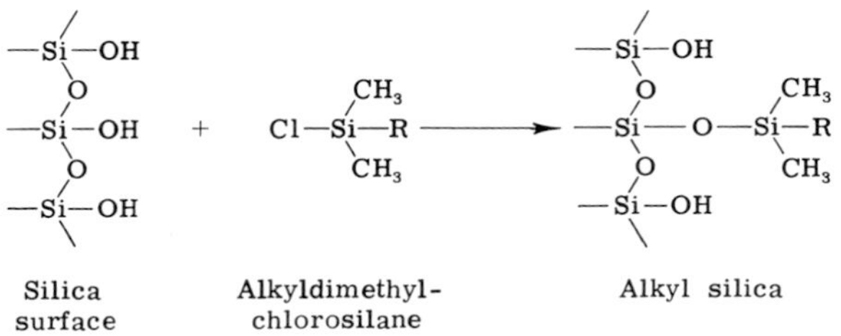

| Silica surface | Alkyldimethyl-chlorosilane | Alkyl silica |

As each reagent molecule can react with only one silanol group, a covalently bound molecular "brush" or "fur" is formed at the surface as illustrated above. Stationary phases thus obtained are called "monomeric" in order to indicate that the surface is covered only with a monomolecular layer of the hydrocarbonaceous functions which are anchored via siloxane bridges to the silica surface as shown for a C_8 bonded phase in Fig. 6. Increasing numbers of commercially available hydrocarbonaceous bonded phases are of the monomeric type as such stationary phases show satisfactory stability in contact with aqueous eluents and can be prepared in a reproducible fashion.

The most reactive silanizing reagents contain three active chlorines. The product of the reaction with the surface silanol groups is not well defined. There are several possibilities for the reaction to occur depending, among others, on the topology of silica surface, the nature (size) of the hydrocarbonaceous function R and the presence of water or another protic agent in the reaction mixture.

In one limiting case a monomolecular layer may be formed according to the following overall scheme:

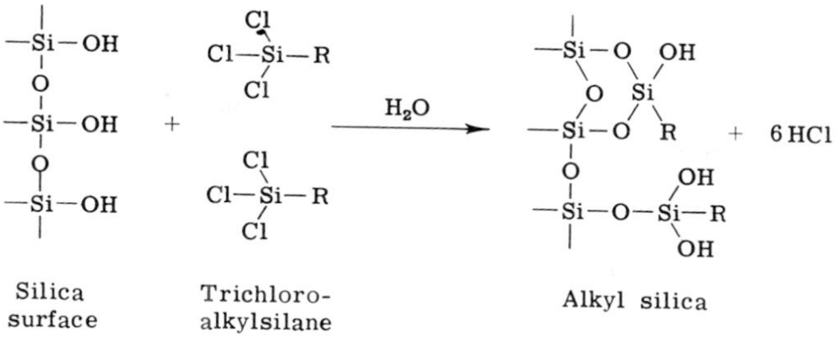

It is seen that the trichlorosilane reacts with the silanol groups to form siloxane bridges. Subsequently the residual chlorines are hydrolyzed. Under carefully controlled reaction conditions it is possible to obtain a product in which the hydrocarbonaceous layer at the surface is similar to that in a corresponding "monomeric" bonded phase. However, the hydrolysis of chlorines that did not react with surface silanols may result in a silanol concentration at the surface that is higher than that in the silica gel proper used as the starting material for the reaction with alkyltrichlorosilanes.

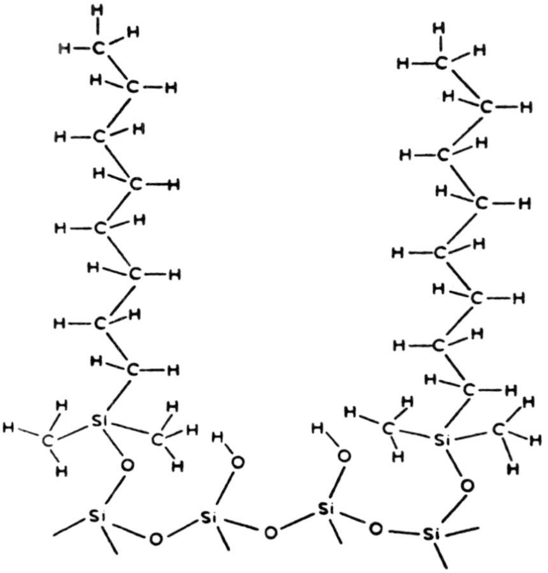

FIG. 6. Surface of "monomeric" octyl silica prepared by the reaction of *n*-octyldimethylchlorosilane with silica gel. Reprinted with permission from Unger (55).

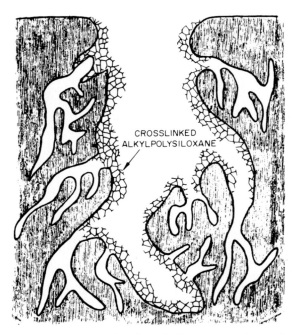

FIG. 7.   Schematic illustration of the alkylpolysiloxane coated pores in a "polymeric" hydrocarbonaceous bonded phase. Reprinted from Horváth and Melander (*129*), *J. Chromatogr. Sci.*, with permission from Preston Publications.

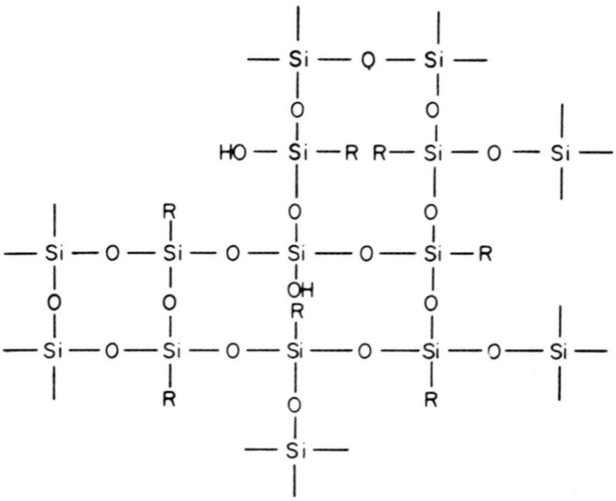

FIG. 8.   Three-dimensional alkylpolysiloxane obtained upon contact of alkyltrichlorosilane with water.

In the other limiting case, a trifunctional reagent can produce a cross-linked alkylpolysiloxane layer at the surface and such stationary phases are called "polymeric." It is believed that the inner walls of the pores in such materials are coated with a silicon rubber layer as illustrated schematically in Fig 7. Whereas the thickness of the silicon rubber layer is exaggerated for sake of illustration, it is believed to be significantly greater than the critical dimension of the hydrocarbonaceous moiety. The reaction conditions in the treatment of microparticulate silica gels are usually so adjusted that the formation of such truly "polymeric" phases is avoided. Figure 8 illustrates the chemical structure of the polymer. Since presence of water is required for the "polymerization" reaction to proceed, control of the water content offers the most convenient means to control the surface properties of the product.

According to the general belief, silanol groups at the surface of hydrocarbonaceous stationary phases interfere with the process of reversed phase chromatography. Indeed, the heterogeneity of the surface often has an untoward effect on the efficiency of the chromatographic system. In stationary phases most widely used in practice, such as phenyl, octyl, or octadecyl silica, the hydrocarbonaceous functions are bulky and have critical dimensions greater than the distance between silanol groups on the surface. Consequently the concentration of residual silanol groups at the surface is significant after the silanization reaction, even when a monochlorotrialkylsilane is used. The most accessible, and therefore most "active" residual silanol groups, however, can be removed by a second silanization reaction in which the hydrogen of the silanols is substituted by the trimethyl silyl function. In the *lingua franca* of HPLC such treatment with trimethylchlorosilane or hexamethyldisilazane $[(CH_3)_3SiNHSi(CH_3)_3]$ is usually referred to as "capping-off" the residual silanol groups. The reaction can be illustrated as follows:

Alkyl silica with residual silanols at the surface	Trimethyl-chlorosilane	Alkyl silica with some residual silanols "capped-off"

It is believed that "capping-off" residual silanol groups is essential to obtain hydrocarbonaceous stationary phases with proper retention behavior and of sufficient stability. Furthermore the batch to batch reproducibility of stationary phase properties can also be improved by this after-treatment.

3. Silanization Procedure

There are numerous methods described in the literature for the preparation of bonded phases (66–76). Generally the pretreated silica is added to a solution of the silanizing agent in a suitable organic solvent such as toluene and the slurry is stirred at elevated temperature, frequently on reflux. When chlorosilanes are used the reaction can be accelerated by adding an amine such as pyridine to the reaction mixture in order to facilitate the removal of HCl. To make a high-quality bonded stationary phase is still an art as expressed by Halász' (73) dictum: "It is not a problem to prepare an RP, but it is extremely difficult to 'cook' a good one."

The physical properties of the silica such as pore diameter play an important role in determining the amount of hydrocarbon which can be bound to the support. The latter is conveniently expressed by the carbon load C_S, i.e., by the weight percent of carbon in the dry stationary phase as measured by elemental analysis. The carbon load of the product usually increases with the reaction time or temperature, as well as with the relative amount of the silanizing agent in the mixture, until a maximum value is reached.

The reproducibility of the process requires rigorous control of the reaction conditions which can vary in a wide range depending on the silica gel and the silanizing reagent as well as on the specifications of the product. After silanization, which may include a second step to "cap-off" surface silanol groups, the product is washed with toluene, methanol, and a methanol–water mixture before filtering and drying to obtain a free flowing powder. The procedure for packing columns is briefly discussed in the next section.

It is also possible to carry out the silanization reaction in situ, so that the silica gel is contacted with the silanizing agent after being packed in the column. A procedure by using trichloroalkylsilanes is described in the literature (77–78). Since corrosive hydrochloric acid was generated in the reaction, the inner wall of the column tubing was coated with Teflon. In situ preparation of hydrocarbonaceous bonded phases has not found widespread use. On the other hand, in situ silanization has been used in the authors' laboratory for rejuvenation of columns. The retentive power of columns that lost hydrocarbonaceous functions from the stationary phase could be restored by perfusion with a solution of octadecyltrimethoxysilane at 120°C.

4. Specifications

In addition to the works (67–76) mentioned above, many attempts have been made to characterize the various hydrocarbonaceous bonded stationary phases available for use in RPC (77–86). A comprehensive treatment of the subject has been given by Unger (55).

The two obvious ways for classification are based on the silica gel support and the chemical nature of the hydrocarbonaceous ligate. The names of commercial stationary phases usually reflect such a distinction and contain the designations of both the silica and the ligate, e.g. Spherisorb ODS and LiChrosorb RP8, where Spherisorb and LiChrosorb are trade names of microparticulate silica gels whereas ODS and RP8 stand for C_{18} octadecyl silica and C_8 octyl silica reversed phase, respectively.

Factors that influence the retentive powers and selectivity of such bonded phases include the surface concentrations of hydrocarbonaceous ligates and free silanol groups. The thermodynamic aspects of solute interactions with the hydrocarbonaceous ligates at the surface, which are hydrophobic interactions in the case of aqueous eluents, are discussed later in this chapter within the framework of the solvophobic theory. In practice, however, solute interactions with surface silanols which may be termed silanophilic interactions can also contribute to retention (71, 75, 93), particularly in the case of amino compounds. Consequently the retention mechanism may be different from that which would be observed with an ideal nonpolar phase. Therefore, increasing attention is paid to the estimation of the concentration of accessible silanols and to their elimination from the surface of bonded phases.

In this section we briefly discuss some significant surface characteristics of silica. The values representing various surface properties are usually derived from three kinds of measurements. First is the evaluation of the specific surface area of the support and/or that of the bonded stationary phase. The use of the BET surface area of silica, S_{Si} (m²/g), is widely accepted although it includes the surface area of micropores that are not explored by most eluites of interest in liquid chromatography. Second is elemental analysis to obtain the carbon load of the bonded phase, C_S (w/w), and third is the measurement of moles of surface silanol groups per gram of column material, m_{OH} [mol/g]. From these data the following specifications are derived most commonly.

a. Surface concentration of silanols, α_{OH}, is used to express a number of micromoles of silanol groups per square meter of BET surface area of the silica and evaluated as

$$\alpha_{OH} = 10^6 m_{OH} / S_{Si} \qquad (1)$$

The magnitude of α_{OH} is greatly dependent on the pretreatment of the silica. For the preparation of bonded phases the surface of the starting material should be saturated with silanols so that α_{OH} is equal to 8–9 μmol/m². This value is in agreement with the findings that the mean area occupied by a silanol group is about 0.2 nm² and the average distance between surface silanols is 0.5 nm (55). Concomitantly, the above α_{OH} value means that there are about four to five silanol groups per square nanometer of silica surface.

The same concept is used in the evaluation of the surface concentration of the silanol groups in bonded phases. In this case m_{OH} is measured with the hydrocarbonaceous phase and the specific BET surface area of the parent silica is used most commonly.

b. Surface concentration of the ligate, α_L, gives the micromoles of ligates per square meter of BET surface area of the silica gel support. It is calculated from the weight of the bound species in grams per gram of adsorbent w, from its molecular weight M, and from the specific surface area of the starting silica corrected by the weight increase upon silanization S_{Si}^* as follows:

$$\alpha_L = 10^6 w / M S_{Si}^* \tag{2}$$

Hydrocarbonaceous bonded phases described in the literature have α_L values ranging from 2 to 4. It is believed that among monomeric phases, those having the highest α_L values are the best for use in RPC. In Table III the pertinent characteristics of some monomeric bonded phases are listed.

c. We introduce the surface carbon equivalent (SCE) in order to express the number of carbon atoms in hydrocarbonaceous bonded phases per square nanometer of BET surface area of the starting silica. It is conveniently calculated from the carbon load and the corrected surface area

$$SCE = 500 C_S / S_{Si}^* \tag{3}$$

For octadecyl silicas the value of SCE may range between 10 and 50.

d. Conversion of surface silanols, X_{OH}, expresses that fraction of the silanol groups at the silica surface which formed a siloxane bond with the silanizing agent. It is a meaningful parameter only in the case of "monomeric" phases prepared with monofunctional silanizing agents. According to the previous discussion it is given as

$$X_{OH} = \alpha_L / \alpha_{OH} \tag{4}$$

where α_{OH} is the silanol concentration at the surface of the parent silica gel. The magnitude of X_{OH} expresses the extent of the surface reaction

TABLE III

Some Characteristics of Alkyl Silica Bonded Phases

Phase	Carbon load % (w/w)	Surface concentration (μmol/m^2)		Hydrocarbonaceous surface area of a ligate (nm^2)
		α_L	α_{OH}	
n-Octyl silica	7.36	2.35	11.70	—
n-Octylmethyl silica	8.43	2.40	6.74	—
n-Octyldimethyl silica	9.24	2.35	3.73	—
Trimethyl silica	4.18	3.37	2.98	1.29
n-Butyldimethyl silica	7.13	2.97	3.68	1.96
n-Octyldimethyl silica	10.43	2.71	3.73	2.86
n-Dodecyldimethyl silica	11.73	2.20	3.81	3.76
n-Hexadecyldimethyl silica	15.67	2.36	4.24	4.65

[a] Reprinted with permission from Roumeliotis and Unger (84).

with a given silanizing agent. In practice there is an upper bound to X_{OH} because the bulky organoligates have a significantly greater surface requirement than the relatively small silanol group. Consequently X_{OH} also depends on the shape and size of the ligate.

In Table IV various specifications of commercial hydrocarbonaceous bonded phases are listed. It is seen that the "conversion of surface sila-

TABLE IV

Approximate Pore Size and Surface Area Data on Some Commercial Alkyl Silica Bonded Phases[a]

Phase	Carbon number	S_{Si}^* (m^2/g)	Pore diameter (nm)	X_{OH}	C_s (%)
LiChrosorb RP2	C_2	350[b]	4[c]	0.67[b]	5[b]
LiChrosorb RP8	C_8	250[b]	6[c]	0.75[b]	12[b]
LiChrosorb RP18	C_{18}	150[b]	10[c]	0.67[b]	20[b]
Spherisorb ODS	C_{18}	200[d]	8[d]	—	8[e]
Partisil ODS	C_{18}	400[b]		—	5[b]
Partisil ODS2	C_{18}	400[b]		—	17[b]
Zorbax ODS	C_{18}	300[e]	6[e]	1.3	18[e]

[a] Considering the rapid developments in the field, commercial products presently available under these trade names may have different specifications.
[b] Ref. (81).
[c] Ref. (89).
[d] Ref. (88).
[e] Ref. (71).

nols" reported for Zorbax ODS (71) is greater than unity because at that time the material was prepared with octadecyltrichlorosilane and a "polymeric" phase was formed. Data for Partisil ODS and ODS 2, which are also "polymeric" bonded phases, are not available. X_{OH} is expected to be greater than unity for Partisil ODS 2 which has a relatively high carbon load.

e. Surface coverage, θ, was introduced by Unger (55) for monomeric bonded phases to express the fractional coverage of the silica surface by covalently bound ligates on the basis of their surface requirements, A_m. The latter parameter may range between 0.2 and 0.8 nm² for hydrocarbonaceous ligates used in RPC. The value of A_m, however, depends on the method used for its determination. By defining the maximum surface concentration of a given ligate in micromoles per square meter as

$$\alpha_{L\,max} = 1/0.6A_m \tag{5}$$

the surface coverage is evaluated by

$$\theta = \alpha_L/\alpha_{L\,max} \tag{6}$$

Since θ depends on the estimate for A_m, it is a useful parameter to compare the quality of the hydrocarbonaceous surfaces only if consistent A_m values are used. According to the general belief the value of θ should closely approach unity for stationary phases in RPC.

f. Estimation of residual surface silanols. The methods used for quantitative determination of silanol groups are very tedious (90). For this reason certain simple tests have been introduced to estimate residual surface silanols in hydrocarbonaceous bonded phases. In the methyl red test (73, 91) the powder is contacted with a solution of methyl red in benzene and it is claimed that a reddish discoloration of the stationary phase suggests the presence of acidic silanols. Certain chromatographic tests for residual polarity of hydrocarbonaceous stationary phases are carried out by measuring the retention factor of small polar eluites such as benzyl alcohol or nitrobenzene with dry n-hexane or n-heptane as the mobile phase (73–75, 81, 92). In the absence of accessible surface silanols no retention of small polar solutes is expected whereas increasing retention suggests increasing surface polarity. Unfortunately, neither of the above tests are reliable; the latter is particularly sensitive to traces of water in the chromatographic system. Another problem with such tests is that the solvation of the hydrocarbonaceous ligates by nonpolar solvents may lead to exposure of silanols that are otherwise inaccessible to eluites under RPC conditions.

Alternative methods measure the relative retention of a polar and nonpolar solute with methanol or acetonitrile as the eluent (93a). Evidently the relative retention of the polar sample component increases with the concentration of accessible silanol groups in the stationary phase. However, the accessibility of the surface silanols to a given eluite, and thus their effect on its chromatographic behavior, depends on the mobile phase and conditions employed.

C. Column Packing

The wide employment of short, high efficiency columns in HPLC is the consequence of the availability of microparticulate stationary phases having uniform particle diameter and the development of techniques to pack powders uniformly into appropriate stainless steel tubes. Most commonly a slurry of the column material is filtered into the column tubing, the bottom of which is closed with a porous frit. The latter is permeable to the suspending liquid but not to the particles. Filtration packing is carried out at relatively high pressure gradients and the column packing is obtained essentially as a cylindrical filter cake of a high aspect ratio (94–103).

1. The Tubing

Stainless steel tubes having $\frac{1}{4}$ in. o.d. are most commonly employed. The tube wall must be thick enough to withstand pressures up to 12,000 psi (816 atm) during the packing procedure. Most columns are prepared from tubes having 4.6 mm i.d. as such tubing is a standard commercial product and withstands these pressures.

The quality of the tube innerwall can have a profound effect on the efficiency of the column. Smooth-walled tubes yield best results and as a rule of thumb the roughness of the surface should not be greater than the particle diameter of the column material. In most instances No. 316 stainless steel is used because this material withstands aqueous solutions in a broad pH range. Nevertheless the presence of halides, particularly in acidic eluents, causes corrosion of this material.

The tubing has to be carefully cleaned and equipped with the proper fitting. At the bottom a sintered stainless steel frit is placed; it acts as a filterplate in the course of the packing procedure and as a packing retainer during the lifetime of the column. The quality of the column may greatly depend on the use of the appropriate fittings and frits as improper distribution of flow at the entrance and exit of short columns can significantly increase overall bandspreading and thereby decrease column efficiency. The role of "plumbing" in HPLC is very important and should not be underestimated.

2. Preparation of the Slurry

When the use of microparticulate stationary phases commenced, the suspensions for column packing were isopycnic, i.e. the densities of the liquid and the column material were equal (94–98). As the bulk density of silica is about 2.5 g/cm³, liquids of high density such as tetrabromo- and tetrachloroethane were employed. These solvents, however, are toxic and chemically unstable. In order to avoid handling of such hazardous chemicals, isopycnic, also called "balanced-density," slurries have been progressively replaced by suspensions made with solvents the density of which is not matched to that of the solid.

The suspending fluid for hydrocarbonaceous bonded phases is usually a blend of different solvents selected from the group of isopropanol, methanol, acetone, cyclohexane, and chloroform. Some solvent mixtures, e.g., isopropanol and tetrachloromethane (104), are corrosive and caution should be exercised in the use of halogen-containing solvents in general. Addition of a suitable electrolyte, e.g., sodium acetate, and water to the slurry can give superior results by facilitating the dissipation of static electric charges. Such electric charges can build up upon stirring the suspension in a nonconducting fluid and lead to agglomeration of the particles. Many workers who pack their own columns have their own formulation with some "magic" ingredients. The role of such additives is mainly to affect surface properties of the particles so that agglomeration is minimized and, if possible, the rate of settling is reduced. The use of highly viscous solvents such as paraffin oil or cyclohexanol has also been recommended (102, 103). Alternately the mobile phase was also employed for the preparation of the slurry (104). The optimum concentration of the solid depends on the composition of the suspending fluid and the packing technique employed. In many cases 10% (w/v) of solid appears to be a suitable concentration for the preparation of the slurry. When using the proper fluid there is no need for high shear stirring in order to disperse the particles. In fact vigorous stirring can disintegrate the particles with the formation of undesirable fines.

3. Packing Procedure

In the two limiting cases filtration can be carried out either in the constant flow or in the constant pressure mode. Under most practical conditions, however, neither the flowrate of the suspending fluid nor the inlet pressure of the column remain constant during the packing procedure. Nonsteady conditions, such as strong pressure waves applied intermittently, have been reported to improve the quality of the packing (71). More than often the packing is performed with a specially designed apparatus as shown in Fig. 9. The slurry is filtrated from a reservoir into the

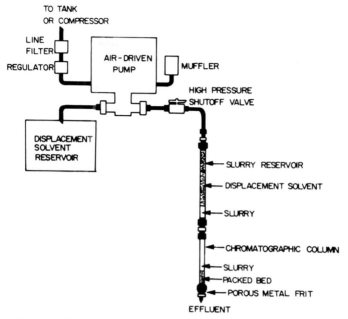

TO TANK
OR COMPRESSOR

LINE
FILTER

REGULATOR

AIR-DRIVEN
PUMP

MUFFLER

HIGH PRESSURE
SHUTOFF VALVE

DISPLACEMENT
SOLVENT
RESERVOIR

SLURRY RESERVOIR

DISPLACEMENT SOLVENT

SLURRY

CHROMATOGRAPHIC COLUMN

SLURRY
PACKED BED
POROUS METAL FRIT

EFFLUENT

FIG. 9.   Apparatus for slurry packing of columns. After the properly fitted column tube is attached to the bottom of the reservoir, both are filled up with the slurry of the microparticulate stationary phase. Thereafter a displacement liquid is pumped into the reservoir by the constant pressure pump, e.g., Haskel Model DST-100, which is driven by pressurized air. Upon displacement, the slurry from the reservoir is filtered over the porous metal frit at the bottom of the column tubing which becomes densely packed with the particles. By intermittently operating the liquid shut-off valve between the pump and the reservoir pressure waves can be generated in order to further compact the column packing. Reprinted from Bakalyar *et al.* (*105*) with permission from Spectra-Physics.

column tubing by pumping a suitable displacement fluid at high pressure into the reservoir. In most instances the slurry enters the top of the vertically placed tubing so that the packing starts to build up at the bottom. Yet "upward" filtration packing has also been recommended as a superior technique with anisopycnic slurries (*106*).

Although efficient columns have reportedly been obtained by packing at pressures not exceeding 2000 psi (136 atm), the use of higher pressures appear to be necessary to optimize not only the efficiency but also the stability of the column. It is not unusual to pack columns at 10,000 to 12,000 psi (680 to 816 atm) final inlet pressure. With silica based bonded phases the packing density is about 0.59 ml of dry column material per milliliter of column volume.

After packing, the column is extensively washed with methanol and/or another suitable solvent to remove the slurry fluid and contaminants from the stationary phase. Subsequently the inlet end of the column is equipped with a stainless steel frit which acts as a retainer, protector, and fluid distributor. In many respects the present column technology appears to be primitive in comparison with the highly advanced instrumentation and further progress in column engineering is essential to develop the full potential of HPLC.

D. Stationary Phase Properties and Chromatographic Behavior

The selection or design of an appropriate stationary phase for a given chromatographic problem is greatly facilitated by adequate knowledge of the relationship between the properties of column materials and the retention behavior of various types of eluites. The retention of an eluite is conveniently expressed by the dimensionless retention factor k which is calculted from the retention time of an unsorbed tracer and the eluite. The retention factor depends on the thermodynamic equilibrium constant K_0 for the reversible binding of the eluite by the stationary phase as

$$k = \phi K_0 \qquad (7)$$

where the proportionality constant ϕ is the so-called phase ratio, which is essentially a property of the column as long as the molecular dimensions of the eluites are commensurable (63). With hydrocarbonaceous bonded phases having alkyl ligates it may be convenient to define ϕ as the ratio of the accessible surface area of the hydrocarbon ligates to the accessible mobile phase volume, both per unit column volume.

The thermodynamics of solute interaction with nonpolar ligates of the stationary phase will be treated later in this chapter within the framework of the solvophobic theory (107–108). According to this theoretical approach the equilibrium constant for the reversible binding of a given eluite to the hydrocarbonaceous ligates at fixed eluent properties and temperature can be approximated by the relationship

$$\log k = a\Delta A + f(V_{\text{lig}}) + C \qquad (8)$$

where ΔA is the contact area between the eluite molecule and the hydrocarbonaceous ligate at the stationary phase surface, $f(V_{\text{lig}})$ is a function of the molecular size of the ligate, V_{lig}, and expresses the effect of van der Waals interactions involved in the chromatographic process (107) (vide infra); a and C depend on the solute and mobile phase properties. It is important to recognize that ΔA for a given solute depends on the shape and the surface area of the ligate which is accessible to the solute as well as on the particular way of eluite binding. Consequently, not only the size and the shape of the stationary phase ligate but also its position at the surface in contact with a given eluent will have an effect on the chromatographic equilibrium constant. Pertinent dimensions of some n-alkyl ligates are listed in Table V in addition to the data presented in Table IV.

In the following discussion we attempt to shed light on some of the stationary phase properties, which in view of Eqs. (7) and (8) affect the magnitude of the retention factor in RPC.

1. Effect of Pore Size and Volume

Although some characteristics of bonded phases are known (cf. Table V), more than often the lack of data makes it difficult to compare different stationary phases in a meaningful fashion. The magnitude of pore volume of the stationary phase, which is filled up with the mobile phase, is related to the intraparticulate porosity of the stationary phase and affects the magnitude of the phase ratio. The pore size distribution also has an effect on the phase ratio as relatively large eluite molecules cannot explore the interior of the small pores (109). In fact, size exclusion effects are ubiquitous in liquid chromatography. They may hinder the precise measurement of the retention factor k, and the interpretation of relative retention data when the molecular dimensions of the eluites are widely different.

The efficiency of the column also can be influenced by the pore size of the stationary phase as the rate of eluite equilibration between the two phases depends on the diffusivity of the eluite inside the pores. Upon attaching the ligates to the pore inner surface the effective pore diameter for eluite transport decreases. The higher the carbon number of the alkyl ligate at a given surface coverage the greater is the constriction of the pores and the concomitant reduction of intraparticulate diffusivity of the eluite.

In practice a compromise has to be reached when optimizing the design of a particular stationary phase. The specific hydrocarbonaceous surface area which can be explored by the eluite molecules without serious diffusion limitations in the course of the chromatographic process should be maximized so that relatively short columns can bring about a desired chromatographic separation. Table IV shows pertinent data on some commercial reversed phase packing materials. It is seen that in the LiChrosorb RP series, which encompasses alkyl silica stationary phases with C_2 to C_{18} ligates, the pore diameter of the silica support increases with the carbon number of the ligates. In Table V some of the critical dimensions of alkyl ligates are listed. Comparing these values with the pore dimensions of the silica given in Table IV we can see that the variation in the effective pore cross section available for intraparticulate diffusion of the eluite is relatively moderate for these column materials due to the

TABLE V

**Approximate Dimension of the Ligates in Alkyl Silica Bonded Phases
Prepared with Alkyldimethylchlorosilanes**

Alkyl function	Symbol	Length (nm)	Cross-sectional area (nm^2)
Methyl	C_1	0.3	0.5
n-Butyl	C_4	0.7	0.6
n-Octyl	C_8	1.2	0.6
n-Decyl	C_{10}	1.7	0.7
n-Hexadecyl	C_{16}	2.3	0.7
n-Octadecyl	C_{18}	2.5	

matching of the pore diameter of the silica and the chain length of the ligate.

The effect of pore size on the retentive capacity of various octadecyl silica bonded phases is illustrated in Fig. 10. Exclusion effects become appreciable when bulky octadecyl groups (see Table V) are attached to the pore wall in 6 nm pore diameter silica (Si-60). As a result, the stationary

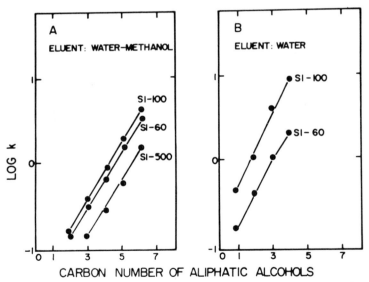

FIG. 10. Effect of the pore size of the parent silica on the retention capacity of stationary phases with octadecyl ligates. The symbols Si-60, Si-100, and Si-500 denote silicas having 6, 10, and 10 nm average pore diameter. The eluites are n-aliphatic C_1–C_6 alcohols and water (A) or (1:1) water–methanol (B) in the mobile phase. From the work by Karch et al. (73) with permission.

phase thus prepared has lower retentive capacity than that made from 10 nm pore diameter silica (Si-100) which has a lower specific surface area, but relatively large pores. However, retention factors obtained under otherwise identical conditions are significantly lower with the bonded phase prepared from 50 nm pore silica (Si-500) with octadecyl ligates due to the low specific surface area of this wide pore silica support.

Stationary phases having wide pores are needed for interactive chromatography of macromolecules in order to avoid undue intraparticulate mass transfer resistances and to reduce size exclusion effects. Thus separation of a wide range of biopolymers by RPC would require silica gel supports having 20 to 50 nm pore diameter. Chromatographic systems appropriate to obtain practically acceptable retention factors and column efficiencies with proteins and nucleic acids have yet to be developed, however, in order to expand the scope of RPC to this field. Unfortunately the surface area of the large pore hydrocarbonaceous bonded phases is relatively small because it decreases with increasing pore diameter in a quadratic fashion. Besides thermodynamic considerations (*110*) the relatively low stationary phase surface area per unit column volume constitutes an impediment to the development of macromolecular separations by interactive chromatography.

2. Nature and Configuration of the Hydrocarbonaceous Ligates

We have already distinguished between three types of hydrocarbonaceous ligates: alkyl, aromatic, and other, such as adamantyl or cholestanyl. Most attention has been paid to the properties of alkyl silicas because this type of stationary phase, particularly octyl and octadecyl silica, is the most popular in RPC. It is relatively easy to prepare such phases and in most studies which attempted to shed light on the role of the bonded function in the chromatographic properties of stationary phases alkyl silicas were employed. A definitive interpretation of data available in the literature, however, has not yet been given for several reasons. First, stationary phases investigated were not prepared from the same type of silica and the variance in the morphology of the support alone can account for substantially different chromatographic behavior. Second, the employment of different types of silanizing agents gives rise to various bonded layers and the structural difference between ''polymeric'' and ''monomeric'' phases can easily mask the effect of the nature of the alkyl ligate. Consequently the use of alkylmethyldichlorosilane or alkyldimethylchlorosilane instead of alkyltrichlorosilane for the treatment of silica can result, under certain conditions, in dramatic differences in selectivity of the stationary phases thus prepared. Moreover, the effect of the above parameters may not be the same when the surface concentration of the ligates or the carbon load changes.

The results of earlier investigations employing bonded phases prepared with alkyltrichlorosilanes are particularly difficult to interpret. The increasing use of alkyldimethylchlorosilanes, however, is expected to lead to the development of a family of stationary phases prepared with such monofunctional reagents under controlled conditions. The results obtained with such materials should then facilitate characterization and definition of hydrocarbonaceous bonded phases.

According to the solvophobic theory, which is discussed in detail later in this chapter, the thermodynamic equilibrium constant for the reversible binding of a given solute from a particular mobile phase to the stationary phase increases with the magnitude of the contact area between the solute and the ligate at the surface, cf. Eq. (8). The theory allows a semiquantitative interpretation of the effect of the ligate on the retention behavior as long as the mechanism of the chromatographic process does not involve specific eluite–stationary phase interactions or such interactions are about the same for the eluites used as molecular probes. It should be noted that the accessibility of the ligates depends on their location in the pores and probably on the mobile phase composition as well. As shown in Eq. (7) the magnitude of the retention factor is also proportional to the phase ratio in the column since the chemical nature of the ligate may affect both the equilibrium constant for the process and the phase ratio of the column. A rigorous interpretation of experimental data is beset with great difficulties.

As stated before the role of surface silanol group in the chromatographic process can be significant particularly when the eluent has a high organic solvent concentration. The magnitude of shielding of the silanol groups by hydrocarbonaceous ligates depends not only on the surface concentration but also on the shape and size of the ligate besides the eluent composition. In view of the interferences attributed to surface silanols in RPC, this secondary effect of the ligate also merits consideration. In the following discussion, however, we consider the chromatographic process to occur as a result of solvophobic interactions without interference due to silanol groups.

According to Eq. (7), the chemical nature of the ligate may affect the retention factor k through its effect on either the phase ratio or the equilibrium constant K_0, or both. In one limiting case, therefore, the retention factor depends linearly on the size or surface concentration of the ligate. This is the case when the phase ratio is proportional to these values and the magnitude of the equilibrium constant remains the same. In the other limiting case the logarithm of the retention factor linearly depends on the size of the ligate—at least in a certain carbon number range. This may occur when the phase ratio is essentially unchanged and the contact surface for solute binding increases with the size of the ligate so that the logarithm of the equilibrium constant is a linear function of the carbon number of the ligate. Under practical conditions, the properties of the eluite, particularly its molecular dimensions and three-dimensional structure, can play an important role in determining which kind of dependence is observed.

As the carbon load of a particular type of stationary phase increases we expect both the phase ratio and the contact surface area for a given eluite probe to reach a plateau so that the dependence of k or log k on the carbon load is likely to be hyperbolic over a sufficiently wide range. A key point in this treatment is the accessible hydrocarbonaceous surface area, which is difficult to quantify and may strongly depend on the molecular architecture of the eluite used as a probe. Furthermore, the involvement of the

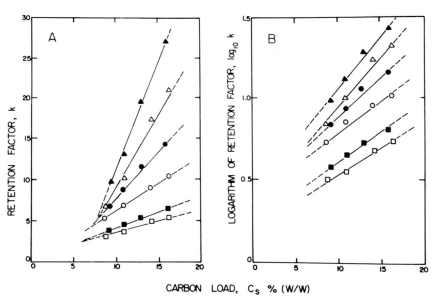

FIG. 11. Dependence of retention factor (A) and the logarithm of the retention factor (B) on the carbon load of various alkyl silicas according to the data of Hemetsberger *et al.* *(80)*. Four different ligates, C_8, C_{11}, C_{15}, and C_{18}, were bound to 10 μm LiChrosorb Si-100 having a BET surface area of 282 m²/g. The open and closed symbols represent data obtained with bonded phases prepared with alkyltrichlorosilane and alkylmethyldichlorosilane, respectively. The mobile phase was 27.6% (v/v) water in methanol at 23°C. The eluites are 1,4-dibromobenzene ($\square$, $\blacksquare$), 1,2,3,4-tetrachlorobenzene ($\bigcirc$, $\bullet$), and 4,4'-dibromodiphenyl ($\triangle$, $\blacktriangle$).

silanol groups in the retention mechanism also can vary with the size and surface concentration of the ligates and the interpretation of results becomes even more difficult.

For reasons enumerated above, experimental results given in the literature do not lend themselves to unambiguous interpretation. The graphs in Fig. 11 illustrate the dependence of the retention factor on the carbon load of stationary phases prepared by C_8, C_{11}, C_{15}, and C_{18} ligates. It is seen that both the direct and the logarithmic plots of k versus C_S are linear, i.e., the increase in the retention factor with the carbon load can be due to an increase in either the phase ratio or the contact surface.

Comparison of the two graphs can serve as a caveat that data obtained in a relatively narrow range of conditions—as usual in practice—may lend themselves to different interpretations. In the present case the measured value is roughly proportional to its logarithm so that the data are unsuitable for model discrimination.

The hyperbolic relationship between the logarithm of the retention factor and the carbon load of octadecyl silica stationary phase is illus-

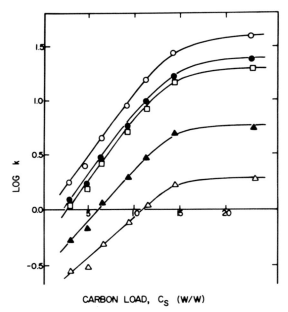

CARBON LOAD, C_S (W/W)

FIG. 12. Graph illustrating the dependence of the logarithm of retention factor for aromatic hydrocarbons on the carbon load of octadecyl silica bonded phases prepared from Partisil with octadecyltrichlorosilane. Mobile phase: methanol–water (70:30); eluites: △, benzene; ▲, naphthalene; □, phenanthrene; ●, anthracene; ○, pyrene. Reprinted with permission from Hennion et al. (76).

trated in Fig. 12. The interpretation of this relationship is straightforward because it is vindicated by either model.

As mentioned earlier [cf. Eq. (7)], the logarithm of the retention factor is linearly dependent on the molecular contact area between the eluite and ligate according to the solvophobic theory. For a given solute, therefore, log k is expected to increase with the size of the ligate until the molecular contact area in the "complex" reaches a maximum. Barring changes in the phase ratio, further increase in the ligate dimensions will not change the magnitude of retention. This hypothesis implies that in a system with n-alkyl silicas and eluites containing n-alkyl chains of increasing lengths, the contact surface for short-chain eluites can reach a maximum when binding to short-chain ligates. On the other hand eluites having long

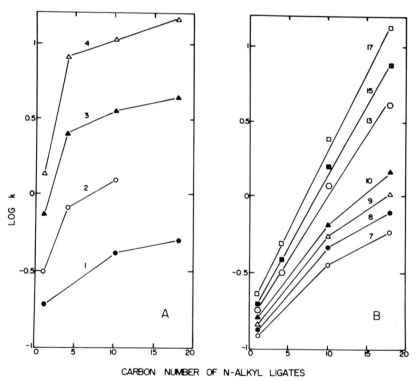

CARBON NUMBER OF N-ALKYL LIGATES

FIG. 13.   Effect of the chain length of n-alkyl ligates in siliceous bonded phases on the retention of (A) short-chain aliphatic alcohols and (B) long-chain alkanes. The number on the curve indicates the carbon number of the solute. The stationary phases were prepared from silica gel having 10 nm average pore size with n-alkyltrichlorosilanes. Mobile phase: (A) methanol; (B) water. Replotted from data by Karch et al. (73).

chains would require long-chain ligates for maximum contact. Consequently, in plots of log k vs the length of ligate for eluites having different chain lengths, the carbon number of the ligate where the curve plateaus should increase with the carbon number of the eluite. Graphs obtained by replotting data of Karch *et al.* (73), shown in Fig. 13, appear to support this hypothesis. As seen log k values for C_1–C_4 aliphatic alcohols show a hyperbolic dependence on the carbon number of *n*-alkyl ligates. Similar behavior is observed with C_7–C_{10} alkanes, whereas C_{13}–C_{17} alkanes exhibit a linear dependence of log k on the chain length of the ligates. The latter finding suggests in view of the above hypothesis that the contact area for the reversible binding of long-chain eluites to the ligates under investigation does not reach a maximum.

The effect of chain length in alkyl silica stationary phases on retention behavior is conveniently interpreted by using a suitable model for the structure of the surface layer. Figure 14 illustrates schematically three possible configurations of covalently bound alkyl chains at the surface. A very dense layer of the alkyl chains, which is termed "picket fence" structure, hypothetically could be formed and this configuration of the bonded phase surface is illustrated in Fig. 14a. In such a case the binding of the eluites would occur only at the tip of the alkyl functions. In a sufficiently dense structure of the closely aligned alkyl chains, eluite molecules of dimensions usually encountered in liquid chromatography would not fit between the alkyl chains and the chromatographic effect of silanol groups at the silica surface proper would be negligible. Since eluites would reach only the top of the three-dimensional "picket fence," the na-

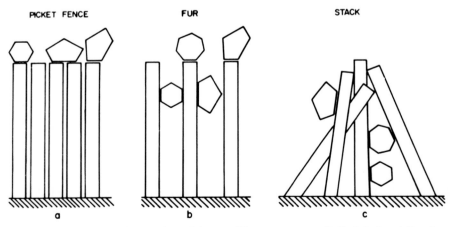

FIG. 14.   Schematic illustration of the possible arrangement of alkyl chains at the stationary phase surface and the bonding of eluites having different shapes and sizes.

ture of the stationary phase surface accessible to the eluite would be very similar for alkyl ligates having different chain lengths. Both the hydrocarbonaceous surface area and the intraparticulate (stagnant) mobile phase volume would decrease upon increasing the chain length of the ligates by using a given silica. Therefore the change in the phase ratio would be relatively small upon increasing the carbon number of the alkyl functions.

In stationary phases commonly used in RPC the "picket fence" configuration is not likely because of the relatively low surface concentration of the ligates, $\alpha_L < 5$ μmol/m^2. This value is much smaller than the surface concentration in compressed monolayers of compounds containing alkyl chains with small polar terminal groups (hydroxyl, carboxyl, amino groups) that was found to be 8 μmol/m^2 at the water surface (111).

Consequently the assumption of a "fur" configuration of the ligates, which is depicted in Fig. 14b, appears to be more realistic. This model implies that the distance between the alkyl chains is sufficiently large for certain solute molecules to bind to the chains laterally. The distance between the chains would permit a sideward movement so that the interligate space would be variable. This structure of the hydrocarbonaceous layer would correspond to a lower carbon load but a higher phase ratio than the "picket fence" configuration in the same silica as the accessible hydrocarbonaceous surface area would be greater due to the greater exposure of the chains to the eluites. The "fur" configuration would also be more selective due to "exclusion" effects by the interligate space. A nearly linear increase in the phase ratio would be expected upon increasing the chain length in the "fur" type bonded phase for solutes having critical dimensions commensurate with the size of the smallest ligand investigated. As long as the "fur" configuration prevails, the phase ratio would increase with the carbon load a given type of stationary phase provided the same silica support is used.

In the "stack" structure, which is illustrated in Fig. 14c the alkyl chains are not perpendicular to the surface and are in close contact with each other. In fact, they appear to form greasy patches on the silica surface. A given stationary phase may have "fur" or "stack" configuration depending on the nature of the pore fluid. When the liquid is capable of wetting the alkyl chains the solvated ligates may form a fur. In contradistinction, in contact with a mobile phase which is incapable of solvating the ligates, they may form stacks as illustrated in Fig. 14c. It is likely that the dependence of phase ratio on the chain length for this configuration is intermediate between those discussed for the "picket fence" and "fur" models.

Although highly speculative, this pictorial representation of the possible stationary phase configurations can convey the complex nature of the problem. An ideal stationary phase would have a very dense "picket fence" type layer. In this case the retention could be readily interpreted on the basis of the solvophobic theory because the hydrocarbonaceous stationary phase layer could be considered a slab and invariant with the mobile phase composition as solvent molecules would not be able to explore the interligate space. Furthermore, residual silanol groups at the surface would not be accessible to eluite molecules. Therefore, the preparation of such a bonded phase could be highly desirable. However, experimental observations accumulated with commercial and homemade hydrocarbonaceous bonded phases suggest that the "fur" and "stack" models are more representative for the arrangement of ligates at the stationary phase surface. It has been generally found that the retention factor of a given eluite increases with the alkyl chain length. The magnitude of the effect depends on the molecular dimensions of the eluite and this can be explained by the dependence of the hydrocarbonaceous surface of the ligates available for eluite binding on the size of the eluite as illustrated in Fig. 14b and c. The frequently observed effect of silanol groups also preclude the existence of a "picket fence" type layer.

Our discussion so far has been restricted to linear alkyl ligates. Of course, silanizing agent containing branched alkyl functions can also be employed and the properties of the stationary phases so obtained can be interpreted in a similar fashion. The properties of a "phenyl" or "benzyl" silica with aromatic ligates at the stationary phase surface can be significantly different from the commonly used octyl or octadecyl silicas. First of all, the benzene ring is relatively small in comparison to the above-mentioned alkyl chains and as a result phenyl bonded phases are generally less retentive toward long-chain aliphatic compounds such as fatty acids than are octadecyl silica bonded phases. Second, the benzene ring is planar as opposed to the cylindrical alkyl chains. Consequently, the geometrical configuration of the hydrocarbonaceous surface layer is expected to be quite different from that illustrated in Fig. 14. Third, the aromatic moiety is readily polarizable and can enter into $\pi-\pi$ interactions. Consequently, the selectivity of aryl silica stationary phases is different from that of alkyl silicas. A "phenyl" bonded phase should, for example, be more retentive for aromatic substances than for aliphatic compounds having comparable size and substituents.

Quite generally a hydrocarbonaceous bonded stationary phase is expected to bind preferentially eluites having the same nonpolar moity as that at the surface. Thus, selective stationary phases can be prepared by using, for instance, adamantyl or cholestanyl functions. According to preliminary experiments in our laboratory cholestanyl bonded phases preferentially retain steroids in RPC. At the present stage of development it appears that the selectivity of the chromatographic system is best adjusted by manipulating the properties of the mobile phase. Nevertheless bonded phases with special hydrocarbonaceous ligates may have significant utility particularly in certain biochemical separations and in separating molecules having rigid three-dimensional structure, e.g., conformationally locked species (*112*).

3. Column Efficiency

The efficiency of properly prepared hydrocarbonaceous stationary phases is comparable to other column materials used in HPLC on the basis of the optimum reduced plate height. The sample loading capacity is at least as high as that of the silica gel used for the preparation of the bonded phase.

With respect to the properties of the support, the particle size, particle shape, and the pore structure of the silica have the greatest influence on the efficiency as discussed in the literature (*71, 73, 108*).

Covalent binding of the ligates to the surface affects the efficiency in at least two ways. First, surface properties change drastically with concomitant effect on the thermodynamics and kinetics (*63*) of the reversible solute binding process. A surface fully and uniformly covered with covalently bound hydrocarbon chains is expected to yield rapid sorption kinetics. Such nonpolar phases are eminently suitable for linear adsorption chromatography because the sorption isotherm, and therefore the chromatographic process, remains linear at relatively high solute concentrations, due to their high specific surface area and high degree of surface uniformity. As a result, close to symmetrical peaks are readily obtained. In fact, in columns packed with particles smaller than 10 μm in diameter, peak asymmetry of nonhydrodynamic origin frequently arises from relatively slow kinetic phenomena which take place in the mobile phase rather than at the surface of the stationary phase.

The other factor which has to be considered with regard to the efficiency of bonded phases is the possibility of pore constriction by the ligates and the resulting reduction in the intraparticulate diffusion rate of the eluite. Untoward consequences can be avoided by selecting sufficiently large support pore dimensions (*89*). For instance, 6 nm diameter pores can be clogged by 2-nm-long octadecyl chains bound to the surface, cf. Fig. 10B. On the other hand, efficient short chain alkyl silica stationary phases can be made for RPC of relatively small sample molecules by using a silica support having 4 to 6 nm average pore diameter.

With "polymeric" phases containing a relatively thick silicon rubber layer inside the pores not only the reduction in pore diameter available for solute diffusion but also the mass transfer resistance in the alkyl polysiloxane layer proper can decrease column efficiency. The mass transfer resistance in the layer depends on the degree of swelling, and consequently on the composition of the eluent. Their frequently poor column efficiency is one of the reasons why polymeric phases lost in popularity.

The interplay of some of the above-mentioned phenomena in determining the effect of carbon load on the efficiency of octadecyl silica stationary phase prepared from Partisil with octadecyltrichlorosilane is illustrated in Fig. 15. As seen from the plot of reduced plate height against carbon load there is an optimum column efficiency at 15% (w/w) carbon content. The minimum of the reduced plate height can be explained as follows. At low carbon load the sorption kinetics on the heterogeneous, partially alkylated silica surface is poor and it also manifests itself in asymmetrical peaks. With increasing carbon load the surface becomes more hydrocarbonaceous and uniform; consequently the reduced plate height decreases. After reaching an optimum value the reduced plate height increases because the efficiency deteriorates due to an increase in the thickness of the hydrocarbonaceous layer in the pores. Diffusion resistances increase with the layer thickness so that the rate of eluite transfer between the two phases is reduced with a concomitant decrease in the column efficiency.

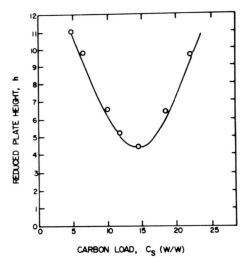

FIG. 15. Graph illustrating the dependence of the reduced plate height on the carbon load of octadecyl silica stationary phase, prepared from Partisil with octadecyltrichlorosilane. Eluent, methanol–water; eluite, polycyclic aromatic hydrocarbons; retention factor, 4. From the data of Hennion *et al.* (76).

4. Column Stability

Aqueous eluents destroy silica gel over an extended period of time particularly when the pH is above neutrality and/or the temperature is high. Whereas the greasy stationary phase layer protects the silica to a large extent, methods for improvement of column stability are usually needed in practice.

In practice there are several approaches to prolong the longevity of the column.

(a) Fines are very carefully removed from the stationary phase prior to column packing and disintegration of the particles during the packing procedure is avoided. Fine particles reduce not only column permeability so that the column inlet pressure becomes unseemingly high at usual flowrates, but also the stability of the column. In a newly made column fine particles constitute an integral part of the packing structure. Upon exposure to eluent flow however, they are subject to relatively rapid dissolution with the result that column efficiency decreases, due to channeling and other structural changes in the column packing.

(b) Silanization is carried out so that a dense layer of the ligates, which approaches the "picket fence" configuration, is formed and the residual silanol groups are capped off. A strong hydrophobic shield can considerably protect the structural silica from hydrolysis with the cleavage of the ligates from the surface.

(c) A silica gel precolumn which saturates the eluent with "dissolved stationary phase" before entering the column has been found remarkably efficient in increasing of column life and in facilitating operation with alkaline eluents (*113*).

(d) Evidence is being accumulated showing that by using large ions as background electrolyte and buffer components in the eluent column life can be prolonged. We can envisage that even in a stationary phase having high surface coverage with long-chain alkyl ligates, small ions such as Na^+, K^+, and NH_4^+ (as well as OH^-) can readily reach the surface silanol groups and accelerate the hydrolytic decomposition of the structural silica and the siloxane bridges of the ligates. However, large cations such as triethylamine would not be able to approach the silica surface so close so that the stationary phase could be more stable in contact with an eluent of alkaline pH. Indeed experimental data show that the octadecyl silica column can be operated with triethylamine buffer, pH 12, for several days without loss of efficiency (*114*). On the other hand, there are indications that long-chain quaternary alkyl amines in the eluent may have detrimental effects on column stability. It is possible that such detergents facilitate "washing-off" the hydrocarbonaceous ligates from the stationary phase.

Silica and bonded phases decompose also in contact with aqueous solution of pH < 2. Therefore it is not advisable to use buffers having pH less than 2. Most commonly phosphoric acid buffers are used in this pH range. Perhaps buffer made with glucose-1-phosphate ($pK_a = 1.1$) and a bulky cation could make possible the long-term use of columns with eluents of pH < 2.

E. Other Nonpolar Stationary Phases

1. Bonded Phases

As discussed by Majors in the first volume of this series (57), modern liquid chromatography employs bonded stationary phases not only with hydrocarbonaceous, but also with polar ligates. Some bonded phases of intermediate polarity (115) can be used with either polar or nonpolar mobile phases. With a sufficiently polar eluent the technique falls into the category of RPC as this chromatographic method by definition employs a mobile phase more polar than the stationary phase so that retention increases with decreasing polarity of eluites having similar molecular dimensions.

On the other hand, nonpolar bonded phases with nonhydrocarbonaceous ligates can also be prepared. For instance, upon replacing hydrogen by fluorine in hydrocarbon molecules, fluorocarbons are obtained and they are considered to be even more nonpolar than the corresponding hydrocarbons. Due to the high chemical stability of such compounds, fluorocarbonaceous bonded phases could play an important role in RPC. Remarkably, enough such stationary phases have not yet received much attention in the literature and only one reference mentions the use of a perfluoroheptyl phase (116).

Among bonded phases of intermediate polarity the so-called cyano phases enjoy the greatest popularity. Some silanizing agents which can be used for the preparation of cyano phases are listed in Table VI. Cy-

TABLE VI

Silanizing Agents for the Preparation of Nonhydrocarbonaceous Bonded Phases for RPC

Phase	Silane	MW	BP (°C/mm Hg)
Cyano	2-Cyanoethyltrichlorosilane	188.5	
Cyano	2-Cyanoethylmethyldichlorosilane	168.1	111–113/40
Cyano	Cyanopropyldimethylchlorosilane	161.7	108–109/15
Cyano	Bis(cyanopropyl)dichlorosilane	235.2	173–177/1
Fluorocarbon	4-Hydrooctafluorobutyltrichlorosilane	335.5	90/195

anoethyl ligates are probably most widely used in such stationary phases, which are prepared from silica by a process very similar to that used in the manufacture of hydrocarbonaceous phases. Under circumstances, the cyano functions impart to these phases selectivity that it is not found in hydrocarbonaceous phases. In contact with a given eluent, the retentive capacity of such phases is usually much lower than that of octyl or octadecyl silicas of commensurate surface coverage and specific surface area. This may be the main reason for the limited use of cyano phases in the practice of RPC.

2. Carbon

Mention was made at the beginning of this chapter that in the 1940s and 1950s charcoal was one of the most popular solid stationary phases. With the advent of instrumentation in chromatography various carbonaceous sorbents found employment again although their role in practical analysis has been rather limited. Graphitized carbon black with high specific and rather uniform surface area has been successfully used in gas chromatography (117). Thereafter the first reversed phase HPLC was probably carried out with such sorbent in pellicular form to separate long-chain fatty acids (118). Pyrocarbon, a recently developed microparticulate carbonaceous sorbent, has received attention as a potential useful stationary phase (119) as well as a tool to gain understanding of the reversed phase chromatographic process (86). So far C_8 and C_{18} hydrocarbonaceous bonded phases proved to be far superior to carbonaceous sorbents in terms of efficiency and peak symmetry. This is partly due to the favorable pore structure of the siliceous support, partly to the homogenous, "soft," greasy, hydrocarbonaceous surface as opposed to the energetically non-uniform, "hard," often highly active surface of carbonaceous sorbents. Results obtained with a column packed with diamond power as the stationary phase in RPC (120) demonstrated that even this material has an energetically inhomogenous surface which can result in pronounced peak asymmetry. Other efforts to prepare a carbonaceous sorbent for RPC include the removal of fluorine from finely dispersed fluorocarbon polymers (121) or carbonization of organic substances on silica support (122).

The great interest in the possibility of replacing silica based bonded phases by carbon in RPC is understandable because the carbon is expected to be more stable toward aqueous eluents than the silica-supported hydrocarbonaceous phases that are used almost exclusively today. Even if a carbonaceous sorbent with uniform surface and favorable porosity would be available its stability may not live up to this expectation, however. The carbon surface is readily oxidized and can undergo other chemical transformations with concomitant changes in its retention properties.

As no controlled comparison has been made we cannot close out the possibility that hydrocarbonaceous bonded phases will prove more stable than carbonaceous sorbents under conditions of RPC.

3. Rigid Porous Polymers

Porous nonpolar polymers are widely used in gel permeation chromatography and are also potentially useful stationary phases in RPC. Macroreticular polyaromatic particles have been successfully used also in gas chromatography and their employment in HPLC has also been advocated (*123*). Experimental results have demonstrated (*124*) that columns packed with rigid porous polymers prepared from styrene and divinylbenzene such as Amberlite XAD (Rohm and Haas) can be used in RPC albeit their efficiency is significantly lower than that of appropriate bonded phases having about the same mean particle diameter. However, since these phases withstand aqueous eluents of high pH, they are far superior to siliceous stationary phases as far as stability in contact with alkaline eluents is concerned.

Macroreticulate nonpolar polymers are usually prepared in spherical form by suspension polymerization in water. The particle size of commercial products ranges from about 100 μm to 3 mm for use in gas chromatography and industrial applications, respectively. The organic phase in the polymerization mixture consists of droplets containing the monomer, such as styrene, a relatively high amount (15–30% v/v) of a cross-linking agent such as divinylbenzene, and an inert diluent (50–80% v/v), such as hexane, which is a good solvent for the monomers but is not a solvent for the polymer formed in the reaction. Thus, upon polymerization the inert diluent will intersperse and occupy a large part of the particle volume. After removing the diluent a rigid, macroporous sorbent particle having relatively high specific surface area and an aromatic hydrocarbonaceous surface is obtained (*125*). Recently spherical microparticles having 10 μm diameter have been introduced for gel permeation chromatography under the trade name μ-Styragel (Waters Associates). Columns packed with this material, however, found no wide use in RPC.

The major technical problems encountered with the use of macroparticular hydrocarbonaceous problems are due to difficulties of packing efficient columns with such materials and the deterioration of efficiency upon changing eluent composition, i.e., under conditions of gradient elution. In contact with organic solvents such materials are expected to swell somewhat despite the high degree of cross-linking. Even if no significant macroscopic swelling occurs, the lesser cross-linked regions at the pore walls are likely to swell in a nonuniform fashion due to a gradient of cross-linking normal to the pore wall. Solute molecules of relatively small dimensions can explore the swollen pore wall region where the average diffusion rate is low. Consequently, the rate of phase exchange and, as a result, the efficiency of the column are relatively low. This is one of the reasons why the column efficiency obtained with such stationary phases is generally inferior to that found in columns packed with rigid siliceous particles of commensurable size, shape, and pore structure. Another possible explanation for the poor efficiency may rest with the insufficient wetting of the particle interior by the polar mobile phase used in RPC. It is recalled that the surface of a nonpolar bonded phase made of silica gel is usually covered with residual silanol groups which facilitate wetting, so that interfacial mass transfer resistances do not impair column performance.

The usual decrease in column efficiency upon changing solvent suggests, however, that the attendant shrinking and swelling of the particles is large enough to cause substantial channeling of the column packing and concomitantly maldistribution of the eluent flow.

The use of Amberlite XAD in RPC, however, has been advocated (123, 124, 126, 127). In order to make this product useful for HPLC, it must be crushed and size-graded in order to obtain particles having uniformly 5 to 10 μm diameter in a fashion similar to the manufacture of irregularly shaped microparticulate silicas. The results obtained with such columns are generally inferior to those obtained with good quality hydrocarbonaceous bonded phases and column stability upon varying eluent composition has not been demonstrated. Nevertheless in one way or another rigid macroporous polymers may find a place as stationary phases in RPC when predominantly aqueous alkaline eluents have to be used.

III. THE MOBILE PHASE

A. Physicochemical Properties of the Solvents

Eluents used in reversed-phase chromatography with bonded nonpolar stationary phases are generally polar solvents or mixtures of polar solvents, such as acetonitrile, with water. The properties of numerous neat solvents of interest, their sources, and their virtues in reversed-phase chromatography have been reviewed (128). Properties of pure solvents which may be of value as eluents are summarized in Table VII. The most significant properties are surface tension, dielectric constant, viscosity, and eluotropic value. Horváth et al. (107) adapted a theory of solvent effects to consider the role of the mobile phase in determining the absolute retention and the selectivity found in reversed-phase chromatography.

TABLE VII

Solvent Properties*

	MW	BP (°C)	n^a	UV^b (nm)	ρ^c (g cm^{-3})	η^d (cP)	ϵ^e	μ^f (Debye)	γ^g (dyn cm^{-1})	E^h
Acetone^i	58.1	56	1.357	330	0.791	0.322	20.7	2.72	23	0.56
Acetonitrile	41.0	82	1.342	190	0.787	0.358	38.8	3.37	29	0.65
Dioxane	88.1	101	1.420	215	1.034	1.26	2.21	0.45	33	0.56
Ethanol	46.1	78	1.359	205	0.789	1.19	24.5	1.68	22	0.88
Methanol	32.0	65	1.326	205	0.792	0.584	32.7	1.66	22	0.95
Isopropanol	60.1	82	1.375	205	0.785	2.39	19.9	1.68	21	0.82
n-Propanol	60.1	97	1.383	205	0.804	2.20	20.3	1.65	23	0.82
Tetrahydrofuran	72.1	66	1.404	210	0.889	0.51	7.58	1.70	27.6	0.45
Water	18.0	100	1.333	170	0.998	1.00	78.5	1.84	73	

* Reprinted from Horváth and Melander (*129*), *J. Chromatogr. Sci.*, with permission from Preston Publications.

^{a} Refractive index at 25°C.

^{b} UV cut-off; the wavelength at which the optical density of a 1-cm-thick neat sample is unity as measured against air.

^{c} Density at 20°C.

^{d} Viscosity at 20°C.

^{e} Dielectric constant.

^{f} Dipole moment.

^{g} Surface tension.

^{h} Elutropic value of alumina according to Snyder (*130*).

^{i} Not suitable for use with UV detector.

The theory, outlined in detail in Section V, anticipates that retention will be governed by the bulk surface tension and the dielectric constant of the solvent in the simplest case. As a first crude approximation, the greater the surface tension of the eluent, the greater the retention of a given eluent will be under otherwise identical conditions. On the other hand, high dielectric constant favors interaction between polar groups of the eluite and the mobile phase and therefore will reduce the capacity factor. Based on quite a different model, Eon (*131*) makes similar predictions regarding the effect of surface tension in an attempt to evaluate the role of solvent interactions in governing retention on silica gel. The result of this work is a general expression for the relationship between interfacial tension, which can be related to surface tension, and the eluent strength which is measured by the eluotropic value. This expression has since been used also to estimate eluotropic strength in RPC with good agreement with experiment (*86*). Traditionally, the eluting power of various solvents has been measured by eluotropic values measured experimentally on an alumina column. Such values have been tabulated by Snyder (*130*) for most solvents of chromatographic interest. As the eluent strength so defined increases with the polarity of the solvent, such an ordering of solvent strength is approximately reversed from that observed in reversed phase chromatography.

The role of the viscosity is quite significant in actual operation due to two effects. At a fixed flow velocity of the eluent, the pressure drop across a column, ΔP, is proportional to the viscosity, η, of the mobile phase according to Darcy's law

$$v_0 = \frac{B^0}{\eta} \cdot \frac{\Delta P}{L} \tag{9}$$

where L and B^0 are the column length and specific permeability coefficient respectively, and v_0 is the superficial velocity defined as the ratio of volumetric flowrate to column cross-sectional area. According to Eq. (9) the pressure drop at a given flowrate is directly proportional to the viscosity and if the pressure drop is held constant, the flowrate is inversely proportional to the viscosity. As a consequence, for example, a change in the eluent from acetonitrile to water will nearly treble the column inlet pressure (see Fig. 17). In addition to determining the column inlet pressure, the viscosity of the medium can have an effect on the efficiency of the separation process. Band spreading of a chromatographic peak arises from a variety of sources. (*63, 109, 132*) If the effect of longitudinal diffusion is neglected, as it can be in all but some exceptional situations in liquid chromatography, the plate height of a given column can be related directly to the reciprocal diffusion constants of the eluite in the mobile and stationary phases at otherwise fixed conditions. The diffusion constant is related to the reciprocal of the viscosity of the medium [cf. the Wilke–Chang equation (*133*)]. Thus, column efficiency decreases with increasing eluent viscosity, i.e., broader peaks can be anticipated when more viscous mobile phases are used under otherwise identical conditions.

Water is the most polar solvent of those listed in Table VII as well as the one with the greatest surface tension. As a consequence it is the weakest eluent, whereas acetonitrile and methanol are the most frequently used stronger eluents. Both solvents are completely miscible with water and have relatively low viscosities. Most importantly they are optically clear far into the ultraviolet region of the spectrum so that spectroscopic monitoring of the column effluent at low detector wavelength settings is facilitated. Most experimental situations require an eluent which is stronger than water and weaker than methanol or acetonitrile. Such eluents of intermediate eluotropic strength are usually obtained by mixing one of these organic solvents with water.

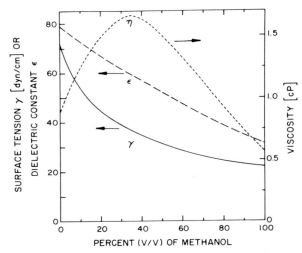

FIG. 16. Dependence of solvent properties pertinent to RPC on composition of water–methanol mixture at 25°C. Surface tension γ data were obtained from Timmermans (134); the viscosity η and dielectric constant data ϵ were taken from Carr and Riddick (135) and Åkerlof (136), respectively. Reprinted from Horváth and Melander (129), *J. Chromatogr. Sci.*, with permission from Preston Publications.

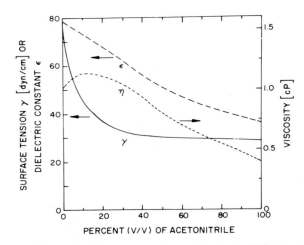

FIG. 17. Dependence of solvent properties pertinent to RPC on composition of water–acetonitrile mixtures at 25°C. Surface tension γ data were obtained from Timmermans (134); the viscosity η and dielectric constant ϵ data were taken from Timmermans (134) and Douhéret and Morenas (137), respectively. Reprinted from Horváth and Melander (129), *J. Chromatogr. Sci.*, with permission from Preston Publications.

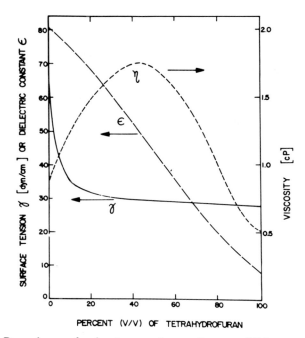

FIG. 18. Dependence of solvent properties pertinent to RPC on composition of water–tetrahydrofuran mixtures at 25°C. Surface tension γ data were obtained from Timmermans (*134*); the viscosity η and dielectric constant ϵ data were taken from Hayduk *et al.* (*138*) and Critchfield *et al.* (*139*).

Figures 16–20 show the variation of various physical properties such as viscosity, density, dielectric constant, and surface tension of such aqueous mixtures when the organic component is methanol, acetonitrile, tetrahydrofuran, *n*-propanol, or ethanol, respectively. In all cases, the surface tension, dielectric constant, and density decrease monotonically with increasing concentration of the organic component. As expected, the eluent strength of the mixture also increases with the concentration of the organic solvent component. It is seen that with increasing organic solvent concentration the viscosity of the eluent first increases to reach a maximum and then monotonically decreases. This effect is particularly marked in the case of methanol and the maximum viscosity obtained with methanol–water mixtures is nearly twice that of pure water. Consequently, the use of methanol–water mixtures is particularly liable to give rise to problems associated with high viscosity when using gradient elution.

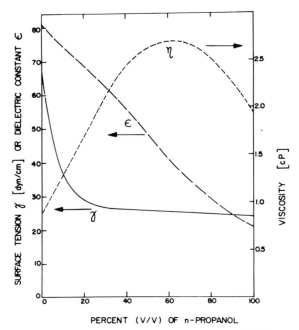

FIG. 19. Dependence of solvent properties pertinent to RPC on composition of water–n-propanol mixtures at 25°C. Surface tension γ data were obtained from Timmermans (134); the viscosity η and dielectric constant ϵ data were taken from Timmermans (134) and Åkerlof (136), respectively.

B. Eluent Strength

The concept of eluotropic strength has been invoked here without a rigorous definition. Snyder (130) developed a series of eluotropic values for solvents by using retention values measured on alumina columns. Colin and Guiochon (86) used a definition similar to that of Snyder to evaluate eluotropic strengths of methanol–water mixtures on various column surfaces. The eluotropic strength, ϵ^0, was calculated by using the equation

$$\log \frac{k_{ij}}{k_{il}} = A_i(\epsilon_l^0 - \epsilon_j^0) + \log \frac{V_j A_l}{A_j V_l} \tag{10}$$

where A and V are the respective molar surface area and volume of eluite i as well as eluents j and l. The retention factors of eluite i in solvents j and l are given by k_{ij} and k_{il}, respectively. The molar volume of water–methanol solvent mixtures, V_{mix}, is given by

$$V_{mix} = V_w \cdot x_w + (1 - x_w) \cdot V_{MeOH} \tag{11}$$

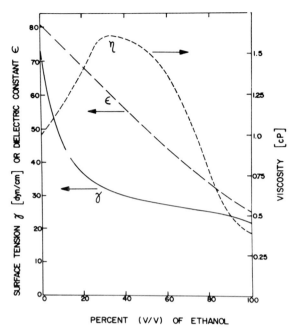

FIG. 20. Dependence of solvent properties pertinent to RPC on composition of water–ethanol mixtures at 25°C. Surface tension γ data were obtained from Timmermans (*134*); the viscosity η and dielectric constant ϵ data were taken from Salceanu (*140*) and Åkerlof (*136*).

where x_w is the mole fraction of water and V_w and V_{MeOH} are the molar volumes of water and methanol, respectively. A convenient rule for estimation of molar surface area of a given compound is given by

$$A_i = V_i^{2/3} N^{1/3} \qquad (12)$$

where N is Avogadro's number (*141*). The results of such calculations by Colin and Guiochon, who did not state the method used for evaluating A_i in solvent mixtures, are shown in Fig. 21. As seen, the eluotropic strength of the neat organic solvent component is arbitrarily set equal to zero. The retention factors for polar and nonpolar eluites were obtained on columns packed with Partisil ODS-2, μBondapak-fatty acid (benzyl silica), and pyrocarbon coated silica. Except for the data obtained on the octadecyl silica stationary phase, the eluotropic strength of various water–methanol mixtures does not appear to depend in a major way on the polarity of the eluite. There is a significant difference, however, in eluotropic strengths calculated for different stationary phases. At first sight this is troubling as

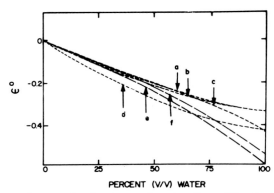

PERCENT (V/V) WATER

FIG. 21.   Eluotropic strengths of water–methanol on three different columns with polar and apolar solutes. The data were taken on a phenyl phase (μ-Bondapak, fatty acids) with (a) polar and (b) apolar solute, on Partisil ODS-2 with (c) polar and (d) apolar eluites, and on pyrocarbon with (e) polar and (f) apolar solutes. Reprinted with permission from Colin and Guiochon (*86*).

the eluotropic values should represent only mobile phase properties. Yet the calculation did not take into account activity coefficients of the eluent components; i.e., ideal solvent behavior was assumed, and this may explain the discrepancy. Despite this shortcoming, such calculations may be useful for crude predictions of solvent effects. A less extensive treatment of eluotropic strength of some acetonitrile–water mixtures has also been given using data obtained by chromatography on graphitized pyrocarbon, (*141*) the results of which have been calculated as stated above and are shown in Fig. 22. As was observed in the case of methanol–water mixtures, the eluotropic values calculated for polar molecules with a given eluent are different from those obtained for nonpolar molecules at the

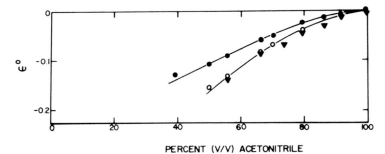

PERCENT (V/V) ACETONITRILE

FIG. 22.   Eluotropic strength of water–acetonitrile on pyrocarbon. The eluites used are ($\blacktriangledown$) 2,3,4,6-tetramethylbenzene, ($\bigcirc$) 1,3,4-trimethylbenzene, and ($\bullet$) 3,4,5-trimethylphenol. Reprinted with permission from Colin *et al.* (*141*).

TABLE VIII

Eluotropic Strength of Various Solvents[a]

Solvent	Log V^0/A^0	A^0 ($\times 10^4$ m²)	ϵ^0				
			Alkyl-benzene	Methyl-benzene	Methyl-phenol	Methyl-naphthalene	Fused aromatic
Methanol	−7.312	9.957	0	0	0	0	0
Acetonitrile	−7.275	10.180	0.038	0.048	0.025	0.039	0.039
Ethanol	−7.259	12.707	0.049	0.051	0.058	0.051	0.051
n-Hexane	−7.142	21.749	0.123	0.128	—	0.091	0.086
Ethyl acetate	−7.184	17.932	0.095	0.105	0.080	0.091	0.091
n-Heptane	−7.125	23.474	0.137	0.142	—	0.110	0.119
Butyl chloride	−7.174	18.727	0.132	0.141	—	0.115	0.112
n-Octane	−7.077	25.159	0.151	0.149	—	0.138	0.139
Tetrahydrofuran	−7.211	15.835	0.133	0.141	—	0.139	0.139
Methylene dichloride	−7.240	13.458	0.127	0.137	—	0.132	0.133
n-Nonane	−7.097	26.796	0.157	0.181	—	0.161	0.161
Chloroform	—	—	—	—	—	—	—
Benzene	−7.198	16.821	—	—	—	0.200	0.204
m-Xylene	−7.231	20.802	—	—	—	—	0.240

[a] The values were obtained from chromatographic retention on pyrocarbon stationary phase. A^0 and V^0 are molar area and volume of the solvent. Reprinted with permission from Colin et al. (141).

[b] At 15°C.

TABLE IX

Eluotropic Strength of Several Solvents with Octyl
or Octadecyl Silica Stationary Phase[a]

	Stationary phase	
Compound	C_8	C_{18}
Methanol	1.0	1.0
Acetic acid	2.7	—
Ethanol	3.2	3.1
Acetonitrile	3.3	3.1
2-Propanol	8.4	8.3
Dimethylformamide	9.4	7.6
Acetone	9.3	8.8
n-Propanol	10.8	10.1
Dioxane	13.5	11.7

[a] The values were determined as the ratio of retention time of the indicated compound relative to that of methanol with water as eluent. Reproduced with permission from Karch et al. (143).

same solvent composition. The authors also calculated the eluotropic strengths of various solvents from the retention of various aromatic substances used as chromatographic probes and the results are presented in Table VIII. Whereas the eluotropic strength calculated from the retention of nonpolar species does not depend very much on the chemical nature of the eluite, as seen in Table VII, different values are obtained with polar eluite. More recently Colin et al. (142) also investigated the effect of temperature on eluotropic strength measured with pyrocarbon as the stationary phase. The thermal effect was relatively small, amounting to ~25% increase in ϵ^0 with a 50° temperature rise.

Alternative methods for the establishment of an eluotropic series in RPC have also been used. Karch et al. (143) determined the retention of several solvents which might prove of use in RPC relative to that of methanol. The basic concept behind this ranking is that the eluent strength of a given solvent increases with its retention factor and when it is used the eluent strengths of the aqueous mixtures having the same organic solvent concentration increase with that of the organic component. The ranking based on data taken with octyl silica and octadecyl silica phases is given in Table IX. It is qualitatively similar to the eluotropic series given by Colin et al.(141) although several inversions of order are noticable.

An earlier attempt to rank solvents (144), particularly hydroorganic mixtures, was made by the comparison of the retention factors of 1,4-

TABLE X

Effect of Organic Modifier on Retention of
1,4-Dimethyl-9,10-anthraquinone[a]

Modifier[b]	k	$k_\alpha{}^c$
Isopropanol	2.5	1.0
Dioxane	3.6	1.44
Acetonitrile	6.0	2.40
Ethanol	11	4.40
DMSO	26.6	10.6
Methanol	27	10.8
Ethylene glycol	139	55.5

[a] Reprinted from Schmidt et al. (144), J. Chromatogr.
Sci., with permission from Preston Publications.
[b] 30% modifier in water at 50°C.
[c] Relative retention with isopropanol as reference.

dimethyl-9,10-anthraquinone measured with aqueous eluents containing 30% organic modifier. The results are presented in Table X as retention factors relative to that obtained when isopropanol was used as the modifier. A qualitatively similar result was also obtained for aqueous mixtures containing 75% organic modifier (145).

More recently Snyder et al. (146) reviewed the literature to evaluate the solvent strength of various organic solvents used in hydroorganic eluents. They have assumed that the dependence of logarithm of the retention factor on the volume fraction of organic component, ϕ_0, in the mobile phase is given by

$$\log k = \log k_w - S\phi_0 \tag{13}$$

where k_w is the retention factor obtained in water. The values of the parameter S, which is obtained as the negative slope of plots of $\log k$ vs volume fraction, are listed in Table XI. From the data a solvent strength ranking may be made and we find that the eluent strength of solvents commonly used in RPC follows the order methanol $\leq$ acetonitrile $<$ ethanol $\simeq$ acetone $\simeq$ dioxane $<$ isopropanol $\leq$ tetrahydrofuran.

In the course of the investigation (146) it was also examined whether Eq. (13) should be replaced by a quadratic relationship between $\ln k$ and ϕ_0, but no marked improvement was found in the fit of retention factor composition data.

Before trying to use the S values in Table XI one should be aware that they are at best semiquantitative. Wide variation in values of S was observed with solute structure (146). However, no obvious correlation of S value with structure emerged except that it increased with the carbon

Wayne R. Melander and Csaba Horváth

TABLE XI

Summary of Solvent Strength Values for Different Organic Solvents in RPC (25°C)[a]

			Solvent				
Meth-anol	Aceto-nitrile	Eth-anol	Ace-tone	Di-oxane	Isopro-panol	Tetrahy-drofuran	Reference
2.4							92,107,143,147–154
3.5	2.9					4.2	155
2.7		3.4			4.1		156
(3)[b]	4.1					4.7	157
	(2.5)[b]		3.4	3.5	4.2	4.4	145
3.0	3.1	3.6	3.4	3.5	4.2	4.4	"Best"[c]

[a] Reprinted with permission from Snyder et al. (146).

[b] Assumed value for calculation of S for other solvents (k versus ϕ data not provided).

[c] The "best" value was determined by Snyder et al. (146) from a critical review of the literature.

number of the solute. Furthermore, S values varied from column to column. The average values of S were 2.65 for Waters' μ-Bondapak C_{18}, 2.94 for Hypersil C_{18}, 3.29 for DuPont's Zorbax C8, and 2.90 and 3.29 for two different DuPont's Zorbax ODS columns. The first of the two Zorbax ODS columns had been prepared by reaction with trichlorosilane (polymeric bonded phase) and the second prepared with monochlorosilane (monomeric bonded phase). This variation in S values between columns is comparable to the variation in S values among solutes on a given column. Thus, these values may only be used as a rough guide in choosing a solvent system for a separation, and the predictions of retention factor obtained by use of Eq. (13) and the values of S in Table XI, or other similarly averaged values, must be regarded as qualitatively useful.

C. Effect of the Composition of Mixed Solvents

1. Effect on Retention

An important question is how increasing organic modifier concentration in the eluent changes the retention factor of a given solute. A number of workers have observed that plots of the logarithm of the retention factor are linear in volume fraction of methanol for methanol–water mixtures. This is shown in Fig. 23, which illustrates the retention factors of butanol, pentanol, and phenol as a function of eluent composition on butyl and octadecyl silica stationary phases (143). Figure 24 shows data taken by other investigators for the retention of n-hexanol and n-octanol in methanol–water mixtures on octadecyl silica stationary phases (148). Similar results

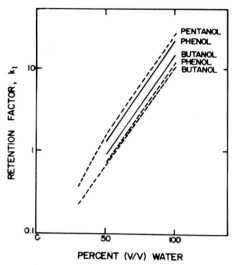

FIG. 23. Dependence of retention factor on solvent composition using methanol–water mixture as the eluent. The data, plotted on a logarithmic scale, were obtained with butanol, pentanol, and phenol as the eluite on Si-100 which had been treated with trichlorobutylsilane (– –) or with trichlorooctadecylsilane (——). Reprinted with permission from Karch *et al.* (*143*).

were also obtained in RPC of aromatic acids (*107*). Exceptions to the apparent linear behavior of log k with volume fraction of methanol can occur due to a flattening of the curve at high methanol concentrations and a somewhat steeper curve observed at very low methanol concentrations.

Similar data have been obtained for the other very popular mixed solvent system, water–acetonitrile. Abbott *et al.* (*152*) reported linear relationships between the logarithm of the retention factors of several aromatic compounds and the acetonitrile volume fractions. These results are shown in Fig. 25. Horváth *et al.* (*107*) also observed linear behavior in the chromatography of aromatic acids over the entire range of solvent composition with minor steepening of the curve at the ends. On the other hand, as shown in Fig. 24, Karger *et al.* (*148*) found marked deviation from linearity with pronounced flattening of the curve in acetonitrile-rich mobile phases.

The quasi-linear relationship between the logarithm of the retention factor and volume fraction organic cosolvent in the mobile phase seems to be the general rule in RPC. However, special effects can occur to cause this rule to be violated. Marked deviation from linearity was observed by Melander *et al.* (*158*) with retention data of poly(ethylene glycol) deriva-

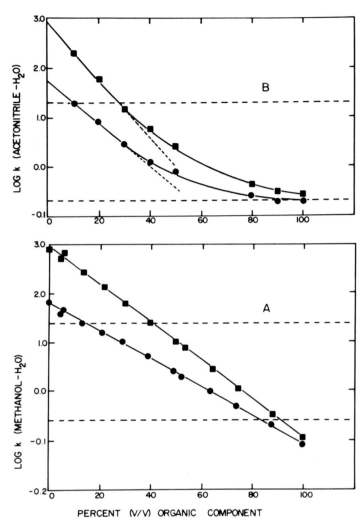

FIG. 24. Logarithm of retention factors of *n*-hexanol (●) and *n*-octanol (■) on octadecyl silica stationary phase at different mobile phase compositions. In (A), the eluents were methanol–water mixtures of various volume fraction. In (B), the eluent was acetonitrile–water in varying volume fraction. Retardation factors greater than 20 or less than 0.2 were obtained by extrapolation of data for lower or higher homologs. The dashed lines indicate $k = 20$ and $k = 0.2$. Reprinted with permission from Karger *et al.* (*148*).

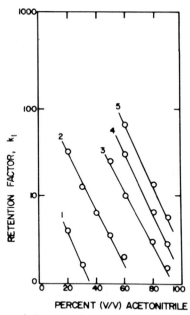

FIG. 25. The logarithm of the retention factor of several fused ring systems with acetonitrile–water as the eluent. The eluent composition is given in volume percent at 35°C. The data were obtained on MicroPak CH-10 column with phenol (1), benzene (2), anthracene (3), benz[a]anthracene (4), and benzo[a]pyrene (5) as eluites. Reprinted with permission from Abbott *et al.* (*152*).

tives obtained by using hydroorganic eluents containing 10–30% of acetonitrile or tetrahydrofuran. The nonlinearity has been explained by solvent mediated formation of a conformer of the eluite which binds to the stationary phase differently than the conformer present in neat water. As a result, the logarithm of the observed retention factors is not a linear function of solvent composition even if the corresponding plots of the individual molecular forms are (*158*). In addition to effects on eluite behavior in the mobile phase, interactions with the stationary phase can also promote deviations from the rule of linearity. For instance, the retention of several crown ethers, antibiotic amines, and basic drugs showed pronounced minima in the corresponding retention factor vs solvent composition plots. The compositions of the eluent in which the minimum occurred with a given eluite were similar, but not identical, for different column packings. These findings have been attributed to interaction of the eluite with the accessible silanol groups at the surface of the stationary phase (*93*).

2. Effect on Selectivity

The effect of eluent composition on group selectivity has also been examined. Group selectivity is defined as the relative retention of a compound bearing a group of interest with respect to a homologous parent compound without that group. Thus, methylene group selectivity can be expressed as the ratio of the retention factors for hexanol and pentanol, whereas the selectivity for the nitro group can be measured by that for nitrobenzene and benzene, for example. Using this sort of approach, the effect of eluent composition on the methylene group and polar group selectivity of reversed phase chromatographic systems has been investigated by Karger *et al.* (*148*). As discussed above, selectivities have been defined by the relative retentions of eluites—one with, the other without an additional methyl or polar group on the molecule, respectively. Changing eluent composition affects the methylene group selectivities differently from the selectivities for polar substituents (*92, 157*).

Selectivity for the methylene group has been mostly the focus of interest and was determined by a host of investigators. Data obtained by Karger *et al.* (*148*) are shown in Fig. 26. It is seen that the selectivity is a linear function of the solvent composition, when water–methanol mixtures are used as the mobile phase, and decreases with the water content of the eluent. The linearity characteristic of methanol mixtures is absent in the data for water–acetonitrile and water–acetone mixtures but again

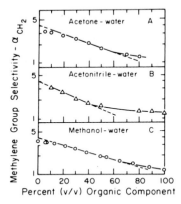

FIG. 26. Methylene group selectivity, α_{CH_2}, of several hydroorganic mobile phases when octadecyl silica stationary phase is used. The selectivity is the ratio of the retention factor of a member of a homologous series to that of another member which differs in having one less methylene group. The solvents shown here are (A) acetone, (B) acetonitrile, and (C) methanol. The data were taken at ambient temperature and the selectivity values are plotted on a logarithmic scale. Reprinted with permission from Karger *et al.* (*148*).

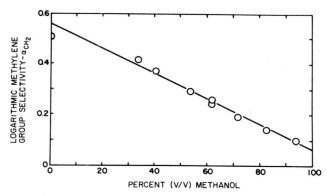

Fɪɢ. 27. Logarithmic methylene group selectivity of alkanes, carboxylic acids, and alcohols chromatographed on octadecyl silica in water–methanol as the eluite. The logarithm of the selectivity is nearly linear with composition, in volume fraction, over the entire range. Data replotted from Tanaka and Thornton (*159*).

the selectivity does decrease as water content decreases. Tanaka and Thornton (*159*) examined the chromatographic behavior of a variety of alkanes, carboxylic acids, and alcohols and found that the selectivity for methylene group was essentially identical for each class of compounds and that it varied with the composition of methanol–water mixtures used as eluents. Their data have been replotted and are shown in Fig. 27. Behavior similar to that observed by Karger *et al*. (*148*) is seen with some evidence for flattening of the curve in the water-rich region. Hoffman and Liao (*160*) examined the methylene group sensitivity in RPC by using hydroorganic mixtures which contained methanol, acetonitrile, ethanol, isopropanol, dioxane, or tetrahydrofuran. Their aim was to establish the "sensitivity" of retention factor to the water content of the eluent. The sensitivity was defined as the change in the logarithm of methylene group selectivity per mole percent water in the hydroorganic eluent. Thus it is a measure of the rate at which water addition increases methylene group selectivity. By this criterion they have ranked the solvents in increasing order of "sensitivity" as tetrahydrofuran, isopropanol, dioxane, acetonitrile, ethanol, and methanol. This is essentially the order of decreasing eluent strength. The results obtained on Partisil ODS and ODS-2 columns were qualitatively similar despite the fact that the carbon contents of the two octadecyl silica stationary phases are quite different.

In order to investigate the effect of eluent composition on polar group selectivity, Tanaka *et al*. (*92*) determined compositions of various hydroorganic eluents at which the observed methylene selectivity in RPC is

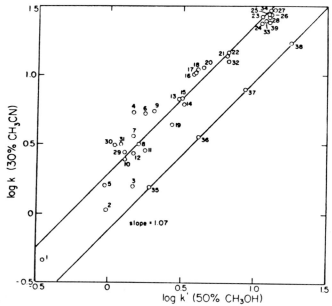

FIG. 28. Plot of logarithm of retention factor in acetonitrile–water (30:70, v/v) vs the logarithm of the retention factor in methanol–water (50:50, v/v) with octyl silica (5 μm Hypersil) capped with trimethylchlorosilane as the stationary phase. The lower line is drawn through the data for four homologous n-alcohols and serves to calibrate changes in hydrophobic character, i.e., changes in the number of methylene groups between the two solvents. The upper line was drawn parallel to the lower one through the point for benzene to determine deviation from solvophobic behavior in the aromatic solutes used. The eluites were (1) benzamide, (2) benzyl alcohol, (3) 2-phenylethyl alcohol, (4) p-dinitrobenzene, (5) phenol, (6) m-dinitrobenzene, (7) benzonitrile, (8) acetophenone, (9) nitrobenzene, (10) p-nitrophenol, (11) p-cresol, (12) m-nitrophenol, (13) anisole, (14) methyl benzoate, (15) benzene, (16) o-nitrotoluene, (17) p-nitrotoluene, (18) p-nitrochlorobenzene, (19) p-chlorophenol, (20) m-nitrotoluene, (21) toluene, (22) chlorobenzene, (23) naphthalene, (24) o-xylene, (25) ethylbenzene, (26) p-xylene, (27) m-chlorotoluene, (28) m-xylene, (29) benzaldehyde, (30) m-nitrobenzaldehyde, (31) p-nitrobenzaldehyde, (32) ethyl benzoate, (33) isopropyl benzoate, (34) p-dichlorobenzene, (35) 1-pentanol, (36) 1-hexanol, (37) 1-heptanol, (38) 1-octanol, and (39) cyclopentane. Reprinted with permission from Tanaka et al. (92).

identical as methanol–water (50:50), acetonitrile–water (30:70), and tetrahydrofuran–water (25:75). In order to study polar selectivities, logarithms of retention factors obtained in one solvent system were plotted against those obtained in another system. As shown in Figs. 28 and 29 a line predicting the retention factors obtained with the two eluents was drawn through the point representing log k for benzene with the slope equal to that found for the corresponding plots for a series of homologous

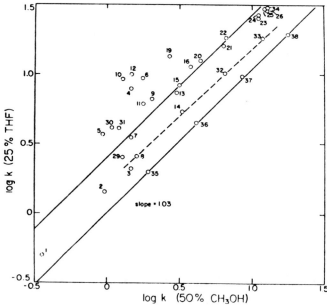

FIG. 29.   Plot of the logarithm of retention factor in THF–water (25:75, v/v) versus the logarithm of the retention factor in methanol–water (50:50, v/v) with octyl silica (5 μm Hypersil) capped with trimethylchlorosilane as the stationary phase. Points lying above the upper curve imply the solute interacts more strongly with methanol–water than it does with THF–water whereas data points below the curve imply stronger interactions in THF–water. Conditions and solutes are same as those given in Fig. 28. Reprinted with permission from Tanaka *et al.* (92).

alcohols. Thus points above this line imply preferential polar interactions of the corresponding eluites with the solvent indicated on the abscissa. Data taken for the retention factors of various aromatic derivatives in acetonitrile–water and methanol–water are shown in Fig. 28 whereas the corresponding plot of tetrahydrofuran–water and methanol–water data is given in Fig. 29. With the exception of two alcohols, the data in Fig. 28 conform well to the expected behavior. On the other hand, great deviation from the calibration line is seen in Fig. 29. In comparison to that observed with aqueous methanol as the eluent, retention in THF–water is greatly enhanced for phenols, dinitrobenzene, and *m*- and *p*-nitrobenzaldehyde and reduced for alkylbenzoates. The graph serves as a convenient means to evaluate the polar selectivity of the mobile phase. The effect of the difference in polar selectivity is dramatically shown by the chromatogram taken from Tanaka *et al.* (92) and depicted in Fig. 30. As seen, the elution

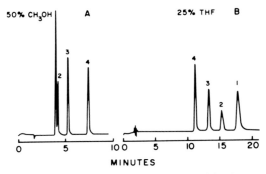

FIG. 30.   Chromatograms illustrating difference in selectivity due to organic components in the mobile phase. The data were obtained on capped octyl silica (5 μm Hypersil) with (A) methanol–water (50:50, v/v) or (B) tetrahydrofuran–water (25:75, v/v) as the mobile phase. The solutes were (1) p-nitrophenol, (2) p-dinitrobenzene, (3) nitrobenzene, and (4) methyl benzoate. Flowrate, 1 ml/min; column 150 × 4.6 mm i.d. Reprinted with permission from Tanaka *et al*. (*92*).

order for four aromatics is reversed when the methanol–water (50:50) eluent was replaced by THF–water (25:75).

Bakalyar *et al*. (*157*) also examined the selectivity of different mobile phases for various aromatics and found it to be markedly different for the eluents methanol–water (50:50), acetonitrile–water (40:60), and THF–water (37:63), compositions of which were chosen to give constant retention factor for benzene at 35°C. The results given in Fig. 31 are in general agreement with those of Tanaka *et al*. (*92*) and serve to show the rather large effect of polar functional groups in eluite molecule on the selectivity in RPC.

D. Gradient Elution

Peak capacity of a given chromatographic system can be increased by the use of gradient elution (*152*) which usually finds application in the separation of multicomponent samples. In RPC, the elutropic strength of the solvent is increased by decreasing the polarity of the eluent in the course of gradient elution. Thus the appropriate solvent gradient is produced by increasing the concentration of the organic solvent in the hydroorganic mobile phase during the chromatographic run.

Gradient shape, i.e., the eluent composition as a function of time, is usually selected *ad hoc* to satisfy the requirements of the particular separation. In most cases linear gradients are satisfactory, although certain difficult separations may require exponential or more complex gradients.

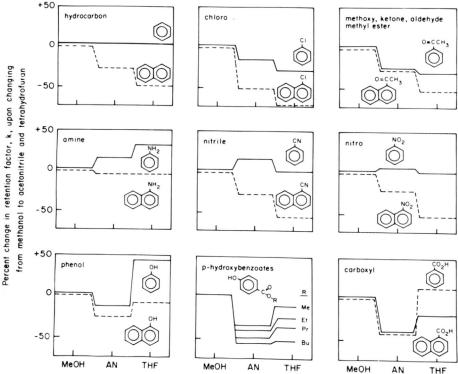

Fig. 31. Solvent selectivity for different functional groups. The data were taken on the eluites indicated in the figure in 50% methanol, 40% acetonitrile or 37% tetrahydrofuran, all given as volume percent at 35°C in water. These conditions were chosen in order to have no difference in benzene retention between solvents. Data were taken on 10 μm LiChrosorb RP-8, 250 × 4.6 mm i.d. at 35°C. Reprinted with permission from Bakalyar *et al.* (*157*).

Abbott *et al.* (*152*) developed a treatment for the prediction of retention time in gradient elution under the assumption that the retention factor as determined under isocratic conditions is a log-linear function of solvent composition, i.e., Eq. (13) is obeyed. The method for the calculation of retention time with linear or exponential gradients was given in addition to means of access to the requisite computer programs. The predicted retention times agreed well with those observed under gradient elution. Schoenmakers *et al.* (*156*) treated the problem more fully in the case of linear gradient. The relationships between retention volume and solvent composition were assumed log-linear, linear, or log-quadratic functions. Analytic expressions for the retention time as a function of the gradient slope and the coefficients relating the isocratic retention factor to eluent composition were developed together with a graphical method for the estimation of elution time, which can be used if the relationship between retention factor and eluent composition is known from experiment. For the case that it is unknown, the authors introduced an expression for the parameters of the quadratic dependence of the logarithm of the retention factor on the solvent composition as a function of the appropriate solubility parameters. The solvent dependence of retention predicted by the use of these correlations agrees fairly well with that observed experimentally.

Engelhardt and Elgass (*161*) found that if the gradient volume, defined as the product of the flowrate and the solvent program time, are held constant and the initial and final compositions of the eluent are fixed, each component of a sample is eluted at a given solvent composition. This result is in agreement with the behavior expected from the expression of Schoenmakers *et al.* (*156*) in the case of linear gradient and eluites for which the logarithm of the retention factor is linear in solvent composition.

Snyder *et al.* (*146*) derived a rather simple relationship between the elution time of an eluite and the rate of change of solvent composition in gradient elution. They found the elution time, t_e, is related to t_0, the column dead-time, and an experimental parameter b by

$$t_e = (t_0/b) \log(2.31k_0 b + 1) + t_0 \tag{14}$$

where k_0 is the retention factor that would be obtained in isocratic elution with the mobile phase composition used at the beginning of the gradient. The parameter b is defined as

$$b = \dot{\Phi} S t_0 / 100 \tag{15}$$

where $\dot{\Phi}$ is the rate of increase in the concentration of the solvent component having eluent strength S and given as volume percent of organic solvent component/min. Average values of S for several solvents are given in Table XI. By using this approach with the optimal value of b given in the literature (*146*) it is possible to calculate the optimal gradient steepness with a given hydroorganic mixture. The result of such calculations is given in Table XII. A companion paper (*155*) showed that this analysis correctly predicts elution time and is generally useful in developing an analytical scheme. Schoenmakers *et al.* (*162*) reexamined the solvent dependence of the logarithm of retention factor and found it to be quadratic for numerous eluties when aqueous methanol, acetonitrile, or tetrahydrofuran mixtures are used as the mobile phase. The authors agreed with Snyder *et al.* (*146*) in declaring linear acetonitrile gradients preferable to convex ones. On the other hand, when gradient elution with

TABLE XII

Optimal Rate of Gradient Development in Hydroorganic Eluent for Maximal Peak Resolution in RPLC[a]

Organic component	$\dot{\Phi}$ (%/min)			
	t_0 (10 sec)	t_0 (30 sec)	t_0 (1 min)	t_0 (2 min)
Methanol	40	13	6.7	3.3
Acetonitrile	39	13	6.5	3.2
Acetone	35	12	5.9	3.0
Dioxane	34	11	5.7	2.8
Ethanol	33	11	5.6	2.8
Isopropanol	29	10	4.8	2.4
Tetrahydrofuran	27	9	4.5	2.3

[a] The elution time of an unretained substance is t_0. The rate of gradient change $\dot{\Phi}$ is given in units of volume percent increase in organic component per minute in hydroorganic eluent. Reprinted with permission from Snyder et al. (146).

methanol–water or THF–water is used, convex gradient shape yields constant peakwidth and therefore high peak capacity.

Changes in eluent viscosity due to changing composition may lead to technical problems in gradient elution. As seen in Figs. 16–20 the viscosity of a hydroorganic solvent first increases to a maximum with increasing organic content. The viscosity decreases with further increase in content of organic cosolvent. As a consequence, in the course of gradient elution, the viscosity of the eluent may be expected to increase with time until the composition of maximum viscosity has been passed. This will be manifested as increased column inlet pressure or as decreased flowrate with instruments operated at constant flowrate or constant pressure, respectively. More generally, when gradient elution with hydroorganic mobile phase is used, systematic variations in the flowrate are expected under conditions of constant pressure operation and systematic variations in the operating pressure will be found when constant flowrate is used. The compressibility of the solvent changes with composition and consequently deviance in the mobile phase composition and the flowrate from that predicted without consideration of compressibility may occur when syringe pumps are used (163–164). This effect can be largely circumvented by the use of a device to generate constant backpressure, such as that described by Abbott et al. (152). Fluctuations in ambient temperature can lead to irreproducible retention times if the column is not thermostated, and this problem becomes especially acute in gradient operation (165).

A frequent problem in gradient elution is the occurrence of "ghost peaks" due to solvent impurities. Specially distilled, and purified solvents to circumvent this problem are commercially available. Water represents a special problem since deionization by ion exchangers can leach out organics from the resin. Storage may lead to additional contamination from leaching of wall components of the storage vessels. The organics may accumulate and eventually elute as a sharp band in the gradient. Engelhardt and Elgass (166) have suggested the use of gradient elution without sample to identify ghost peaks from various sources. Majors (167) discussed the removal of various contaminants from water. The chromatographic significance of water treatment has been considered and a method has been given (168). Water, specially purified for use in HPLC, is commercially now available. An octadecyl silica column of appropriate capacity can be used for purification of water in the laboratory. A cartridge packed with this material and placed between the pump of the aqueous eluent component and the gradient mixer often suffices to remove interfering trace contaminants.

E. Special Considerations

Bakalyar et al. (169) have discussed the effect of dissolved gases in HPLC. The appearance of gas bubbles due to outgassing of the solvent in gradient elution can occur because the solubility of many gases is lower in water and hydroorganic mixtures than in plain organic solvents. In addition, oxygen can form ultraviolet absorbing species with solvents including acetonitrile, methanol, and THF. As a consequence changes in dissolved oxygen concentration upon changing solvent composition can lead to baseline shifts when photometric detection in UV is used. Oxygen can also cause fluorescence quenching and thereby shift baseline when the effluent is monitored by fluorometric detector. Helium sparge is recommended to reduce problems associated with dissolved oxygen.

When sample components having ionizable groups are chromatographed the use of a background electrolyte and control of the eluent pH with an appropriate buffer are mandatory. It is advisable to maintain a fairly high concentration of buffer in the medium in order to rapidly reestablish protonic equilibria and to thereby avoid peak splitting or asymmetrical peaks due to slow kinetic processes. Acetic acid, phosphoric acid, and perchloric acid and their salts have been used for the control of pH.

Perchloric and phosphoric acids are used especially at pH values less than 3. Whereas sulfuric acid has found some use, other common inorganic acids are generally avoided because halide ions have a detrimental effect on No. 316 stainless steel, the favored structural material for wetted parts of chromatographic instruments. At higher pH in the acidic range acetate and phosphate buffers are used most commonly. However, if eluite detection by optical absorption below 240 nm is desired, the use of acetate or any other carboxylic buffer should be avoided. Phosphate buffers are preferred over acetate buffers because of the superior column efficiency often obtained with phosphate in the eluent. The reason for the favorable chromatographic properties of phosphate is not known but it may be due to its propensity to facilitate proton exchange or transfer or to form chromatographically "well-behaved" complexes with various sample components. The presence of trifluoracetic or phosphoric acid in the eluent has been reported to be essential in reversed phase chromatography of larger peptides (170). The use of trialkylammonium phosphate buffers has also been described. In particular, the use of triethylammonium phosphate prepared from phosphoric acid and triethylamine has been reported to give high resolution and high recovery in RPC of peptides and proteins (171).

F. Selection of the Appropriate Mobile Phase

Typical mobile phases used in RPC are mixtures of water or an aqueous buffer and one or more organic solvents. The number of eluents is theoretically infinite although the solubility of the buffer salt frequently limits the buffer concentration in hydroorganic eluents. The selection of an eluent by random testing of various solvent combinations is impractical and rapid methods for choosing the mobile phase appropriate to various separation problems are necessary. Of course, the simplest approach is to reproduce methods given in the literature; however, the mobile phase compositions used by others may not give acceptable results if the stationary phase used is slightly different from that cited. Although from the theoretical point of view little intrinsic difference is evident in stationary phases most commonly used in reversed phase chromatography, as determined by their respective selectivities for alkyl groups (172–173), in practice significant differences in retention behavior might be observed due to variations in the phase ratio from one to another (173). As a consequence, mobile phases having different concentrations of the organic cosolvent may be required to elute certain sample components with commensurate retention factors from columns packed with alkyl silicas of different provenance. Rapid optimization of mobile phase composition may therefore be necessary even if methods to carry out the analysis under similar conditions have been described already.

Many samples contain ionogenic substances which may undergo ionization in an aqueous eluent and as a result severe peak broadening may occur. The effect is reduced or eliminated by the use of buffer solutions which serve as proton sources or sinks and thus maintain the pH in the vicinity of the eluite in the column constant, and which may manifest felicitous kinetic effects. In practice, a buffer concentration between 10 and 100 mM seems to be sufficient in the pH region of peak buffering capacity. This rule suggests that no buffer should be used at pH values where the absolute value of the difference between the pH and the pK_a, the negative logarithm of the acid dissociation constant, is greater than 1. Conventional phosphate buffer is commonly used in pH ranges 2–3 and 6–8. In a recent study Melander *et al.* (*174*) proposed the use of novel amine phosphate buffers to span also the pH range where phosphate buffers poorly. Such buffers prepared with 2-(2-aminoethyl)piperidine (pK_a = 3.94) or with methylpiperazine (pK_a = 5.76) possess other attributes desirable for chromatography as they are optically clear in the ultraviolet and compatible with commonly used organic cosolvents.

In many cases economic considerations or the operating wavelength of the UV-detector mandate the organic solvent component of the eluent. The two most commonly used solvents are methanol and acetonitrile, each of which can be obtained in a specially purified form for HPLC. The optimal concentration of organic cosolvent may be determined by trial and error using isocratic elution with a number of mobile phases differing in fraction of organic cosolvent in order to find the most advantageous eluent composition. An alternative, and more rapid, method is to make a gradient run with increasing concentration of the organic component of the eluent. The concentration at which the elution of a sample component occurs is close to the optimal eluent composition for isocratic elution and can serve as a guide for optimization. Relationships developed by Snyder *et al.* (*146*) may be useful in making this decision. Generally either of the above two organic solvents can be employed in the aqueous eluent. In certain cases, however, the hydroorganic mobile phase chosen on the basis of the experiments suggested above may prove to be unsuitable in practice. For example, the viscosity of the mixture may produce an excessively large pressure drop under the desired operating conditions of flowrate, temperature, etc. An alternative solvent may be selected by use of solvent strength data of the sort presented in Section III,B. For example, by reference to Table XI, one can estimate that the volume fraction of acetonitrile required to obtain a given retention is 0.97 times the volume fraction of methanol which produces the same retention because the reciprocal of their solvent strength ratio is given by 3.0/3.1 = 0.97.

One can also look for special effects with solvents. If the eluotropic strength needed to elute the sample components in a practical range of the retention factor, k, is known, ternary mixtures of the same strength can be prepared under the assumption that the effects are additive in volume fraction. This selectivity can be enhanced by special solvent interactions. In many cases the choice of the solvent mixture will be dictated by knowledge of the chemistry of the eluites under investigation. For instance, the selectivity of tetrahydrofuran toward polar molecules has been exhaustively investigated (92) and similar data can serve as a guide in the final solvent choice (157).

The use of a hetaeron, a complex forming agent, which can selectively interact with certain sample components, stands as a special case of eluent manipulation to augment chromatographic selectivity. The essence of the technique, discussed in Section VII, is the addition to the mobile phase of a species which will form a complex with certain sample components. In "ion-pair" chromatography, the hetaerons are lipophilic ions. They interact with sample components having opposite charges to increase their retention. In another example the retention of substances having double bonds decreases when silver ions are added to the eluent due to the formation of a charged complex. By such manipulation of the eluent composition, it is possible to separate species which might otherwise cochromatograph. The scope of this approach is illustrated by the many types of analyses described later (Section VII, Table XVII). It has been long suspected that buffers also interact with certain eluites and act as hetaerons. For example, amine buffers can be used to advantage in reducing tailing and changing retention factor (174). The effect may be due to an interaction of the buffer amine with acidic silanol groups at the stationary phase surface (93) so that eluite interaction with residual silanols is eliminated.

The choice of the mobile phase is perhaps more an art than an exact science. However, the above guidelines are based on sound, rational grounds. We know that usually a buffer should be used in an analysis, the choice of which can be made according to the particular requirements. An organic cosolvent is usually required in order to carry out the chromatographic run within a reasonable time. The choice can be rapidly made on the basis of a single gradient analysis. Alternative choices of mobile phase can be approximated from considerations of eluotropic strength. Finally, a third solvent or a complexing agent can be added to the mobile phase to impart greater selectivity to the chromatographic system.

IV. EFFECT OF TEMPERATURE

A. Retention Time and Resolution

The effect of temperature on retention has been studied less extensively in RPC than the effect of solvent composition. In fact the use of aqueous eluents and siliceous stationary phases limit the practical column temperature to a range from 5 to 100°C. Over that temperature range one can expect a roughly tenfold decrease in retention factor if the enthalpy of eluite binding to the stationary phase is -5 kcal/mol, a typical value encountered in RPC. On the other hand, a retention change of similar magnitude typically occurs upon increasing the fraction of organic component in the solvent by about 30%. Consequently the effect of temperature on retention may not appear to be important at the first glance.

Nevertheless, temperature effects can be of importance in determining column inlet pressure, analysis time, and resolution. The pressure drop across a column is related to flowrate by Darcy's law according to Eq. (9). The flowrate can be increased with no change in operating pressure if the viscosity is decreased. This can be accomplished by increasing temperature of the column. The dependence of eluent viscosity upon temperature and composition of methanol–water and acetonitrile–water mixtures are shown in Figs. 32 and 33, where solvent compositions are given in volume percent at 20.5°C. Inspection of the graphs shows that the viscosity at 55°C is approximately one-half that at 15°C. Therefore increasing the operating temperature from 20 to 60°C will lead to a roughly 50% decrease in the inlet pressure needed to maintain a given flowrate. Thus, a longer column or a higher flowrate to facilitate faster separation can be used at fixed inlet pressure. Further decreases in analysis time may be obtained due to a reduction in retention factors caused by the effect of increasing temperature on the interaction of eluite with the stationary phase as discussed in Section IV,B.

Temperature changes can also affect the efficiency of the chromatographic system since the diffusion coefficient of a solute increases with the temperature. In eluents used in RPC solute diffusivity increases by a factor of about 2.7 over the temperature range 15–72°C. As a result the plate height may decrease as much as 30% over this temperature range.

In most cases selectivity is not affected strongly by changing temperature in RPC. Colin et al. (142) found a small decrease in selectivity with increasing temperature on pyrocarbon columns. The relative retention of the pair octylbenzene–3,4,5-trimethylphenol decreased from ~ 1.08 to 1.05 over the range 20 to 70°C whereas the relative retention of 1,2,3- and 1,2,4-trimethylbenzenes decreased from 1.24 to 1.2 over the same range. This finding is hardly surprising as enthalpies for eluite binding by the sta-

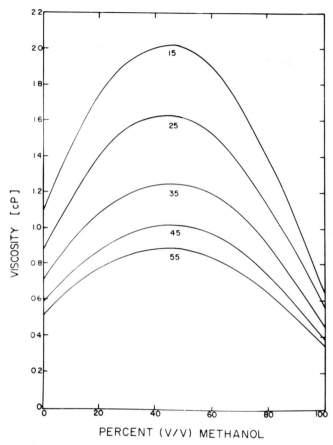

FIG. 32. Variation of solvent viscosity with changing temperature and composition of water–methanol mixtures. The composition is expressed as percent methanol (v/v) at 20.5°C and the viscosity is in centipoise. The temperature at which the viscosity was measured is indicated below each curve. The data are taken from Colin *et al.* (*142*).

tionary phase are usually small and the difference is often less than 1 kcal for closely related substances.

Gant *et al.* (*175*) examined the effect of temperature on resolution and on selectivity, retention factors, and plate number, which determine the magnitude of resolution. They found that these data can be used together with the temperature dependence of solvent viscosity to optimize analysis rate with required resolution. This is of particular interest when RPC is used for automated repetitive analyses of large numbers of samples.

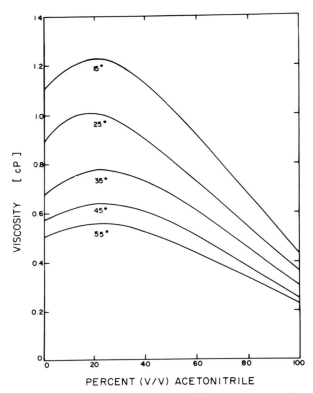

FIG. 33.   Variation of solvent viscosity with changing temperature and composition of water–acetonitrile mixtures. The composition is expressed as percent acetonitrile (v/v) at 20.5° and the viscosity is in centipoise. The temperature at which the viscosity was measured is indicated below the curve. The data are taken from Colin et al. (142).

The need for proper temperature control was shown in the data of Perchalski and Wilder (176) who have examined the effect of heating the mobile phase to the nominal column temperature in a precolumn heater. Some of their data are reproduced in Figs. 34 and 35. They found the peak shape to be non-Gaussian and the resolution between peaks to be poorer when the temperature of the mobile phase was significantly lower than that of the column. The effect was less pronounced at low flowrates. They interpreted the results by assuming that a stream of cold mobile phase passes through a heated column annulus. Peak asymmetry is believed to arise from nonuniform flow profile, and differences in equilibrium constants and diffusion rates in various column regions.

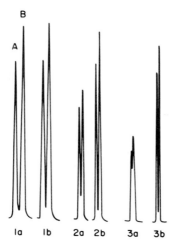

Fɪɢ. 34. Effect of thermal pre-equilibration of the mobile phase. Chromatograms of identical samples, containing (A) 3-bromo-*N*-methyl cinnamamide and (B) clonazepam run at 40 ml/h (1), 80 ml/h (2), and 120 ml/h (3) with unheated (a) and heated (b) precolumns on a 5-μm Spherisorb ODS column. Reprinted with permission from Perchalski and Wilder (*176*), *Anal. Chem.* Copyright © 1979 by the American Chemical Society.

B. Enthalpy of Binding

The effect of temperature on retention is largely determined by the enthalpy of eluite interaction with the stationary phase. The enthalpy is evaluated from the slope of plots of log k versus the reciprocal of the absolute temperature that are called van't Hoff plots. When the equilibrium for eluite binding by the stationary phase can be written in the terms of a single eluite species and a single bound species, then the enthalpy of this particular binding process is obtained. If, however, there is more than one kind of eluite species, such as when the eluite is partly ionized, and both are present in significant quantities, van't Hoff plots yield the overall retention enthalpy which is a weighted average of the enthalpies for the binding of the species. An example of such a system, which may be encountered in RPC, is an acid chromatographed at a pH value within one unit of its pK_a value. The relationship between the enthalpy observed in the chromatographic experiment and enthalpies for ionization and for binding of the species by the stationary phase is given by Eq. (19) in Section IV,C. With more than one species present, we can expect curvature in the van't Hoff plot. Of course, departure from linearity can also occur if the heat capacities of the bound and free forms of the eluite are different. In either case the value of the enthalpy obtained from the slope of the

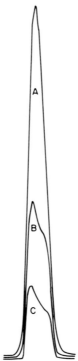

FIG. 35. Effect of thermal pre-equilibration of the mobile phase. Chromatograms of identical samples of 3-bromo-*N*-ethyl cinnamamide run at (A) 58 ml/h, (B) 115 ml/h, and (C) 230 ml/h on a 5-μm Spherisorb ODS with unheated precolumn. Reprinted with permission from Perchalski and Wilder (*176*), *Anal. Chem.* Copyright © 1979 by the American Chemical Society.

van't Hoff plot can be used for calculations at the particular temperature used in its determination only. The source of nonlinearity can be determined from extrachromatographic experiments.

In most cases, however, van't Hoff plots of chromatographic data are linear and allow facile determination of the enthalpy according to the general experience (*142, 177, 178*). Examples of such plots are shown in Figs. 36 and 37 which present data taken on octadecyl silica and on pyrocarbon, respectively. One can readily see that the enthalpies, which are directly proportional to the slope, do not differ greatly among the several solutes although relatively large differences in retention factors, and in their logarithms, are observed.

In many physicochemical interactions which are subject to the same intrinsic mechanism, the free energy change is proportional to the enthalpy.

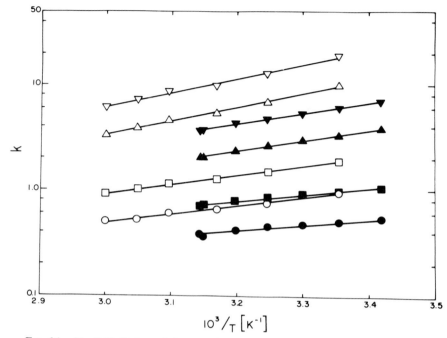

FIG. 36.   Van't Hoff plots of the retention factors of aromatic acids in reversed-phase chromatography using octadecyl silica as the stationary phase and neat aqueous 50 mM NaH_2PO_4 buffer (pH 2.0) (open symbols), or the same buffer containing 6% (v/v) of acetonitrile (closed symbols) as the eluent. Column: 5 μm Spherisorb ODS, 250 × 4.6 mm. Eluites: 3,4-dihydroxymandelic acid (O, ●); 4-hydroxymandelic acid (□, ■); 4-hydroxyphenylacetic acid (▽, ▼); 3,4-dihydroxyphenylacetic acid (△, ▲). Reprinted with permission from Melander et al. (177).

If that were true in RPC, one would expect the logarithm of retention factor to increase linearly with the enthalpy and the slope of a plot of this sort to be approximately equal to $1/RT$ when natural logarithm is used. Indeed there is ample evidence in the literature that the logarithm of the retention factor is linearly related to the enthalpy in RPC (142, 177, 178). The increase in logarithm of retention factor with the enthalpy, however, is much greater than expected. This is due to changes in binding enthalpy which are accompanied by corresponding changes in the binding entropy caused by structural modifications of the eluite molecule or changes in eluent composition. Indeed, linear relationships have been found in RPC between the enthalpy and the entropy for reversible eluite binding to hydrocarbonaceous bonded phases (177) and pyrocarbon (142). The effect is called enthalpy–entropy compensation, and it is an example of a linear

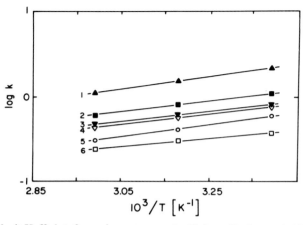

FIG. 37.   Van't Hoff plots for various compounds. Eluites: (1) cinnamyl aldehyde, (2) estragole, (3) cinnamyl alcohol, (4) cinnamyl acetate, (5) methyleugenol, and (6) safrole. The data were obtained on pyrocarbon, 400 × 2.17 mm i.d., with acetonitrile as mobile phase. Reprinted with permission from Colin *et al.* (*142*).

TABLE XIII

Enthalpies of Binding of Eluite to Stationary Phases

Solutes	Solvent system		$-\Delta H$ (kcal/mol)	Ref.
Catechol derivatives	0.05 M KH$_2$PO$_4$ in H$_2$O	Partisil ODS	1–3	*180*
Aromatic acids	0.05 M KH$_2$PO$_4$ in H$_2$O, pH 2.1	5 μm Partisil ODS-2	3.5–5.7	*107*
Aromatic acids, alcohols, and esters	CH$_3$CN	Pyrocarbon	1.1–1.8	*142*
Alkyl benzenes and phenols	Methanol–H$_2$O (80:20)	Pyrocarbon	3.5–6.5	*142*
Aromatic acids	0.05 M KH$_2$PO$_4$ in H$_2$O, pH 2.1 or 6% in this buffer	5 μm Partisil ODS-2	2.5–6	*177*
Substituted hydantoins, allantoin, and phenylacetic acid	0.05 M KH$_2$PO$_4$ in H$_2$O, pH 2.1 or 30% CH$_3$CN in this buffer	5 μm Partisil ODS-2	1–8	*177*
Benzene derivatives	Water–methanol (60:40, v/v)	Permaphase ODS	2.9–5.8	*178*

free energy relationship (*179*). The slope of the plots has the dimension of temperature which is called the compensation temperature. Comparison of "compensation temperatures" obtained from such data can be used to investigate whether the intrinsic mechanism of retention in a given chromatographic system is identical to that found on another (*177*).

The enthalpy of binding of different solutes does not range very widely. A sample of enthalpy values obtained from the literature is given in Table XIII. They range between -3 to -8 kcal/mol. If we assume that -5 kcal/mol is an appropriate value for an average eluite, we find that the retention factor will decrease by 36% upon increasing the temperature from 20 to 60°C.

C. Secondary Equilibria in the Mobile Phase

So far only a few studies have shown curvature in van't Hoff plots of data measured in RPC. As mentioned above, curvature may be found when the eluite is present in more than one form and an analysis of such systems is presented in this section.

The simplest case in which secondary equilibria can lead to curvature is an isomerization of eluite. Consider the process

$$A \xrightleftharpoons{K} B \tag{16}$$

where A and B are isomers and K is the appropriate equilibrium constant. If both forms can bind to the stationary phase the observed retention factor, k, can be written as

$$k = \frac{k_A + Kk_B}{1 + K} \tag{17}$$

where k_A and k_B are the retention factors of species A and B, respectively. The retention factors and equilibrium constant are assumed to depend on temperature as follows:

$$k_A = \phi \, \exp(-\Delta H_A^\circ/RT) \, \exp(\Delta S_A^\circ/R) \tag{18a}$$

$$k_B = \phi \, \exp(-\Delta H_B^\circ/RT) \, \exp(\Delta S_B^\circ/R) \tag{18b}$$

$$K = \exp(-\Delta H^\circ/RT) \, \exp(\Delta S^\circ/R) \tag{18c}$$

where ΔH_A°, ΔH_B°, and ΔH° are the standard enthalpies of binding of A and B by the stationary phase and the enthalpy of the isomerization, respectively, whereas ΔS_A°, ΔS_B°, and ΔS° are the respective entropy changes. The phase ratio defined in Section II is given by ϕ.

The enthalpy observed at any temperature can be obtained by combining Eqs. (18a)–(18c) and differentiating with respect to reciprocal temperature. The result is given by

$$\Delta H_b^{\circ} = \frac{k_A \, \Delta H_A^{\circ}}{(k_A + Kk_B)} + \frac{k_B K \, \Delta H_B^{\circ}}{(k_A + Kk_B)} + \frac{\Delta H^{\circ}(k_B - k_A)K}{(1 + K)(k_A + Kk_B)} \tag{19}$$

where ΔH_b° is the observed retention enthalpy for the overall chromatographic process at temperature T.

Three conclusions can be drawn from the relationship given in Eq. (19). The first is that if secondary equilibria occur in the mobile phase, the enthalpy observed will be a weighted mean of the enthalpies associated with each process. This result is general and will be true if the equilibria involved in the chromatographic retention process are more complex than those of this example. The second conclusion is that van't Hoff plots will be nonlinear unless one works under conditions such that the formation of all isomers but one is suppressed or a most felicitous combination of enthalpies and entropies is found. The third conclusion is that changes in the sign of the enthalpy may be seen in the temperature range under investigation as can be demonstrated by a simple example as follows. Suppose the binding enthalpy of each form is identical but the entropies are different in such a fashion that the ratio of the retention factors k_B/k_A is 3:1. Suppose, furthermore, that the enthalpy for isomerization is opposite in sign to that of binding but four times greater in absolute magnitude. Then at a given temperature, at which the equilibrium constant for isomerization is unity, the enthalpy will be zero, and at higher and lower temperatures, the enthalpies will have opposite signs.

Increased temperature has been found to give enhanced retention under some circumstances. For instance, in RPC of certain aryl oligoethylene glycols the sign of the enthalpy has been found to depend on both the oligoethylene glycol chain length and the solvent (158). The chromatographic data were consistent with the assumption that the oligoethylene glycol moiety can have at least two conformations and that the energetics of the binding for the two forms of the eluite are different. Evidence from extrachromatographic experiments gave strong support to such interpretation of the observed retention behavior (158).

D. Multiple Retention Mechanisms

Alternative mechanistic schemes can also be devised to account for nonlinear van't Hoff plots. A particularly simple case occurs when more than one binding site on the stationary phase surface is involved in the retention of eluite. The retention process can formally be modeled for the case of two sites by the use of Eqs. (17)–(19) with the following modifications: $\Delta H° = 0$ and $K = 1$.

If the retention enthalpies of the two sites differ, curvature may be observed in the plots. Moreover, if the enthalpies are opposite in sign, a minimum will occur in the van't Hoff plot at a temperature where the ratio of the retention factors for the two mechanisms equals the absolute value of the reciprocal of the ratio of the corresponding enthalpies. Most frequently, however, less dramatic curvature would be expected. Such behavior may be anticipated in the RPC of amines with large nonpolar moieties which could be retained by silanophilic interactions with surface silanols and by solvophobic interactions with nonpolar ligates of a reversed phase with low surface coverage. Recently an analysis of this behavior has been reported (93).

The examples given in Section IV,C and here serve to demonstrate that nonlinear van't Hoff plots can be expected whenever one of the following three conditions hold: (i) the eluite exists in two or more forms having different retention factors; (ii) the eluite is retained by two or more mechanisms due to the heterogeneity of the stationary phase surface containing more than one type of binding site; (iii) eluite exists in more than one form and the surface is heterogeneous. We may expect that such phenomena will be increasingly encountered upon detailed investigation of the effect of temperature on retention.

V. THEORY OF REVERSED-PHASE CHROMATOGRAPHY

A. Hydrophobic and/or Solvophobic Effect

Since the stationary phase is basically nonpolar in RPC, it is not expected to effect solute retention by ionic attraction, hydrogen bonding, formation of charge transfer complexes, or by any of the other strong noncovalent interactions familiar to the chemist. The only attractive force between the stationary phase and the eluite would seem to be van der Waals forces. However, these forces also act between the eluite and the mobile phase so that the net effect is generally not sufficient to account for the strong retention often observed with nonpolar compounds in RPC.

The apparent paradox has been resolved by the recognition that the so-called "hydrophobic effect" very likely plays a dominant role in determining the overall energetics of the equilibrium distribution under conditions of RPC. The hydrophobic effect has been the object of intensive theoretical and experimental study since its importance in maintaining protein structure was demonstrated by Kauzmann (*181*). This Section (V,A) briefly reviews the hydrophobic effect and outlines the "solvophobic theory" which has been adopted to interpret solvent effects in RPC. Section V,B considers in detail the process of solute transfer from gas to liquid on which the original formulation of the solvophobic theory is based. The extension of the theory to chromatography, and molecular associations in general, is made in Section V,C whereas some of the general predictions of the theory for chromatography are presented in Section V,D. Those readers who find mathematically couched arguments soporfic or those who prefer the vistas at journey's end to the pleasures of the journey itself are advised to fly over Section V,B and C.

Whereas the consequences of the hydrophobic effect itself have long been recognized, e.g., in the adage "you cannot mix oil with water," an understanding of the physicochemical nature of the incompatibility probably dates to an observation of Frank and Evans (*182*). They found that the entropy of mixing apolar solutes with water is anomalous when compared to the entropy of mixing the same solutes with most other solvents. Furthermore, they also found that the enthalpies, although small, are such as to favor mixing. They summarized their results by stating, "The nature of the deviations [of entropies of vaporization] found for non-polar solutes in water, together with the large effect of temperature on them, leads to the idea that the water forms frozen patches or microscopic icebergs around such solute molecules, the extent of iceberg increasing with the size of the solute molecule. Such icebergs are apparently formed also about the non-polar parts of the molecules of polar substances such as alcohols and amines dissolved in water . . . " (p. 507). The concept of the "iceberg," which was not intended to imply the occurrence of any known ice structure, was used to explain the unusual entropy. The most important feature of this picture is the conclusion that the average solvent molecule in the vicinity of the solute is in an energy state differing from that of the bulk molecules.

A statistical thermodynamic treatment of the water structure as an incompletely hydrogen-bonded network was presented by Nemethy and Scheraga (*183*). In companion papers they showed that the thermodynamic properties of nonpolar substances in water can be duplicated using this framework with the assumption that in the vicinity of a solute the average number of hydrogen bonds to water is increased over that in the bulk solvent. When two apolar solutes come together to form a "hydrophobic bond," the driving force for the reaction is derived from a decrease in the number of water molecules coordinating to the solute, and the concomitant relaxation of the water released to a lower energy state.

This molecular picture has gained widespread popularity in part because its qualitative features are easily grasped.

Eyring and his co-workers (184) proposed a somewhat different picture of water structure, suggesting that liquid water can be regarded as a mixture of the following significant structures: monomeric water, ice-I-like molecules and ice-III-like molecules. Hermann (185) adopted this model and combined it with the assumption that the water molecules coordinating the solute have properties of those at the air–water interface. He also took into account the number of solvent molecules around the solute and found that the free energy of solution is proportional to the number of water molecules surrounding the solute, i.e., to the surface area of the solute. This result was tested by others (186, 187) for alkanes and aromatic hydrocarbons as well as alcohols and acids. In each case good correlation was obtained between the activity of solute in dilute solutions and its molecular surface area. The surface area was taken as that scribed by the center of a sphere of radius 1.5 Å, i.e., the van der Waals radius of a water molecule rolling over the surface of the solute molecule. Alternatively, the surface was measured by the number of spheres of the same diameter which could be close-packed around the appropriate space-filling molecular model. The results suggested that the hydrophobic effect is related to the nonpolar surface area of a solute and the strength of a hydrophobic bond would therefore be proportional to the decrease in molecular surface area that occurs upon formation of a complex.

While informative and suggestive, the above-cited treatments of the hydrophobic effect were not seen by us as appropriate in treatment of data obtained in RPC. By their nature, the calculations of energies in these treatments is a somewhat laborious process which frequently requires detailed molecular information about the solute or solvent or both. In addition the results obtained focus on the properties of aqueous solutions. As chromatographers most commonly use less polar solvents or mixed solvents, not even the qualitative features of the above-mentioned theoretical approaches could be used with equanimity. In order to circumvent the difficulties, recourse was made to a theory of solvent effects on chemical equilibria developed by Sinanoğlu and his co-workers (188–194). Since this theoretical analysis makes no special assumptions about the nature of solvent but is appropriate to not only water but also to any other solvent, it is called "solvophobic theory." Another prominent feature of the theory is that it uses only bulk properties of both solute and solvent in the calculation of the "solvophobic effect." Electrostatic and any other special interactions are explicitly included in the theory although ionic effects were not considered in the original treatment. As will be seen, the solvo-

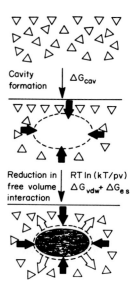

FIG. 38. The process of going into solution is treated as two steps in the solvophobic theory. The first step is the creation of a cavity of size and shape suitable for the incoming solute. In the second step the solute enters the cavity and interacts with the surrounding solvent. Reprinted from Horváth and Melander (*129*), *J. Chromatogr. Sci.*, with permission from Preston Publications.

phobic theory preserves several important features of the above-mentioned treatments of the hydrophobic effect.

The theory is formulated in terms of the free energy change required for dissolving a solute from a hypothetical gas phase containing the solute at atmospheric pressure. The process is decomposed into two steps for the purpose of calculating the energy of the process which is portrayed in Fig. 38. In the first step, a cavity is prepared in the solvent of size and shape appropriate to the incoming solute. This is accompanied only with changes in the solvent; the molecules at the face of the cavity are assumed to have properties similar to those of condensed molecules at the air–liquid interface. In the second step, the solute is placed in the cavity and allowed to interact with the surrounding solvent molecules via van der Waals and electrostatic interactions.

B. Energetics of Solvent Effect

1. Free Energy Change due to Cavity Formation

The energies associated with the interactions are readily calculated, at least in principle, if certain properties of solute and solvent are known. As a result of cavity formation, the surface area of the solvent increases by A, the molecular surface area of the solute. Consequently the energy should increase by an amount proportional to the surface tension γ and the increase in area. However, the cavity surface has a very small radius of curvature, and therefore the surface tension of the cavity will be different from that of the bulk by a proportionality factor $\kappa^e(r)$. Thus the free energy of cavity formation, ΔG_c, is given approximately by

$$\Delta G_c = \kappa^e(r)A\gamma N \tag{20}$$

where N is Avogadro's number. The coefficient $\kappa^e(r)$ is dependent upon the size of the cavity itself. Although the dependence on cavity size has not been closely examined for polar liquids, it is believed to be functionally similar to that for nonpolar liquids (192). Thus

$$\kappa^e(r) = 1 + (\kappa^e - 1) A_s/A \tag{21}$$

where κ^e is the corresponding parameter for the creation of a cavity the size of a solvent molecule and A_s is the molecular surface area of the solvent. This result can be combined with Eq. (20) to yield the following expression for the energy of cavity formation

$$\Delta G_c = A\gamma N + (\kappa^e - 1) A_s\gamma N \tag{22}$$

Equation (22) has been found to be somewhat more useful than Eq. (20) for the evaluation of the free energy change related to cavity formation when more than one solute species is present. In the form given by Eq. (22) the cavity term can be calculated if macroscopic surface tension, γ, κ^e, and molecular surface area of both the solute and solvent are known. The latter values may be calculated for spherical or quasi-spherical molecules as

$$A_i = 6.78 \times 10^{-16} V_i^{2/3} \tag{23}$$

where V_i is the molar volume of pure species i. The κ^e term is determined from the heat of vaporization and the surface tension of the solvent at the temperature of interest in addition to the temperature derivative of the surface tension and the thermal coefficient of expansion. All these properties are known for solvents of chromatographic interest or can be reliably estimated (107).

It should be noted that the expression of the cavity term in Eq. (22) differs from that given by Halicioğlu and Sinanoğlu (191, 192) who presented a more exacting treatment of the thermodynamics of cavity formation. However, the difference between the energy calculated by the rigorous formulation and by the approximation in Eq. (22) is only a few percent and seldom exceeds 0.4 kcal/mol.

2. Free Energy Change due to Solute–Solvent Interaction

The calculation of the energy for the interaction between the solvent and solute is more complicated. Formally, the free energy for the process, ΔG_{int}, is composed of two chemical parts and an essentially entropic term. The chemical terms are associated with van der Waals interactions ΔG_{vdw}, and electrostatic effects, ΔG_{es}, between solute and solvent. The entropic term measures the "free volume" i.e., the volume a molecule explores before encountering another, which is assumed to be proportional to the molar volume of the solvent. Thus, the free energy change associated with solute–solvent interactions at temperature T is given by

$$\Delta G_{int} = \Delta G_{vdw} + \Delta G_{es} + RT \ln(RT/PV) \tag{24}$$

where P is one atmosphere and R and V are the gas constant and the molar volume of the solvent, respectively.

a. Van der Waals Interactions. The evaluation of van der Waals interaction energies is possible provided the appropriate interatomic potentials are known or can be reliably estimated. Whereas in the usual treatment of the problem it is assumed that the molecules are in the gas phase, i.e., the potential is not modified by the presence of other molecules, Sinanoğlu developed a method for the calculation of van der Waals potential in condensed media (*190, 191*). The potential between two molecules in the absence of any perturbations by other molecules is given by

$$V_{AB} = C_{AB} \left(\frac{\bar{\sigma}^6}{\bar{\rho}^{12}} - \frac{1}{\bar{\rho}^6} \right) \tag{25}$$

where C_{AB} is the London parameter defined by

$$C_{AB} = 2.3\alpha_A \alpha_B I_A I_B / (I_A + I_B) \tag{26}$$

α_A and α_B are the polarizability of species A and B and I_A and I_B are the corresponding ionization potentials. The values of $\bar{\sigma}$ and $\bar{\rho}$ are determined from the molecular volumes and acentric factors of the individual species. The values for the molecular diameter of species i, R_i, and its Kihara parameters ℓ_i and σ_i are given by

$$R_i = 1.74 \, (3v_i/4\pi)^{1/3} \tag{27}$$

$$\ell_i = R_i \frac{0.24 + 7\omega_i}{3.24 + 7\omega_i} \tag{28}$$

$$\sigma_i = R_i \frac{2.67}{3.24 + 7\omega_i} \tag{29}$$

where v_i and ω_i are molecular volume and acentric factor, respectively.

Arithmetic mean values of the diameter, $\bar{R}$, Kihara diameter, and $\bar{\ell}$ and $\bar{\sigma}$ are used in the evaluation of the interaction between two molecules A and B. These values, and an associated parameter t, are defined as

$$\bar{R} = (R_A + R_B)/2 \tag{30}$$

$$\bar{\ell} = (\ell_A + \ell_B)/2 \tag{31}$$

$$\bar{\sigma} = (\sigma_A + \sigma_B)/2 \tag{32}$$

$$t = \bar{\ell}/(\bar{R} - \bar{\ell}) \tag{33}$$

Finally the core-corrected intermolecular distance, $\bar{\rho}$, is defined as

$$\bar{\rho} = r - \bar{\ell} \tag{34}$$

where r is the center-to-center intermolecular distance.

The effective potential V_{eff} which accounts for the perturbation of the intermolecular gas phase potential by intervening molecules in the condensed phase, is related to the gas phase potential, V_{AB}, by

$$V_{\text{eff}} = V_{AB} \left[1 - \frac{1}{n} \frac{\phi_{AB}}{1 - (\bar{\sigma}/\bar{\rho})^6} \right] \tag{35}$$

where n is an integer, 1, 2 or 3, corresponding to solute–solute, solute–solvent, or solvent–solvent interactions, respectively, and ϕ_{AB} is a function dependent upon the polarizability and dielectric of the medium as well as the ionization potentials of the molecules of interest and their distance of separation. These are given explicitly by Sinanoğlu (190) who provided a very detailed treatment of this approach (191). The integration of V_{eff}, which implicitly treats a discrete solvent layer followed by a solvent continuum, yields the appropriate free energy, ΔG_{vdw}, through the following expressions:

$$\Delta G_{\text{vdw}} = \frac{9}{4\pi} C_{AB} (1 - x) (Q' + Q'') \tag{36}$$

where for situations likely to be encountered in RPC (190) the value of the parameter x is 0.436.

$$Q' = V_S \int_R^\infty (\bar{\sigma}^6/\bar{\rho}^{12} - 1/\bar{\rho}^6) r^2 dr \tag{37a}$$

$$= V_S \left[\frac{\bar{\sigma}^6}{(R - \bar{\ell})^9} \left(\frac{t^2}{11} + \frac{t}{5} + \frac{1}{9} \right) - \frac{1}{(R - \bar{\ell})^3} \left(\frac{t^2}{5} + \frac{t}{2} + \frac{1}{3} \right) \right] \tag{37b}$$

and

$$Q'' = V_S \int_R^\infty \frac{\phi_{AB}}{n\bar{\rho}^6} r^2 dr \tag{38}$$

Due to the nature of the modifications introduced above, the potential Q'' is not an analytical function. Since the potential perturbation function is known, however, the function can be evaluated numerically. Furthermore, Sinanoğlu (193) has found a relatively simple relationship for the estimation of the effective potential in the solvent from the gas phase potential which can be calculated from Eqs. (25) and (26). The approximation is given by

$$V_{\text{eff}} \approx V_{\text{AB}} \left[1 - \frac{1}{n} \cdot \frac{3}{2} \cdot \frac{\mathscr{D}}{1 + \mathscr{D}} \right]$$ (39a)

where

$$\mathscr{D} = (n_{\text{B}}^2 - 1)/(n_{\text{B}}^2 + 2)$$ (39b)

and n_{B} is the refraction index of the solvent. By substituting values appropriate for water and various alkanes or aromatic solutes into Eqs. (37)–(39b) and integrating numerically, we obtain the relationship

$$Q'' \approx -0.125 \, Q'$$ (40)

for solutes when the solvent is water. For other solvents the numerical coefficient is somewhat larger and has a value of about 0.15.

As the second integral Q'' is small in comparison to the first integral Q', it can be neglected in a simplified calculation of V_{eff}. While less rigorous and much less esthetically satisfactory, this assumption frequently is adequate for qualitative predictions (193–195).

b. *Electrostatic Interactions.* (i) *Dipoles.* The evaluation of the electrostatic terms is somewhat less tedious. Two kinds of effects need to be considered; that of a dipole and that of a monopole (ion) in a condensed medium. In the present theory, dipole effects are treated according to the Onsager reaction field approach (196) and the corresponding free energy change is expressed by

$$\Delta G_{\text{es}} = \frac{-N}{2} \frac{\mu_i^2}{r_i} \mathscr{D} \bigg/ \left(1 - \frac{\mathscr{D} \bar{\alpha}_i}{r_i^3} \right)$$ (41a)

where μ_i, $\bar{\alpha}_i$, and r_i are the dipole moment, polarizability, and molecular radius of the solute, respectively, and ϵ is the dielectric constant of the solvent. The value of $\mathscr{D}$ is given by

$$\mathscr{D} = 2(\epsilon - 1)/(2\epsilon + 1)$$ (41b)

The ratio $\bar{\alpha}/r^3$ can be estimated in some cases by using the Clausius–Mosotti function as

$$\frac{\bar{\alpha}}{r^3} = \frac{4\pi N\alpha}{3V} = \frac{n_A^2 - 1}{n_A^2 + 2} \qquad (42)$$

where n_A is the refractive index of the solute. The function $\mathcal{D}$ is rather insensitive to variation in dielectric constant and it has the value of 0.981 and 0.954 for water ($\epsilon = 78.5$) and for methanol ($\epsilon = 32$), respectively. However, with a nonpolar liquid, having $\epsilon = 3$, the value of $\mathcal{D}$ falls to 0.5.

(ii) *Monopoles.* The corresponding term for an ionic species (or monopole) can be estimated from conventional electrostatic theories. The energy for placing an ionic species into a medium can be estimated by the Debye–Hückel treatment as

$$\Delta G_{es} = \frac{Z^2 e^2 N}{\epsilon} \left(\frac{1}{b} - \frac{\kappa}{1 + \kappa a} \right) \qquad (43a)$$

where Ze is the electronic charge on the ion, b is its ionic radius, and a is the distance of closest approach, usually taken as 3 Å. The value of the Debye screening parameter κ is calculated from the relationship

$$\kappa^2 = 4\pi e^2 \, NI/\epsilon RT \qquad (43b)$$

where e is the electronic charge and I the ionic strength of the medium. The Debye treatment is limited to very low ionic concentrations but is generally considered valid up to an ionic strength of 0.01.

At higher ionic strength the limiting Debye–Hückel treatment no longer adequately describes the effect of the ionic atmosphere on the activity of an ion, and this deficiency has been the subject of much theoretical investigation.

The activity coefficient, y, is a measure of the change in free energy of a component at fixed concentration x in a mixture due to nonideal behavior. In this case, nonideality is due to electrostatic effects. The general expression for free energy of a component including this effect is

$$\Delta G^\circ = \Delta G^\circ(0) - RT \ln y - RT \ln x \qquad (44)$$

where $\Delta G^\circ(0)$ is the free energy of the corresponding ideal solution at infinite dilution.

Two semiempirical approaches which to date seem to account best for the effects observed at high ionic strength, will be discussed here. The first, due to Lietzke *et al.* (197), views the solution as a mixture of two components: one has the limiting Debye–Hückel character and the other component exhibits the behavior of fused salts. The activity coefficient for fused salts is given by

$$\ln y = BI^{1/3} + CI \qquad (45)$$

where I is the ionic strength; B and C are experimentally determined constants appropriate to the salt. The fraction of the solution which exhibits Debye–Hückel behavior is given by

$$f = \exp(-aMV/1000) \tag{46}$$

where a is an empirically determined value, and M and V are the salt molarity and the molar volume of dry salt, respectively. The volume fraction exhibiting fused salt behavior is $1-f$. Combining Eqs. (45) and (46), we obtain the relationship between the activity coefficient, and thus the free energy, and the ionic strength as

$$\ln y = (\mathscr{S}\sqrt{I}/(1 + 1.5\sqrt{I}) + BI^{1/3} + CI) \exp(-aMV/1000)$$
$$+ BI^{1/3} + CI \tag{47}$$

where $\mathscr{S}$ is the Debye–Hückel coefficient evaluated from Eqs. (43a) and (43b) as

$$\mathscr{S} = \frac{e^3}{(\epsilon kT)^{3/2}} \left(\frac{2\pi N}{1000}\right)^{1/2} \tag{48}$$

In water at 25°C, $\mathscr{S}$ is equal to 1.18. The parameters a, B, and C have been determined for several salts and regularities have been discovered in a and B. The magnitude of a is related to charge type since the observed values for 1–1, 1–2, or 2–1 and 1–3 electrolytes were clustered into these groups with narrow ranges of intragroup values. Lietzke et al. (197) showed that B should be dependent upon charge type and found experimentally that the data did conform to the expression derived and, in an addendum, noted that the parameter could be calculated on the basis of Madelung constant and charge type, i.e. it can be regarded as not adjustable. Similarly, the value of a is not adjustable either.

The second semiempirical approach to the evaluation of the activity coefficient is the use of the Davies equation which modifies the Bronsted extension of the limiting Debye–Hückel expression and is given by

$$\ln y = \mathscr{S}\sqrt{I}/(1 + 1.5\sqrt{I}) + BI \tag{49}$$

where B is an experimentally defined parameter (197).

The methods described above are appropriate for "simple" ions, but not for the calculation of the activity coefficients of more complex compounds such as zwitterions, i.e., those which bear more than one functional group, have a low molecular weight, which is arbitrarily put at less than 500, and are approximately spherical in shape so that both the quasi-spherical assumption used in the van der Waals integral and the present definition of cavity area are satisfied. Many substances of interest in reversed phase chromatography do not meet these criteria. Amino acids fall into the latter category and they have two ionizable groups separated so widely that solvent molecules can interpose between them and modify their interaction. Two methods are available for the treatment of the energetics of such systems. The first follows the solution thermodynamics of amino acids and polypeptides developed by Kirkwood (198) and Linderstrøm-Lang (199), respectively. Formulas given by those authors for the effect of ionic strength on the activity coefficient of such species can be used directly for our purpose (60). A more general procedure considers each functional group separately and sums the individual contributions to obtain the overall free energy (116).

C. Free Energy Change in the Chromatographic Process

In the preceding sections we have considered the role of the solvent in determining the solubility of a molecule taken from the gas phase. However, we are interested in a quantitative evaluation of the solvent effect on phenomena which involve more than one molecule, such as the equilibrium distribution process in reversed-phase chromatography using hydrocarbonaceous bonded phases.

The formulation developed above can be directly used for the analysis of truly liquid–liquid chromatographic systems. In that case the Gibbs free energy of transfer for the partition process is calculated by the difference in the energies of solution, which are evaluated according to the preceding analysis for the two phases. From the free energy change the pertinent partition coefficient is readily obtained.

On the other hand, adsorption on a bonded stationary phase may be regarded formally as a reversible reaction involving the eluite, E, and the binding surface, B, to form a complex, EB:

$$E + B \rightleftarrows EB \tag{50}$$

The chromatographic retention factor, k, is related to the equilibrium constant for adsorption to a solid phase or distribution between liquid mobile and stationary phases by the well-known expression

$$k = \phi K_0 \tag{51}$$

where K_0 is the adsorption or distribution constant and ϕ, the phase ratio, is defined as the ratio of stationary phase surface area to mobile phase volume in the case of adsorption or by the volume ratio of stationary to mobile phase in the case of liquid–liquid chromatography. This can be related to the Gibbs free energy for the process of binding, $\Delta G°$, through the equally familiar relationship between the free energy and equilibrium constant for a process to obtain

$$\ln k = \ln \phi - \Delta G°/RT \tag{52}$$

The equilibrium constant, K_0, is given, in view of Eq. (50), by the appropriate concentrations as

$$K_0 = [EB]/[E][B] \tag{53}$$

Two phenomena have to be taken into account in order to make the previous treatment of an isolated solute applicable to the present situation. First, one must consider the strength of the interaction between E and B in the absence of any solvent effects. Second, an additional entropic term must be introduced because B, and consequently EB, are bonded so that the complex has restricted translational freedom at the surface. We will neglect, however, the latter effect because its contribution to the free energy of binding is expected to be small, i.e., not more than a few hundred calories per mole. Moreover, the entropic contributions for both species B and EB should be nearly identical and cancel when the individual terms are summed.

The energetics of interaction between the associating molecules and the solvent can be evaluated from the van der Waals potential between a solute and a solvent molecule at a distance r apart. For such calculations standard formulas are available and the most useful may be that given by Sinanoğlu (190) since it correctly accounts for the effects of intervening solvent in modifying the potential between molecules. It is given by

$$V_{\text{eff}} = C_{AB} \left[\frac{\bar{\sigma}^6}{(r - \bar{\ell})^{12}} - \frac{1}{(r - \bar{\ell})^6} \right] \left[1 - \frac{1}{2} \frac{\phi_{AB}}{1 - (\bar{\sigma}/r - \bar{\ell})^6} \right] \tag{54}$$

which differs from Eq. (35) in that the coefficient in the denominator of the second term is appropriate to solute–solvent interaction. Equation (54) can be replaced in view of Eq. (39a) by the approximation

$$V_{\text{eff}} \approx C_{AB} \left[\frac{\bar{\sigma}^6}{(r - \bar{\ell})^{12}} - \frac{1}{(r - \bar{\ell})^6} \right] \left[1 - \frac{3}{4} \cdot \frac{\mathscr{D}}{1 + \mathscr{D}} \right] \tag{55}$$

where $\mathscr{D}$ is given by Eq. (39b). This expression is integrated to obtain the final expression of the van der Waals energy for the interaction

$$\Delta G_{\text{vdw}} \approx 0.0404 C_{AB} Q' \left(1 - \frac{3}{4} \cdot \frac{\mathscr{D}}{1 + \mathscr{D}} \right) \tag{56}$$

where C_{AB} and Q' are given by Eqs. (26) and (37b).

Although this approach appears simple, there is certain ambiguity because the nature of the complex together formed by the eluite and the ligand of the stationary phase is unknown. The problem is shown in Fig. 14 which illustrates two of many possible ways of binding the eluite to the stationary phase ligates. One possible mode of binding involves the intercalation of the eluite between the stationary phase ligates, as shown in Fig. 14, which may occur when the configuration of the hydrocarbonaceous ligates at the stationary phase surface is that of a molecular fur (see Fig. 14b). In such a case the van der Waals term can be calculated by evaluating it first for the interaction of the eluite with a single ligate and then multiplying that value by the number of ligates which are in contact with the eluite molecule. The appropriate refractive index for the calculation is that of the free ligate. Alternatively, the process might be considered partitioning between a liquid stationary phase composed of the ligate molecules and the mobile phase in question. In this case the free energy of interaction can be readily evaluated by using the approach described above.

A second mode of binding depicted in Fig. 14 occurs when the surface is densely covered with ligates or the ligates form clumps on the surface although the clumps may be well separated. In this case the eluite may lie on the tips of the ligates which are arranged as those shown in Fig. 14a. The van der Waals term can again be evaluated by using Eq. (54); however, both the geometry of the ligates and the possible interactions between different portions of the ligates and eluite molecules must be considered. For stationary phases with octadecyl groups, for example, this means that the energy of van der Waals interactions between the eluite and terminal methyl as well as between the eluite and the methylene groups must be evaluated and summed. The geometry of the ligates in hydrocarbonaceous bonded phases having high carbon load may be assumed to be close-packed so that the methylene groups probably form an array which is similar to the array of methyl groups but is displaced at a certain distance from it.

In the simplest case one can establish an orthogonal set of axes, x, y, and z such that the z axis is the axis of a ligate molecule and that the eluite binds in a plane parallel to the x–y plane defined by the centers of the methyl group. In this case the interactions with the eluite of each of the methyl groups except the central one are identical due to symmetry. Similar relationships hold for the methylene groups. As a consequence, the calculations of the energies is a relatively simple task.

Electrostatic interactions encompass ionic, dipole–dipole, dipole-induced dipole, or charge-induced dipole interactions. No formal elaboration of the appropriate equations will be given here because the alkyl bonded stationary phases most commonly used in RPC are not expected

to enter into such interactions. However, a polarizable hydrocarbonaceous bonded phase such as phenyl or benzyl silica could be involved in
induced dipole interactions. The chief effect in the calculations of such
electrostatic phenomena is due to the relatively low effective dielectric
constant in the vicinity of the stationary phase. The use of an apparent
dielectric constant given by the mean of the values for bulk eluent and stationary phase proper seems to be often acceptable in practice. The dielectric constant of the stationary phase is taken as that of the corresponding
pure hydrocarbon and can usually be approximated as two.

With the preceding points in mind it is now possible to formulate the
overall binding free energy as

$$\Delta G^\circ = \Delta G^\circ_{EB} - \Delta G^\circ_E - \Delta G^\circ_B + \Delta G^\circ_r \tag{57}$$

where ΔG°_{EB}, ΔG°_E, and ΔG°_B are the energies of complex, eluite, and
ligand interaction with the solvent and ΔG°_r is the energy change due to
reaction as evaluated above. If the solvent interaction is evaluated by
Eq. (24) for each species, one obtains

$$\Delta G^\circ = \Delta\Delta G_c + \Delta\Delta G_{int} + \Delta G_r - RT \ln(RT/PV) \tag{58a}$$

where the net free energy related to the cavity size, $\Delta\Delta G_c$, is evaluated
according to Eq. (22) as

$$\Delta\Delta G_c = N\gamma(A_{EB} - A_E - A_B) - N\gamma A_s(\kappa^e - 1) \tag{58b}$$

In Eq. (58b), A_{EB}, A_E, and A_B are the molecular surface areas of the complex, eluite, and ligate, respectively.

The net interaction energy is calculated from the following relationship:

$$\Delta\Delta G_{int} = \Delta G_{vdw,EB} - \Delta G_{vdw,E} - \Delta G_{vdw,B}$$
$$+ \Delta G_{es,EB} - \Delta G_{es,E} - \Delta G_{es,B} \tag{58c}$$

Thus, according to Eqs. (57) and (58a) the overall free energy change in
binding to a stationary phase is given by the difference between cavity
and electrostatic terms plus the free energy of the association in the absence of any solvent (but the van der Waals potential modified such that
the solvent effect is included) with a contribution from the change in free
volume.

For simplicity the term involving cavity formation can be rewritten as

$$\Delta\Delta G_c = -N\gamma\Delta A - N\gamma A_s(\kappa^e - 1) \tag{59a}$$

where

$$\Delta A = A_E + A_B - A_{EB} \tag{59b}$$

Thus ΔA is the surface area no longer accessible to the solvent when the complex formation occurs, i.e., the contact area between the two species in the complex. This quantity determines the relative magnitude of the cavity term when different eluites are used in a particular solvent.

The evaluation of the interaction term is done with the prescriptions given in Section V,B. The chief difficulty in evaluating electrostatic terms is the indeterminate dielectric constant at the surface which has been estimated as the mean value of the dielectric constants for the solvent and the material forming the stationary phase layer. Therefore, the apparent dielectric at the surface, ϵ_s, may be expressed as

$$\epsilon_s \approx 1 + \epsilon/2 \qquad (60)$$

where ϵ is the dielectric constant of the solvent. This approximation needs to be examined in each system since its validity will depend on the nature of stationary phase material and the proximity of polar groups to the surface. If the polar groups are well solvated even when the complex is bound, the effective dielectric constant may be that of the bulk eluent and hence the contribution of this term will be negligible.

D. Solvophobic Interpretation of Experimental Observations

1. Effect of Eluent

A general prediction of this theory is that the retention factor should decrease with decreasing surface tension of the solvent. The effect has generally been observed in RPC. However, the relationship is not simple insofar as the value of the factor relating microscopic surface tension to bulk surface tension, κ^e, depends on the solvent composition. In the case of neat solvents, the value of κ^e is readily calculated using the approach and formulas of Halicioğlu and Sinanoğlu (*188–194*). However, the problem becomes more difficult if the solvent is composed of more than one component. In fact, a theoretically satisfactory *a priori* method for the determination of κ^e has not been achieved although it is the object of current study (*194*). Various ad hoc approaches have been used to estimate κ^e in mixed solvent systems. All of those attempted to date in this laboratory have used arguments similar to those used to evaluate κ^e for pure solvent. Various combination rules and simplifying approximations were used for the estimation of quantities used in the determination of κ^e, such as solvent heat of vaporization and thermal expansion coefficient of the mixed solvent.

It is interesting to note that in all cases studied, the value of κ^e of mixed solvents showed a dependence on the solvent composition very similar to that illustrated by curve a in Figs. 39A and B. The plots have been obtained from analysis of RPC data with octadecyl silica as the stationary

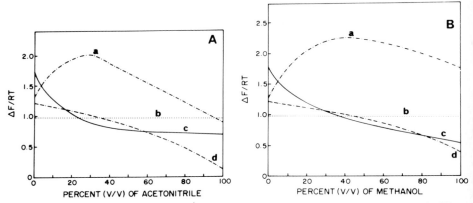

FIG. 39. Graph illustrating the properties of (A) acetonitrile–water and (B) methanol–water mixtures at 25°C as a function of the composition. The properties are expressed in dimensionless energy units and are represented by the curves (a) κ^e, (b) $-\mathcal{D}$, (c) $A\gamma/RT$ with $A = 10$ Å², and (d) ln $(RT/P_0V) - 6$. Reprinted with permission from Horváth *et al.* (*107*).

phase and water–methanol and water–acetonitrile mixtures of varying composition as the mobile phase. As seen from the illustrations, in such hydroorganic mixtures, the addition of organic solvent component caused an increase in the value of κ^e from that of water and it was followed by a decrease in κ^e upon further increase in concentration of the organic component. In the absence of a satisfactory theoretical method for the determination of κ^e in mixed solvents, data from chromatographic experiments may be used for its estimation upon making certain simplifying assumptions. Comparison of κ^e values derived from the retention factors of various solutes at different solvent compositions shows that the κ^e values obtained in this fashion are internally consistent. Indeed, the κ^e values evaluated with a variety of eluites for a given eluent composition agree well and the small deviations observed do not exhibit dependence on molecular properties of the eluites used.

The logarithm of a group selectivity coefficient, τ, is defined (*200*) as

$$\tau_{ji} = \log(k_j/k_i) \tag{61}$$

where k_j and k_i are the retention factors of two eluites whose molecular structures differ only by the presence of the group or substituent in question. A relationship between τ_{ji} and the bulk surface tension, γ, was obtained by Riley *et al.* (*201*) from Eqs. (58a)–(58c) and it is given by

$$\tau_{ji} = \frac{\Delta(\Delta G_{vdw})_{ji} + \gamma N \,(\Delta A_j - \Delta A_i)}{2.3RT} \tag{62}$$

where $\Delta(\Delta G_{vdw})_{ji}$ is the difference in van der Waals interaction energies appropriate for the two compounds, j and i, and $(\Delta A_j - \Delta A_i)$ is the difference in the respective molecular contact areas. They tested the prediction that τ_{ji} is a linear function of the surface tension of the eluent. Appropriate plots of τ values against the surface tension were indeed linear and had a common point of intersection near to a surface tension value predicted from other, related arguments. Their argument minimized the effect of the microscopic correction factor, κ^e, to find that the dependence on surface tension of relative retention data is correctly given by the theory.

The second important macroscopic quantity which appears in the theory is the dielectric constant. The theoretical treatment predicts that for *neutral* species with small dipole moments retention is not affected significantly by changes in the dielectric constant with the composition of hydroorganic eluents. This is due to the occurrence of the dielectric constant in the ratio $2(\epsilon - 1)/(2\epsilon + 1)$ which is nearly invariant over the entire composition range if all solvent components have dielectric constants greater than about 20, a condition which holds for alcohols, acetonitrile, and most other solvents commonly used in RPC. However, one anticipates larger effects if the eluite is *ionic* in character. A consideration of the effect of dielectric constant change as a function of changing solvent composition is given in Section VI in which the effect of using agents in the mobile phase to form a complex with the eluite and thereby modify its retention factor is treated. The theoretical predictions are consistent with the experimentally observed results. In general, the effect of increased dielectric constant is to decrease retention.

2. Molecular Size of the Eluite

The effect of eluite size on its chromatographic retention is not explicitly given in the preceding formulation. It will, however, be manifested by contributions to each of the following three terms: "gas phase interaction," van der Waals interaction, and energy of cavity formation. In reversed-phase chromatography the energy of interaction between the alkyl ligates of the stationary phase and the eluite in a hypothetical gas phase is believed to arise essentially from van der Waals interactions only. It can therefore be evaluated from the appropriate potential energy functions provided they are known. The calculation differs from that used for a purely gaseous phase interaction insofar as the effect of neighboring solvent molecules in modifying the potential need be considered. This can formally be done in a separate calculation which gives rise to a "reduction term" in Sinanoğlu's formulation (*193*). However, it can equally well be included in the calculation directly if it is remembered that this term now includes the effect of modification of the potential by vicinal solvent molecules. The effect is to reduce the interaction energy about 10% from that which would be found in the gas phase for isolated molecules. Formulations for the calculation of the interaction energy of the eluite with solvent have been given (191). The interaction energy calculated for various hydrocarbons in water does not depend linearly upon carbon number but more nearly upon the molecular surface area. Similarly, calculations with branched hydrocarbons showed linear dependence of the interaction energy on the surface area. According to the results of these calculations the energy of van der Waals interactions is about one-half of the corresponding energy related to cavity formation. Since the van der Waals interaction and the cavity term have a similar dependence on eluite surface area, the sum of the two terms will also be determined by the molecular surface area of the eluite in a given solvent although the magnitude of the dependence is difficult to predict.

The approach to calculate the van der Waals and cavity terms from the molecular surface areas can be used for the calculation of partition coefficients. The results show that for the distribution of hydrocarbons between water and *n*-octanol the calculated partition coefficient is linear in carbon number. Qualitatively similar data are obtained for the distribution between other solvents and water and the results can be used to predict the retention in liquid–liquid chromatography. On the other hand, if retention in RPC occurs due to reversible binding at the surface of the stationary phase, the significant parameter is not the total surface area of the eluite but rather the net decrease in the molecular surface area of the stationary phase ligates and that of the eluite upon binding, i.e., the contact area in the complex.

The relationship between change in surface area upon binding and total eluite surface area can be examined by considering the binding of spherical or rod-shaped eluite molecules to rodlike, spherical, or planar surfaces. From geometrical considerations, the relationship between the surface area change and total eluite area in each binding configuration can be determined analytically if characteristic distances, such as radii, of the eluite and the surface ligate are known and if solvent molecules are regarded as hard spheres of known diameter. According to such calculations the surface area change is about 20–25% of the eluite molecular area. The result is largely independent of the binding geometry but it does assume the molecular diameters for eluite and eluent are commensurable with that of the stationary phase or less than it. A consequence of these results, when combined with the relationship between the magnitude of the interaction term and eluite area, is that logarithm of retention factor is linear in molecular surface area. In more useful terms, plots of the logarithms of the retention factor of a number of homologous series would be linear. Indeed, a variety of workers have observed that the logarithm of retention factor is linearly dependent on the carbon number in *n*-alkanes and their derivatives (*143, 159, 160, 202*).

3. Substituents in the Eluite

Linear dependence of log *k* on the number of molecular units in homologs is observed to hold for a variety of substances, such as for oligoalanines (*203*) and oligo-γ-glutamates of folic acid (*204*) as depicted in Figs. 40 and 41, respectively. The analysis of such plots is complex because of the polar functions present in the molecules. Moreover, the effect of ionization must be considered.

Several workers have observed that deuterium- or tritium-containing molecules are retained less than their completely protiated homologs (*202, 205, 206*). This result is explicable in the framework of the solvophobic theory by considering two phenomena which are opposite in their effects on the retention factor. The first is an increase in the molecular surface area upon replacing hydrogens by its heavy isotope and this results in an increase in retention factor. The second is an increase in the Kihara parameter. As a result the van der Waals interaction energy with the solvent will become more favored and a decrease in retention factor will occur. Estimating the magnitude of each term we find that the latter effect is usually slightly greater, i.e., the retention factor of protiated species is

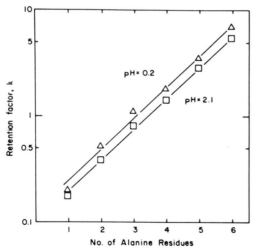

Fig. 40. Plots of the logarithm of the retention factor against number of residues in alanine oligomers. Reprinted with permission from Molnar and Horváth (*203*).

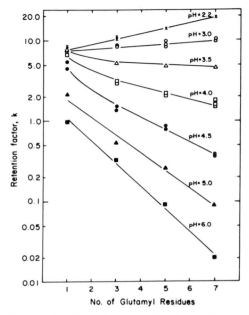

FIG. 41.   Plots of the retention factor (on a logarithmic scale) against the number of the glutamyl residues in pteroyl-oligo-γ-glutamates at different pH values of the eluent. Column, 5 μm Partisil ODS-2; eluent, 0.1 M phosphate solution containing 6% (v/v) acetonitrile; temperature, 45°C. Reprinted with permission from Bush *et al.* (*204*).

expected to be greater than that of the deuterated or tritiated analogs, at least with aqueous methanol as the eluent.

Hoffman and Liao (*160*) observed apparent deviations from linearity in plots of the logarithm of retention factor versus carbon number for normal alcohols which were chromatographed in acetonitrile–water mixtures rich in acetonitrile. The results have been explained in terms of "normal phase" interactions with the otherwise nonpolar stationary phase, i.e., the alcohols were assumed to form hydrogen bonds or otherwise interact with residual silanols at the surface of octadecyl silica stationary phase.

The theoretical treatment outlined above does not involve any attractive interaction between stationary phase and eluite besides that caused by van der Waals force. Consequently, the above observation is not predicted by the theory. Nevertheless, surface silanols can have profound effect on the selectivity of the stationary phase in contact with eluents rich in organic solvent. The phenomenon and its practical importance in RPC of polar substances, particularly those carrying positive charge, have long been appreciated. Yet, only recently has   retention behavior analyzed in the case when the eluite interacts with two different kinds of binding sites at the stationary phase surface (93b).

4. Ionization of Eluite

Electrolyte concentration and pH have profound effect on electrostatic interactions and consequently on the retention behavior of ionogenic sample components in RPC. In agreement with the results of a detailed theoretical treatment (207), which is summarized in Section V,B,2,b, ionization results in a decrease of the retention factor although exceptions

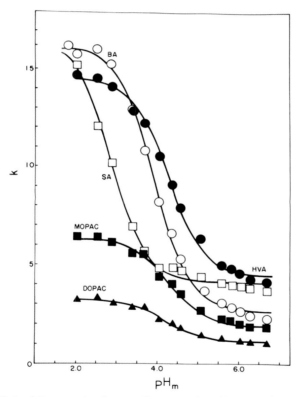

FIG. 42. Plots of the retention factors of monoprotic acid versus the pH of the eluent. The data were obtained on benzoic acid (BA), 3,4-dihydroxyphenylacetic acid (DOPAC), homovanillic acid (HVA), parahydroxyphenylacetic acid (MOPAC), and salicylic acid (SA). The data follow the pH titration curve. The pK_a values determined from these data, as well as those of other acids, are compared to literature values in Table XIX. Column, Partisil ODS, 250 × 4.6 mm i.d.; eluent 1.0 M Na$_2$SO$_4$ in 0.05 M phosphate buffers, 25°C. Reprinted with permission from Horváth et al. (207), Anal. Chem. Copyright © 1977 by the American Chemical Society.

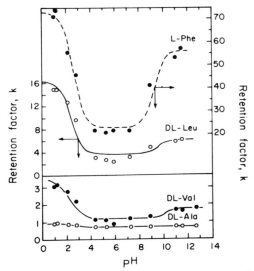

FIG. 43.   Retention factor for several amino acids on XAD-4 as a function of pH. The data conform to Eq. (92) as shown by the curves drawn using parameters given by Kroeff and Pietrzyk. The eluites are L-phenylalanine (L-Phe); DL-leucine (DL-Leu); DL-valine (DL-Val), and DL-alanine (DL-Ala). Data were taken with 0.04 M phosphate buffer at 0.2 M ionic strength with NaCl. Reprinted with permission from Kroeff and Pietrzyk (208), *Anal. Chem.* Copyright © 1978 by the American Chemical Society.

have been reported (209). The effect is shown in Figs. 42 and 43. It is encouraging that the solvophobic theory used in the analysis allows one to estimate physically realistic values for the distance of closest approach or the average dielectric constant of the surface.

The effect of neutral salts on ionized species is especially interesting. In light of the theory, one would expect two different effects. At low salt concentrations, where the limiting Debye–Hückel treatment is valid, the predominant effect of added salt is to enhance the ionic atmosphere about each eluite molecule, and thus reduce electrostatic repulsion between eluites. Thus in this concentration regime the addition of salt causes a decrease in the retention factor. Whereas, at higher salt concentrations further increase in shielding due to salt addition is negligible, the surface tension of the eluent increases with salt concentration and results in increased retention factors according to Eq. (59a). As a consequence, with increasing salt concentration in the eluent, one anticipates an initial decrease followed by monotonic increases in the retention factor of ionized eluites. If the eluite is not ionized, the retention factor should monotonically increase with the salt concentration in the eluent due to the increasing surface tension.

This prediction is supported by the retention factor data shown in Fig. 44 which includes both neutral and ionized eluites. It is interesting to note that the limiting slopes of the log k vs salt concentration plots at high salt concentrations is the same for both neutral and ionized species in Fig. 44. This behavior is expected in this example because the slopes are determined by the area change upon binding the eluite to the stationary phase which is about the same for both types of substances due to their similar molecular dimensions.

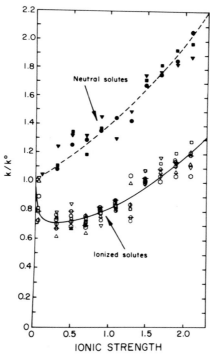

FIG. 44. The dependence of the normalized retention factor on the ionic strength of the eluent for a variety of un-ionized (---) and ionized (——) solutes. $k°$ denotes the retention factor at zero ionic strength. The broken line was calculated for un-ionized acids and the solid line was calculated for ionized species from the solvophobic theory. The chromatographic conditions are: Column, Partisil ODS; eluent, 0.01 M phosphate buffer, pH 4.0 with increasing concentration of KCl; temperature, 25°C. The symbols represent the following solutes: (▼) 3,4-dihydroxyphenylacetic acid; (●) homovanillic acid; (■) vanillmandelic acid; (∇) normetanephrine; (○) metanephrine; (□) 3-O-methyldopamine; (△) dopamine; (○) adrenaline; (θ) tyramine; (φ) paranephrine. Reprinted with permission from Horváth *et al.* (207), *Anal. Chem.* Copyright © 1977 by the American Chemical Society.

The data shown in Fig. 43 indicate that the retention is determined by the degree of ionization and not the total charge. Apolar amino acids may serve as examples for this situation. Although they are fully ionized at pH 5 they carry no net charge. On the other hand, at high and low pH values they are anionic and cationic because the amino and carboxyl groups are not fully ionized. If retention depends on the net charge on the eluite, or its absolute value, retention would be less at the extremes of pH than at pH 5. Yet, minimum retention occurs near pH 6 as seen in Fig. 43.

This finding suggests a fundamental difference between the charge dependence of retention in RPC and electrophoretic migration. In transport of charged species in an electrical field, the charge centers move together. Since the "probe," the electrodes in an electrophoretic cell, is well removed from the molecules, i.e. the electrostatic "interaction" occurs over many molecular diameters, the molecule will usually behave as if all the charge is located at its center. On the other hand, the "probe" used in RPC is the stationary phase surface, i.e. the hydrocarbonaceous ligate, which can be very close to the charge center of the eluite. Contact could require loss of strongly associated solvent from the eluite. In many cases an energetically less costly alternative to the eluite may be to bind with the charges oriented away from the surface. Because this effect will be manifested at distances commensurate with molecular diameters, the interaction with the surface will depend critically on the location of the charge centers. As a consequence the effect is not as if the molecule has its charges summed and located at its center but rather the charges exert individual effects. This point of view was useful in interpreting the pH dependence of poly-γ-glutamyl folates in RPC (204).

E. Mechanism and Kinetics of Retention

1. Binding Mechanism

Two limiting mechanisms for solute retention can be imagined to occur in RPC: binding to the stationary phase surface or partitioning into a liquid layer at the surface. In the previous treatment we assumed that retention is caused by eluite interaction with the hydrocarbonaceous surface, i.e., the first type of mechanism prevails. When the eluent is a mixed solvent, however, the less polar solvent component could accumulate near the apolar surface of the stationary phase. In the extreme case, an essentially stagnant layer of the mobile phase rich in the less polar solvent could exist at the surface. As a result eluites could partition between this layer and the bulk mobile phase without interacting directly with the stationary phase proper.

Scott and Kucera (210) have urged the occurrence of such partitioning mechanism because they found from mass balance experiments that octyl silica binds acetonitrile (among other solvents) in preference to water from hydroorganic mixtures range in composition from 32 to 56% (v/v). However, they observed no sensible increase in acetonitrile concentration upon adsorbtion of acetophenone on octyl and octadecyl silica from acetonitrile–water mixtures in shake flask experiments. Attempts to detect displacement of acetonitrile in frontal analyses from the two stationary phases were similarly unsuccessful. The results have been used to support the proposition that acetonitrile, the less polar solvent component, forms a monolayer on the surface which is not displaced by the eluite in the course of the chromatographic process. Consequently, retention has been assumed to occur due to eluite interaction with the liquid layer at the surface.

Slaats et al. (211) examined the thesis of Locke (212) that selectivities are determined mostly by mobile phase effects. According to the theory in this case, the retention factor should be directly proportional to the activity coefficient, weighted by the ratio of mobile phase to acetonitrile molar volumes in water–acetonitrile mixtures. Plots of the relative retentions on octyl silica against the water content of the eluent were not parallel. On the other hand, plots of retention factor versus the weighted activity coefficient in acetonitrile–water were linear but, contrary to theory, did not pass through the origin. Furthermore, data taken in methanol–water mixtures were scattered; neither linearity or an intercept at the origin were observed. As a consequence of these results, those workers concluded that interactions with the mobile phase change significantly with solvent composition so that activity coefficients do not generally suffice for prediction of retention behavior in RPC.

Colin and Guiochon (86) examined the retention mechanism by comparing isotherms of octylbenzene on both octadecyl silica and n-octadecane. Assuming a partitioning model, they found that the "partition" coefficient calculated for octylbenzene in an octadecyl silica–methanol system was about fivefold greater than that obtained for methanol–octadecane partitioning. This anomaly suggested that a pure partition mechanism cannot operate. Although the results for the binding of octylbenzene to octadecyl silica were consistent with the adsorption model, it was felt that the data were insufficient to conclude that solute binding proceeds strictly by adsorption.

A useful distinction between partitioning and adsorption has also been given by Colin and Guiochon (86, p. 184). They stated that:

One must first define precisely the difference between adsorption and partition. Adsorption means the interactions of the stationary phase with the solute or solvent molecules covering the external molecular layer of the "adsorbent." In the simplest case, adsorption is characterized by a monolayer in which the solute and solvent molecules are in competition to cover the external surface of the packing. More often, the interfacial region between the adsorbent and the bulk mobile phase can be composed of several layers. The separation surface results from an arbitrary choice and is generally selected in a position such that the Gibbs' surface excess for the solvent is zero (213). In a partition process, the sole role of the support is to create a large surface area where the stationary liquid phase will be deposited. The phenomena in the interfacial region are then generally neglected and the solute molecules partition between the two liquid phases. Each phase contains two components: the solvent and the solute for the mobile phase and the solute for the stationary phase.

This definition erases some of the differences between the interpretation of experimental data by different workers. For instance, the conclusions of Scott and Kucera (210), which apparently support a partitioning mechanism, are also consistent with the adsorption model in the view of the above definition.

In light of the solvophobic theory, the chief difference between the two mechanisms is that in partitioning the magnitude of retention has a greater dependence on the molecular surface area of the eluite than in adsorption. As discussed before, the net surface area change upon binding is one-quarter to one-third of the surface area when adsorption prevails. In contradiction the corresponding term is calculated with total molecular surface area of the eluite when partitioning occurs. If we regard water as the reference mobile phase, the corresponding apparent area change is about two-thirds of the total area with most cosolvents. In the initial formulation of solvophobic theory to chromatography, the change in surface area with binding was approximately one-fourth the total surface area (107). Subsequent analyses of various data sets have consistently found the surface area change to total surface area ratio on reversed phase to be about 0.25–0.35. Furthermore, the binding of diphenylguanidium ions on XAD-2 were quantitatively consistent with the formation of a bound cation monolayer of close-packed ions lying flat on the surface (214).

The correlation between partition coefficient and chromatographic retention has been examined by many workers in an attempt to predict log P, the logarithm of water–octanol partition coefficient used in establishing structure–function relationships, from retention data or vice versa. The data (92, 203, 215) seem to imply changes in log k are linear in changes in log P but that log k changes more slowly. Some literature data suggest that the retention is directly proportional to the partition coefficient. This apparent equality disappears, however, if the data are recalculated to identical experimental conditions.

All the evidence is not yet in. On the basis of the available data, however, it appears that the binding mechanism operating in reversed-phase chromatography is adsorption according to the definition given by Colin and Guiochon (86) and that no purely partitioning mechanism exists.

2. *Kinetics of Binding*

The kinetics of eluite binding to hydrocarbonaceous bonded phases has received relatively little attention. Horváth and Lin (109) have examined the effect of kinetic resistances on plate height by taking advantage of a simple relationship between peak width and retention factor which holds if all sample components have similar diffusivities and binding rate constants. The plate height contribution due to kinetic resistances was examined on four reversed-phase and two silica columns. It has been found that with well-packed 5 μm octadecyl silica a substantial part of the measured plate height arises from the slowness of binding kinetics in RPC with plain aqueous eluent.

Activation energies for binding and dissociation were calculated to be ~3 kcal/mol and 7–9 kcal/mol, respectively. Whereas the calculations were too crude to draw conclusions regarding the mechanism, the activation energies are clearly commensurate with those involved in diffusion and rupture of weak chemical bonds, respectively.

Perhaps more important is the observation that the kinetics of binding can contribute significantly to band spreading. If the results presented in that work are general, the kinetic plate height contribution sets a lower limit on the size of particles used in RPC at about 5 μm and no practical gain in column efficiency would be made by using smaller particles.

F. Conclusions

Reversible binding of the eluite to monomolecularly bonded stationary phases in RPC is believed to proceed not through partitioning but rather through adsorption in the general sense of that term. As a consequence, a calculation of the corresponding free energy change and, therefore, the retention factor, requires knowledge of the energies of both the bound eluite and the eluite in the mobile phase. Locke's (212) treatment of RPC is formulated in terms of the eluite activity coefficient in the two phases. The activity coefficient of the bound eluite, however, is not known *a priori* in most cases. Moreover it is likely that the activity coefficients of both adsorbed eluite and eluite in the mobile phase depend on the composition of the mobile phase. Consequently this approach may not be suitable to predict the effect of changing mobile phase composition on retention. Indeed experimental data indicate that the variation of retention with solvent composition does not parallel the corresponding change of activity coefficient in the mobile phase. We may conclude, therefore, that the activity of the eluite on the surface is not independent of solvent composition. The deviation from retention predicted on the basis of mobile phase activities is usually not great, however, and data for eluite solubility in the mobile phase, or other data suitable for calculation of activity coefficient if available, may serve as a reliable guide to retention behavior.

In many respects the solvophobic approach is inferior to that of Locke. In order to use it rigorously, both the dependence of surface tension correction factor on solvent composition and the area change attendant on binding must be known. The latter is equivalent to knowing the structure of the adsorbed species. Nevertheless, these obstacles to quantitative analysis of binding are not particularly important in the qualitative and semiquantitative use of the theory for interpretation of retention in RPC. It should be noted that other data required for the calculations are usually available in handbooks or can be estimated from such data on analogous molecules. Another attractive feature of the theory is that the free energy, i.e., the equilibrium constant, can be easily estimated for different assumed modes of eluite binding to the stationary phase. In this way energetically impossible hypotheses can be revealed and discarded.

The most important prediction of the solvophobic analysis is that the binding energy increases with the eluite surface area. Although the van der Waals contribution to the interaction energy is not negligible, it is apparently proportional to the surface area. Since it is opposite in sign to the contribution due to cavity formation, it can be regarded as simply reducing the apparent surface area. Therefore the retention order of eluites should be in order of size if the polar interactions are the same.

It is important, however, not to neglect other components of the interaction term. Since the energy of desolvation of a polar group is likely to be large, it is expected that upon binding of the eluite molecule most of its polar groups will be oriented toward the mobile phase rather than the stationary phase. This effect will reduce the total area available for binding and may contribute to the reduced retention of polar molecules compared to their apolar analogs in RPC.

The solvophobic theory has been successfully applied to treat the effect of solute ionization as well as the effect of salts on the retention of both neutral and ionized species. There is ample experimental evidence that retention of a species decreases upon ionization according to the theoretical prediction. Addition of salts to aqueous eluents increases surface tension and consequently the retention of neutral eluites on nonpolar stationary phases. With ionized solutes, however, the solvophobic theory predicts a minimum at low ionic strength in plots of retention versus ionic strength and this phenomenon has also been experimentally demonstrated.

The solvophobic theory could be extended to the treatment of other special effects such as hydrogen bonding between eluite and species present in the eluent. The predictive power of the theory may be improved by such extension. In addition a rigorous theory for treatment eluite interaction with surface silanols would be needed.

VI. MODULATION OF SELECTIVITY BY SECONDARY EQUILIBRIA

A. General Considerations

In some cases the selectivity of the chromatographic system can be modified by simply changing the eluent strength of the hydroorganic mobile phase. For example, Colin and Guiochon (86) have demonstrated that with octadecyl silica as the stationary phase the retention of benzene relative to that of phenol increases from 3 to 5 when the methanol content of the aqueous eluent increases from 20 to 50% (v/v). More than often, however, changes in selectivity reached in this fashion, i.e., by changing the volume ratio of the organic component in a given hydroorganic eluent, are accompanied by concomitant changes in the retention of all sample components and an appreciable change in relative retention might be obtained only with retention factors that fall outside the practical range of $0.3 < k < 5$. It is therefore desirable to find ways to change the selectivity of the chromatographic system for closely eluting sample components without drastic changes in the eluent strength at large.

A variety of methods have been found to effect such modulation of selectivity. For instance, changes in the pH of the eluent can often modify the retention of ionogenic substances dramatically while leaving the retention of neutral substances unchanged. At present, however, the most widely heralded method involves the use of alkyl sulfates or long-chain alkyl quaternary ammonium salts in the eluent and the technique is often referred to as ion-pair chromatography. In fact, this approach is a particular example of a general method for modifying chromatographic retention by including in the eluent a modifier that is capable of specific chemical interactions with some sample components.

The essence of this approach for modulating chromatographic selectivity is the formation of a complex that is retained to a different extent than the uncomplexed eluite from a sample component and the modifier of the eluent. By selecting complexing agents of appropriate specificity the selectivity of the chromatographic system can be deliberately altered according to the need imposed by a particular separation problem. In light of this view, we have termed the complexing agent, which is kept in the eluent at a fixed concentration level "hetaeron" from the Greek ἐταιρον (companion). Concomitantly the technique may be called "hetaeric chromatography" (34).

An early demonstration of the power of hetaeric chromatography in bonded reversed-phase systems was made by Wittmer et al. (216) who showed the effect of tertiary and quaternary amines in the mobile phase on the separation of tartrazine, a sulfonic acid dye, and intermediates in its synthesis. Figure 45 reproduces a chromatogram obtained by them with this technique.

Of course, the enhancement of chromatographic selectivity by secondary chemical equilibria is neither new nor confined to reversed-phase systems. Most widespread probably has been the exploitation of protonic equilibria by appropriately adjusting the pH of the eluent so that the degree of ionization of the eluite is altered. Generally the ionized and neutral forms of an eluite are retained differently (207, 208). Formation of metal complexes of certain eluites has also been utilized for modulating retention behavior for higher selectivity.

Other recent applications of suitable secondary chemical equilibria in the mobile phase include the separation of optical isomers by using opti-

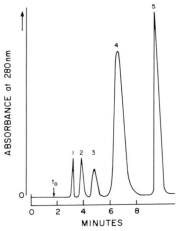

FIG. 45.   An early example of hetaeric chromatography. Tartrazine, synthetic intermediates, and internal standards are separated on μ-Bondapak C_{18}. The mobile phase is water–methanol–formic acid (400:400:1) containing 3×10^{-3} M tetrabutylammonium hydroxide and 4×10^{-4} M tridecylamine. Eluites: (1) sulfanilic acid, (2) pyrazolone, (3) 3-nitrosalicylic acid, (4) tartrazine, and (5) m-chlorobenzoic acid. Flowrate 1 ml/min; column 300 × 4 mm i.d. Detection at 280 nm. Reprinted with permission from Wittmer et al. (216), Anal. Chem. Copyright © 1975 by the American Chemical Society.

cally active hetaerons which can form metal complexes with amino acids and other substances. As the complex eluites containing optical antipodes do not have the same retention, this approach, which is discussed in detail by Karger *et al.* (52), is rapidly becoming an important mode of RPC.

B. Phenomenological Model for the Process of Retention

If an eluite, E, interacts with a hetaeron, $\mathcal{H}$, to form a complex E$\mathcal{H}$, a number of equilibria may need to be considered in evaluating the details of the retention mechanism. For instance the complex may be formed in the mobile phase and subsequently bind to the stationary phase with a retention factor different from that of the free eluite. Alternatively, the hetaeron itself may strongly bind to the stationary phase and thereby promote one or both of two effects. It may convert the stationary phase into a dynamically coated "complex" or "ion"-exchanger, depending on the nature of the interaction. The retention is modified as a result of the interaction between the eluite and the functional groups of the hetaeron which is bound to the stationary phase surface. In contradistinction to classical ligand or ion-exchange chromatography, however, the exchanger is not covalently bound but dynamically coated, i.e, the surface concentration of the functional groups depends on the hetaeron concentration in the mobile phase.

On the other hand, binding of hetaeron to the surface of the stationary phase can also affect retention by occupying space which the complexes formed in the mobile phase could occupy otherwise. Therefore binding of the hetaeron could result in decreased eluite retention unless the complex displaces the hetaeron molecule from the surface or exchange of ligands occurs between the complex formed in the mobile phase and the hetaeron bound to the surface. Exhaustive modeling of all possible phenomena involved in the general case is a burdensome task beyond the aims of this section. Nevertheless, important features concerning the effect of hetaeron concentration on eluite retention can be deduced from considering certain simple limiting cases of the general model.

If the hetaeron binds to the stationary phase surface according to Langmuir isotherm, the surface concentration of hetaeron $[\mathcal{H}_s]$ is given by

$$[\mathcal{H}_s] = \frac{K_1[\mathcal{H}_m][\mathcal{H}_s]^*}{1 + K_1[\mathcal{H}_m]} \qquad (63)$$

where $[\mathcal{H}_m]$, $[\mathcal{H}_s]^*$, and K_1 are the concentration of hetaeron in the eluent, the maximum surface concentration of bound hetaeron, and the equilibrium constant for binding of hetaeron to the stationary phase surface, respectively. The expression given in Eq. (63) assumes that a monolayer will be formed on the surface at sufficiently high hetaeron concentration in the eluent. Accordingly, the concentration of available surface sites, i.e. the sites not occupied by hetaeron, is given by the difference between the maximal surface density of binding sites and the actual surface concentration as

$$[\mathscr{H}_s]^* - [\mathscr{H}_s] = \frac{[\mathscr{H}_s]^*}{1 + K_1[\mathscr{H}_m]} \tag{64}$$

In all cases discussed below we assume that the eluite can bind independently to the stationary phase, according to the following equilibrium:

$$E_m \underset{}{\overset{K_0}{\rightleftharpoons}} E_s \tag{65}$$

In view of Eq. (63) the limiting cases mentioned above can be now individually considered. According to the ion-pairing model, mechanism I, the eluite and hetaeron first form a complex, $E\mathscr{H}_m$, in the mobile phase, and then the complex binds reversibly to empty sites on the surface as $E\mathscr{H}_s$. The corresponding equilibria are given by

$$\mathscr{H}_m + E_m \underset{}{\overset{K_2}{\rightleftharpoons}} E\mathscr{H}_m \tag{66}$$

$$E\mathscr{H}_m \underset{}{\overset{K_3}{\rightleftharpoons}} E\mathscr{H}_s \tag{67}$$

Combining Eqs. (64), (66), and (67) we obtain for the surface concentration of the complex according to mechanism I the following expression

$$[E\mathscr{H}_s] = \frac{K_2 K_3[E_m][\mathscr{H}_m][\mathscr{H}_s]^*}{1 + K_1[\mathscr{H}_m]} \tag{68}$$

The important chromatographic parameter, which can directly be obtained from the chromatogram, is the retention factor k. It is given by the ratio of mass of eluite bound to the stationary phase to the mass in the mobile phase and is conveniently expressed by the corresponding equilibrium concentrations and the phase ratio, ϕ. When a complex formation in the mobile phase dominates, i.e., the chromatographic process can be represented by the first limiting case, the retention factor is obtained by combining Eqs. (51), (65), (66), and (68) to obtain

$$k = \phi \frac{K_0[\mathscr{H}_s]^* + K_2 K_3[\mathscr{H}_m][\mathscr{H}_s]^*}{(1 + K_1[\mathscr{H}_m])(1 + K_2[\mathscr{H}_m])} \tag{69}$$

In mechanism II, which represents another limiting case, the complex is formed directly with a hetaeron already bound to the surface of the stationary phase, according to the equilibrium

$$\mathcal{H}_s + E_m \xrightleftharpoons{K_4} E\mathcal{H}_s \tag{70}$$

where $[\mathcal{H}_s]$ is the concentration of the hetaeron bound to the surface and $[E\mathcal{H}_s]$ is the surface concentration of the complex. The concentration of each species can be found by combining Eqs. (63) and (70) and the surface concentration of the complex can be expressed as

$$[E\mathcal{H}_s] = \frac{K_1 K_4 [E_m][\mathcal{H}_m][\mathcal{H}_s]^*}{1 + K_1[\mathcal{H}_m]} \tag{71}$$

Equations (51), (65), (66), and (71) together yield an expression for the retention factor when mechanism II, i.e., dynamic ion exchange, governs the chromatographic retention and it is given by

$$k = \frac{K_0[\mathcal{H}_s]^* + K_1 K_4 [\mathcal{H}_m][\mathcal{H}_s]^*}{(1 + K_1[\mathcal{H}_m])(1 + K_2[\mathcal{H}_m])} \tag{72}$$

The third possible mechanism, IIIa, proposed here involves a transfer of solute from the complex formed in the mobile phase to the bound hetaeron and complex formation at the surface. The corresponding equilibrium for this metathetical process, which may be called "dynamic complex exchange," can be written as

$$E\mathcal{H}_m + \mathcal{H}_s \xrightleftharpoons{K_5} E\mathcal{H}_s + \mathcal{H}_m \tag{73}$$

The surface concentration of complex is given by

$$[E\mathcal{H}_s] = \frac{K_1 K_2 K_5 [\mathcal{H}_m][E_m][\mathcal{H}_s]^*}{(1 + K_1[\mathcal{H}_m])} \tag{74}$$

and the corresponding retention factor is expressed as

$$k = \phi \frac{K_0[\mathcal{H}_s]^* + K_1 K_2 K_5 [\mathcal{H}_m][\mathcal{H}_s]^*}{(1 + K_1[\mathcal{H}_m])(1 + K_2[\mathcal{H}_m])} \tag{75}$$

An alternative mechanism, IIIb, may be formulated in the case when the eluite–hetaeron complex first binds the surface that is already covered with bound hetaeron so that the following equilibrium represents the binding of eluite by the stationary phase

$$E\mathcal{H}_m + \mathcal{H}_s \xrightleftharpoons{K_6} \mathcal{H}E\mathcal{H}_s \tag{76}$$

and the binding is followed by the exchange process

$$\mathcal{H}E\mathcal{H}_s \xrightleftharpoons{K_7} \mathcal{H}E_s + \mathcal{H}_s \tag{77}$$

The concentration of the complexes bound to the surface is given by

$$[\mathcal{H}E\mathcal{H}_s] = K_6[E\mathcal{H}_m][\mathcal{H}_s] \tag{78a}$$

$$= \frac{K_1 K_2 K_6[E_m][\mathcal{H}_m]^2[\mathcal{H}_s]^*}{1 + K_1[\mathcal{H}_m]} \tag{78b}$$

and

$$[E\mathcal{H}_s] = K_7[\mathcal{H}E\mathcal{H}_s]/[\mathcal{H}_s] \tag{79}$$

or

$$[E\mathcal{H}_s] = K_2 K_6 K_7[\mathcal{H}_m][E_m] \tag{80}$$

The total concentration of the eluite bound to the surface, $[E_s]_T$, by this mechanism is given by

$$[E_s]_T = [\mathcal{H}E\mathcal{H}_s] + [E\mathcal{H}_s] \tag{81a}$$

$$= \frac{K_2 K_6 K_7[\mathcal{H}_m][E_m](1 + K_1[\mathcal{H}_m]) + K_1 K_2 K_6[E_m][\mathcal{H}_m]^2[\mathcal{H}_s]^*}{1 + K_1[\mathcal{H}_m]} \tag{81b}$$

so that mechanism IIIb leads to an expression of the retention factor given by

$$k = \phi \frac{K_0[\mathcal{H}_s]^* + K_2 K_6 K_7[\mathcal{H}_m](1 + K_1[\mathcal{H}_m]) + K_1 K_2 K_6[\mathcal{H}_m]^2[\mathcal{H}_s]^*}{(1 + K_1[\mathcal{H}_m])(1 + K_2[\mathcal{H}_m])} \tag{82}$$

If the surface species formed through the various mechanisms discussed above are chemically distinguishable, i.e., if a Maxwell demon exploring the stationary phase could determine by inspection of the complex how each eluite it encountered had been bound, it would be meaningful to express the retention factor as a function of the hetaeron concentration in the mobile phase with all pertinent equilibrium constants as parameters as given in the following expressions. If retention occurs simultaneously through mechanisms I, II, and IIIa, the retention factor is given by

$$k = \frac{\phi K_0[\mathcal{H}_s]^*}{(1 + K_1[\mathcal{H}_m])(1 + K_2[\mathcal{H}_m]} + \frac{\phi K_2 K_3[\mathcal{H}_m][\mathcal{H}_s]^*}{(1 + K_1[\mathcal{H}_m])(1 + K_2[\mathcal{H}_m]}$$

$$+ \frac{\phi K_1 K_4[\mathcal{H}_m][\mathcal{H}_s]^*}{(1 + K_2[\mathcal{H}_m])(1 + K_1[\mathcal{H}_m])} + \frac{\phi K_1 K_2 K_5[\mathcal{H}_m][\mathcal{H}_s]^*}{(1 + K_1[\mathcal{H}_m])(1 + K_2[\mathcal{H}_m])} \tag{83}$$

However, if the process follows mechanisms I, II and IIIB, the retention factor is given by

$$k = \phi \frac{K_0[\mathscr{H}_s]^*}{(1 + K_1[\mathscr{H}_m])(1 + K_2[\mathscr{H}_m])} + \phi \frac{K_2 K_3[\mathscr{H}_m][\mathscr{H}_s]^*}{(1 + K_1[\mathscr{H}_m])(1 + K_2[\mathscr{H}_m])}$$

$$+ \frac{\phi K_1 K_4[\mathscr{H}_m][\mathscr{H}_s]^*}{(1 + K_2[\mathscr{H}_m])(1 + K_1[\mathscr{H}_m])}$$

$$+ \phi \frac{K_2 K_6 K_7[\mathscr{H}_m](1 + K_1[\mathscr{H}_m]) + K_1 K_2 K_6[\mathscr{H}_m]^2[\mathscr{H}_s]^*}{(1 + K_1[\mathscr{H}_m])(1 + K_2[\mathscr{H}_m])} \tag{84}$$

In order to make use of the preceding expressions in the analysis of experimental data, it is mandatory to maintain the free hetaeron concentration in the mobile phase constant. This requires that the hetaeron concentration in the eluent is sufficiently higher than the mean eluite concentration in the elution band so that no appreciable depletion occurs in the moving zone occupied by the eluite. If this precaution is not observed, a significant fraction of the hetaeron may be bound to the eluite and, as a consequence, the true unbound hetaeron concentration in the vicinity of the eluite may be significantly less than that at the column inlet. In the worst case, eluite binding to the stationary phase will cease to be linear and anomalous band spreading will be observed.

The first term of Eqs. (83) and (84) is a measure of retention due to binding of eluite to the stationary phase with no intervention of hetaeron. The second term is the contribution due to complex formation in the mobile phase, which governs retention according to mechanism I. The third term expresses the retention factor increment that arises from complex formation with the bound hetaeron, according to mechanism II. The fourth term in both Eqs. (83) and (84) expresses the effect of "complex exchange" on the retention as given by mechanisms IIIa or IIIb. If surface bound species formed by two or more mechanisms are indistinguishable, i.e., a Maxwell demon could not ascertain from the structure of the complex the particular mechanism by which it was formed, then the terms which are related to the formation of those species in Eqs. (83) and (84) must be identical and therefore interchangeable. Consequently, in such cases all but one of those terms in Eqs. (83) and (84) should be dropped in order to avoid a multiple counting of the corresponding retention factor increment. The first term of Eqs. (83) and (84) measures retention due to binding of eluite with no intervention of hetaeron, the second term is the contribution due to complex formation in the mobile phase (mechanism I), the third arises from complex formation with the bound hetaeron (mechanism III), whereas the fourth is a consequence of the "complex exchange" mechanism (IIIa or IIIb). If the some of the bound species are indistinguishable, the terms corresponding to those forms in Eq. (83) and (84) are identical.

Two important conclusions can be derived from the foregoing analysis. First, if complex formation in the mobile phase and the adsorption of hetaeron to the stationary phase are both important in affecting the retention of eluite in the range of hetaeron concentration covered experimentally, i.e. $K_1[\mathcal{H}_m]$ and $K_2[\mathcal{H}_m]$ are both of the order of unity or greater at the highest hetaeron concentrations, Eqs. (69), (75), (82), and (84) each imply that the retention factor will first increase and then decrease with the increasing hetaeron concentration. This result is independent of the actual mechanism of complex formation or binding to the stationary phase as can be seen by comparing these equations. The second, third, and fourth terms of these equations have a first-order dependence on hetaeron concentration in the numerator of the expression and a quadratic dependence in the denominator. The first term decreases approximately as the inverse square of the hetaeron concentration and the fourth term, which in the high concentration limit has a zero order dependence on hetaeron concentration, rises to a constant value. Thus a parabolic dependence of retention factor on hetaeron concentration in the eluite is predicted unless "complex exchange," as detailed by Eqs. (76) to (82), occurs in which case the retention rises to a limiting value with increasing hetaeron concentration. Parabolic behavior will be observed also if complex formation in the mobile phase or binding of the complexing agent to the stationary phase, but not both, occurs.

If the hetaeron is a detergent or another amphiphile, we also may assume that at sufficiently high concentrations micelles are formed by agglomeration of n_1 hetaeron molecules and the solute can be distributed between the mobile phase proper and the micelles in the eluent. The corresponding partition equilibrium can be expressed by

$$E_m + n_1\mathcal{H}_m \xrightleftharpoons{K_8} E\mathcal{H}_{n_1,m} \tag{85}$$

The concentration of solute included in the micelles, $[E_{mic}]$ is assumed to depend on the free solute and hetaeron concentrations as

$$[E_{mic}] = K_8[\mathcal{H}_m]^{n_1}[E_m] \tag{86}$$

An approximately parabolic dependence will occur if the hetaeron forms micelles into which the eluite can partition. In that case, the eluite molecules in the mobile phase can be present as free eluite, as eluite bound to hetaeron and bound to micelles. The total eluite concentration in the eluent, $[E_m]_T$, is given by

$$[E_m]_T = [E_m](1 + K_2[\mathcal{H}_m] + K_8[\mathcal{H}_m]^{n_1}) \tag{87}$$

and the factor $(1 + K_2[\mathcal{H}_m])$ in each term of Eqs. (69), (72), (75), (82), and

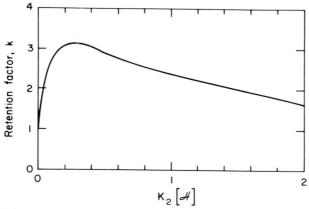

FIG. 46. Schematic representation of the dependence of retention factor on hetaeron concentration. The data were calculated according to Eq. (72) with the following assumed values: K_4, 0; K_1, 1; $\phi K_0[H_s]^*$, 1; K_1/K_2, 10; $\phi K_0 K_2[H_s]^*$, 50. The plot shows that concentration dependence in hetaeric chromatography is, in general, parabolic.

(84) is replaced by $(1 + K_2[\mathcal{H}_m] + K_8[\mathcal{H}_m]^{n_1})$. If such partitioning occurs, the decrease in retention with hetaeron concentration will be much sharper at high hetaeron concentrations than predicted when no micelles are present. However, if the stability constant for hetaeron binding to stationary phase, K_1, is so small that it can be neglected, the retention factor of the eluite is expected to increase with hetaeron concentration. The effect of changing hetaeron concentration on retention factor is shown schematically in Fig. 46.

A second important conclusion from this analysis is that data obtained in a study on the dependence of retention factor on the hetaeron concentration alone do not suffice to ascertain the mechanism of the particular process that governs retention. It is so because for all the four mechanisms I, II, IIIa, and IIIb, which were discussed above, the dependence of retention factor on hetaeron concentration in the eluent is identical. The problems arising from this and the need for extrachromatographic data in attempting to establish the mechanism of retention are discussed further in Section VI,C.

Two assumptions essential to obtain Eqs. (82) or (84) are (i) the hetaeron adsorption has an asymptotic limiting value, and (ii) complexes between the eluite and hetaeron form in the mobile phase. The Langmuir adsorption isotherm is used here because it is simple to manipulate mathematically and because most data on the adsorption of detergents used as hetaerons are reported to obey this relationship as determined directly (207, 209) or from analyses of its effect on chromatographic retention (34).

Furthermore, adsorption data taken from the literature (*201, 217*) were found to conform to this model when appropriately analyzed. The particular choice of a Langmuir isotherm is not necessary to draw the above conclusions regarding the dependence of the retention factor on the hetaeron concentration.

The use of a function that shows a similar "saturation" behavior is required, however, to account for the observed concentration dependence of hetaeron binding to the stationary phase. The need for saturation behavior when combined with the assumption that complex formation occurs in the mobile phase, accounts for the frequently observed decrease in retention upon increasing the hetaeron concentration above a certain level. Any other process which involves an interaction of the sample component with the hetaeron in the mobile phase, including formation of complexes of other than 1:1 stoichiometry or partitioning into micelles, will introduce the qualitative feature of decreased retention at high hetaeron concentrations to the theory.

An especially interesting result can be obtained from Eq. (69) in the special case where the hetaeron does not bind to the stationary phase. This assumption is plausible when it is a small ion, including the proton. In that case the retention factor is given as

$$k = \phi \frac{K_0[\mathcal{H}_s]^* + K_2 K_3 [\mathcal{H}_s]^* [\mathcal{H}_m]}{1 + K_2 [\mathcal{H}_m]} \tag{88}$$

The implications and practical significance of Eq. (88) are discussed in Sections VII,C and D.

C. Protonic Equilibria: pH Control

1. Monobasic Substances

Horváth *et al.* (*207*) considered the effect of ionization on the retention factor of monoprotic acids in RPC by using an argument slightly different from the general form given in the preceding section to arrive at the following expression for retention factor

$$k = \frac{k_0 + k_{-1}(K_{d1}/[H^+])}{1 + (K_{d1}/[H^+])} \tag{89}$$

where k_0 and k_{-1} are the limiting retention factors of the neutral and ionized forms of the acid and K_{d1} and $[H^+]$ are the acid dissociation constant and proton concentration, or more precisely, proton activity, respectively. This expression, which is identical to Eq. (88), predicts that plots of retention factor vs pH will be sigmoidal with a midpoint pH value which corresponds to the pK_a of the acid. The chromatographic behavior of several organic acids as shown in Fig. 42, support these expectations and the pK_a values determined chromatographically agree well with those obtained potentiometrically and given in the literature (see Table XIX, Section VII,D).

A similar derivation was presented for weak monoprotic bases with the retention factor given by

$$k = \frac{k_0 + k_1([H^+]/K_{d1})}{1 + ([H^+]/K_{d1})} \tag{90}$$

where K_{d1} is the protonic dissociation constant for the base and k_1 is the limiting retention factor of the positively charged species.

2. Diprotic and Oligoprotic Substances

Expressions were also derived for the retention factors of other ionogenic species (207). For diprotic acids the retention factors have been given by

$$k = \frac{k_0 + k_{-1}(K_{d1}/[H^+]) + k_{-2}(K_{d2}K_{d1}/[H^+]^2)}{1 + (K_{d1}/[H^+]) + (K_{d1}K_{d2}/[H^+]^2)} \tag{91}$$

where k_0, K_{-1}, and k_{-2} are the retention factors of species with charge 0, -1, and -2, respectively, and K_{d1} and K_{d2} are the respective first and second dissociation constants. In a similar fashion, the retention factor of ampholytes is given by

$$k = \frac{k_0 + k_{-1}(K_{d2}/[H^+]) + k_1([H^+]/K_{d1})}{1 + (K_{d2}/[H^+]) + ([H^+]/K_{d1})} \tag{92}$$

where k_0, k_{-1}, and k_1 are the limiting retention factors of the zwitterionic, anionic, and cationic forms of the ampholyte and K_{d1} and K_{d2} are its first and second acid dissociation constants, respectively. It is perhaps interesting to note in passing that Eq. (92) can be obtained from Eq. (91) by the simple expedient of multiplying each term in the numerator and denominator by $[H^+]/K_{d2}$ in which case the "new" expression is identical to Eq. (91) except that k_0, k_{-1}, and k_{-2} in Eq. (91) correspond to k_1, k_0, and k_{-1} in Eq. (92), respectively. This trivial exercise may quiet the alarms of those who, finding the treatments of these equilibria by different authors "look" different, would leap to the conclusion that one or all such treatments are erroneous.

Kroeff and Pietrzyk (208) developed similar sets of expressions for the retention factors of ionizable substances and have presented data obtained on column packed with porous cross-linked polystyrene, Amberlite XAD. These data, which are reproduced in Fig. 43, demonstrate clearly that the effect of pH on the retention factors of dibasic substances

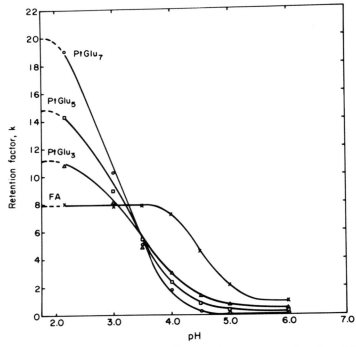

FIG. 47. Retention factors of pteroyl-oligo-γ-L-glutamates as a function of the eluent pH. The solid lines were calculated as discussed in Bush *et al.* (*204*). Data were obtained for folic acid (FA) and oligo-γ-glutamates containing one (Pt Glu$_1$), three (Pt Glu$_3$), five (Pt Glu$_5$), and seven (Pt Glu$_7$) glutamyl residues. The data were obtained on 5 μm Partisil ODS-2 with 0.1 M phosphate buffers containing 6% (v/v) acetonitrile as eluent, 45°C, with detection at 254 nm. Reprinted with permission from Bush *et al.* (*204*).

can be complex. Bush *et al.* (*204*) have examined the effect of pH on the retention behavior of pteroyl-oligo-γ-L-glutamates. Some of the data shown in Fig. 47 could be treated by an extension of Eq. 91, which could be rendered tractable with assumptions specific to the system studied.

The assumptions used for the derivation of Eqs. (91)–(92) are given in Table XIV.

D. Ion-Pair Chromatography

1. Importance of the Technique

The use of amphiphilic ions, such as alkylsulfonate or alkylquaternary ammonium compounds, as hetaerons to augment retention of ionized sample components having charges opposite is quite popular in RPC. By

TABLE XIV

Equilibria and Constants Used in the Derivation of Eqs. (91)–(92)[a]

	Diprotic acids	Zwitterions
Dissociation equilibria in the mobile phase	$H_2A \rightleftarrows HA^- + H^+$ $HA^- \rightleftarrows A^{2-} + H^+$	$H^+BA^- \rightleftarrows BA^- + H^+$ $H^+BAH \rightleftarrows H^+BA^- + H^+$
Acid dissociation constants in the mobile phase	$K_{d1} = \dfrac{[HA^-]_m[H^+]_m}{[H_2A]_m}$ $K_{d2} = \dfrac{[A^{-2}]_m[H^+]_m}{[HA^-]_m}$	$K_{d1} = \dfrac{[H^+]_m[BA^-]_m}{[H^+BA^-]_m}$ $K_{d2} = \dfrac{[H^+]_m[H^+BA^-]_m}{[H^+BAH]_m}$
Equilibria in the solute–ligand binding process	$H_2A + L \rightleftarrows LH_2A$ $HA^- + L \rightleftarrows LHA^-$ $A^{2-} + L \rightleftarrows LA^{2-}$	$H^+BA^- + L \rightleftarrows LH^+BA^-$ $BA^- + L \rightleftarrows LBA^-$ $H^+BAH + L \rightleftarrows LH^+BAH$
Equilibrium constants for the binding process	$K_{LH_2A} = \dfrac{[LH_2A]_s}{[H_2A]_m[L]_s}$ $K_{LHA^-} = \dfrac{[LHA^-]_s}{[HA^-]_m[L]_s}$ $K_{LA^{-}} = \dfrac{[LA^{2-}]_s}{[A^{2-}]_m[L]_s}$	$K_{LH^+BA^-} = \dfrac{[LH^+BA^-]_s}{[H^+BA^-]_m[L]_s}$ $K_{LBA^-} = \dfrac{[LBA^-]_s}{[BA^-]_m[L]_s}$ $K_{LH^+BAH} = \dfrac{[LH^+BAH]_s}{[H^+BAH]_m[L]_s}$
Overall retention factor	$k = \varphi \dfrac{[LH_2A]_s + [LHA^-]_s + [LA^{2-}]_s}{[H_2A]_m + [HA^-]_m + [A^{2-}]_m}$ $k = \dfrac{k_0 + k_{-1}\dfrac{K_{d1}}{[H^+]} + k_{-2}\dfrac{K_{d1}K_{d2}}{[H^+]^2}}{1 + \dfrac{K_{d1}}{[H^+]} + \dfrac{K_{d1}K_{d2}}{[H^+]^2}}$	$k = \varphi \dfrac{[LH^+BA^-]_s + [LBA^-]_s + [LH^+BAH]_s}{[H^+BA^-]_m + [BA^-]_m + [H^+BAH]_m}$ $k = \dfrac{k_0 + k_{-1}\dfrac{K_{d1}}{[H^+]} + k_1 \dfrac{[H^+]}{K_{d2}}}{1 + \dfrac{K_{d1}}{[H^+]} + \dfrac{[H^+]}{K_{d2}}}$
Retention factors of the individual species	$k_0 = \varphi[L]_s K_{LH_2A}$ $k_{-1} = \varphi[L]_s K_{LHA^-}$ $k_{-2} = \varphi[L]_s K_{LA^{-}}$	$k_0 = \varphi[L]_s K_{LH^+BA^-}$ $k_{-1} = \varphi[L]_s K_{LBA^-}$ $k_1 = \varphi[L]_s K_{LH^+BAH}$

[a] Reprinted with permission from Horváth et al. (207), Anal. Chem. Copyright © 1977 by the American Chemical Society.

using this approach the retention of ionized sample components can be increased, the selectivity enhanced and, under circumstances, even the elution order typical in plain RPC can be reversed.

Furthermore, the technique, which originally was called "ion-pair" chromatography, provides a powerful alternative to ion-exchange chromatography. The analogy between these two branches of liquid chromatography is striking because in both cases retention arises primarily from electrostatic interactions, either between the sample component and the amphiphilic hetaeron in ion-pair chromatography or between the eluite and the ion exchange resin, in ion-exchange chromatography. On the other hand, the nonpolar bonded stationary phases employed in RPC are rigid porous and mechanically stable column materials unlike the resinous, cellulosic, and other types of ion exchangers used in ion-exchange chromatography. Since the porosity of the latter stationary phases depends on the degree of swelling in contact with the aqueous eluent, the use of hydroorganic mobile phases is usually precluded in conventional ion-exchange chromatography whereas hydroorganic eluents can readily be used for ion-pair chromatography with bonded-phase columns in RPC. Another advantage of the latter technique is that diffusion rates are much higher in siliceous stationary phase particles than in ion-exchange resins. Consequently, separations more rapid than those obtained in conventional ion-exchange chromatography with resin-packed columns can be obtained even with samples consisting of relatively large molecular weight substances.

Whereas the expansion of the scope of RPC and a supersedure of ion-exchange chromatography by the "ion-pairing" approach in coping with a wide range of separation problems has been quite successful in practice, research to find out what is actually going on in the column has not lead to a generally accepted picture as far as the mechanism of the chromatographic process is concerned. As we shall see below numerous models have been put forward for the mechanism of the process. Yet, its very complexity is likely to preclude a satisfactory interpretation of experimental observations by using a single simple model. Though this type of chromatography has become the wellspring of various theories, none of them has been proven to satisfaction over a sufficiently wide range of conditions. The remarkable galaxy of names arising from this endeavor is unparalleled in chromatographic history. The technique has been described among others as "solvophobic ion," "soap," "solvent-generated (dynamic) ion exchange," "ion association," "detergent-based cation exchange," "surfactant," and "Paired-Ion™ chromatography."

The original name "ion-pair" chromatography has been criticized on the basis that either no ion-pair formation occurs under conditions used in

RPC or the formation of an ion-pair with the oppositely charged hetaeron in the mobile phase, and the enhanced retention of the eluite complex, is not responsible for the observed increase in retention. It has been repeatedly put forward that in actuality the amphiphilic hetaeron ions are absorbed on the stationary phase surface and thus it is converted into a dynamically coated ion exchanger. As the reader may have gathered from the previous discussion of mechanisms I and II an unambiguous interpretation of chromatographic results is difficult because the formal dependence of retention on the hetaeron concentration in the eluent is the same in both cases. Moreover, the parameter values represented by the magnitude of the equilibrium constants and, as a consequence, the actual mechanism of retention may change upon varying conditions such as the organic solvent content of the eluent or the chain length in the aliphatic moiety of the amphiphilic hetaeron. Only the results of appropriate extrachromatographic experiments can yield data on the magnitude of the

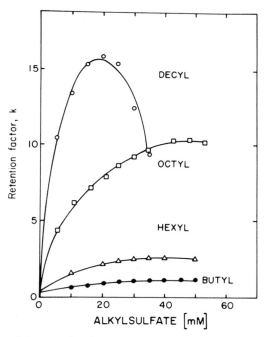

FIG. 48. Plots of the retention factor of adrenaline vs the hetaeron concentration for various *n*-alkylsulfates. Column, Partisil ODS; flowrate, 2.0 ml/min; temp. 40°C; inlet pressure, 400 psi; eluent 0.05 *M* phosphate in water, pH 2.55, containing various concentrations of the hetaeron. Reprinted with permission from Horváth *et al.* (207), *Anal. Chem.* Copyright © 1977 by the American Chemical Society.

various equilibrium constants involved so that the predictions of one or the other postulated mechanism can be substantiated unambiguously.

In the following sections, the data and arguments which support either the ion-pairing or dynamic ion-exchange mechanism, mechanisms I and II, respectively, will be reviewed briefly. Subsequently, a general model will be developed on the basis of mechanisms IIIa and IIIb which were discussed previously and available experimental observations, both chromatographic and extrachromatographic, will be compared to behavior predicted by this model. It will be shown that satisfactory agreement gives support to the model and to the conclusion that the mechanism of the process is likely to change with the chromatographic conditions. Fortunately, in practice the chromatographer can predict the effect of changing parameters or hetaeron concentration and optimize separation without having precise knowledge of the particular retention mechanism.

2. Basic Model for Ion-Pairing and Dynamic Ion Exchange

The elucidation of the retention mechanism in ion-pair reversed-phase chromatography using alkyl amines or alkyl sulfonates as hetaerons has evoked significant interest not only for the great potential of the method in the separation of ionic compounds but also for theoretical reasons.

Several workers have concluded that under conditions used in their study ion-pairing in the mobile phase between amphiphilic hetaeron ions and oppositely charged sample components governed retention. Horváth et al. (34) examined the effect of alkyl sulfates and other alkyl anions on the retention of catecholamines in which both the concentration and the length of the alkyl chains of the hetaerons were varied. The hyperbolic concentration dependence of the retention factor shown in Fig. 48, was found to be similar to that reported by others.

Experimental results were interpreted by these authors (34) who assumed that the physicochemical phenomena underlying the retention process can be described by mechanisms I or II discussed in the previous section. Thus, two limiting mechanism were formulated for enhancement of retention due to eluite–hetaeron interactions. According to the first mechanism the ionized sample component and the oppositely charged hetaeron form ion pairs in the mobile phase and retention is due to the reversible binding of the complex to the stationary phase, the surface of which has no bound hetaeron. This simple case corresponds to mechanism I and consequently the retention factor is given by Eq. (69). According to the other limiting case retention occurs by dynamic ion-exchange due to the interaction of the ionized sample component with the oppositely charged hetaeron bound to the stationary phase (53, 217–219).

Such a physicochemical situation is represented by mechanism II so that the retention factor under such conditions is given by Eq. (72).

What is especially important about Eqs. (69) and (72) is that they describe intrinsically different sets of equilibria, yet they are indistinguishable in the sense that one cannot make a mechanistic choice based upon the interpretation of the dependence of retention factor on the hetaeron concentration from these expressions. Both predict that with increasing hetaeron concentration the retention factor first increases linearly and ultimately reaches a limiting value. However, the hetaeron concentrations which correspond to the left- and right-hand inflection points of the parabolic plot of k against $[\mathscr{H}_m]$ depend on the actual mechanism. They are given by $1/K_1$ and $1/K_2$ according to Eq. (72) when ion-pairing (mechanism II) predominates. On the other hand, the respective parameters are given by $1/K_2$ and $1/K_1$ when Eq. (69) is valid because dynamic ion exchange (mechanism II) governs the retention process at low hetaeron concentrations.

In the case where the retention factor increases to a limiting value with no subsequent decrease at increasing hetaeron concentrations, i.e., the plot of k versus $[\mathscr{H}_m]$ is a rectangular hyperbola because one of the terms in the denominator of Eqs. (69) and (72) vanishes, the relationship between retention factor and hetaeron concentration for either of the two mechanisms takes the simple form

$$k = (A + B[\mathscr{H}_m])/(1 + C[\mathscr{H}_m]) \tag{93}$$

where the parameter A is the limiting retention factor obtained in the absence of hetaeron and C is equal to the reciprocal hetaeron concentration at the inflection point of the curve. On the other hand, the ratio B/C is the other limiting value of the retention factor obtained at saturating hetaeron concentration.

The parameters obtained by analyzing experimental data in view of Eq. (93) can be combined to calculate the so-called retention modulus (116), η, which in our case is given by the ratio of the maximum retention factor at "saturating" hetaeron concentration to the retention factor measured with an otherwise identical chromatographic system in the absence of hetaeron as

$$\eta = B/AC \tag{94}$$

As the modulus is greater than unity under such conditions, it is conveniently called enhancement factor (34) because it measures the maximum possible increase in retention upon ion-pair formation in a given chromatographic system. The behavior of the enhancement factor as a function hetaeron or eluite properties has been analyzed (34) in the light of the information reproduced in Tables XV and XVI and the behavior of each parameter with changes in eluite or hetaeron properties has been estimated by considering the variation of each equilibrium constant as predicted by the solvophobic theory. In this approach it is assumed that the mechanism of the process is described by either the "ion-pair" formation (mechanism I) or the "dynamic ion-exchange" (mechanism II) model and as a result a qualitative treatment leading to conclusions similar to those given in the tables can be made. For example, if ion-pair formation in the mobile phase dominates retention, the parameter B/C, which is the overall equilibrium constant for the binding of the complex to the stationary phase, should depend on the hydrophobic character of the hetaeron. When linear alkane derivatives are the hetaerons, the logarithm of B should increase linearly with carbon number in accord with the general experience in RPC. When the retention process follows this mechanism, the parameter C is the ion-pair formation constant, and it is expected to depend mainly on the charge carried by the hetaeron according to recent investigations (34) although hydrophobic interactions between eluite and hetaeron can be of great significance in augmenting the formation constant of ion pairs when both species are nonpolar (220–222). The retention factor of the eluite in the absence of hetaeron, A, will depend on both the size of the hydrophobic moiety of the eluite and its charge. As a consequence, this approach predicts that when ion-pairing governs retention (mechanism I) the logarithm of the enhancement factor is a linear function

TABLE XV

**Relationship between Hetaeron Properties and the Parameters
of Equation (93) as Predicted for the Two Limiting Mechanisms
in Ion-Pair Reversed-Phase Chromatography[a]**

| Parameter | Ion-pair formation occurs in the | |
	Mobile phase	Stationary phase
A	—	—
B	Hydrophobic surface area (carbon number)	Hydrophobic surface area (carbon number)
C	Charge type ($C = K_2$)	Hydrophobic surface area (carbon number) + charge type ($C = K_1$)
$\eta = B/AC$	Hydrophobic surface area (carbon number)	Charge type

[a] Reprinted with permission from Horváth et al. (34), Anal. Chem. Copyright © 1977 by the American Chemical Society.

TABLE XVI

Relationship between Eluite Properties and the Parameters of Equation (93) as Predicted for the Two Limiting Mechanisms with Ion-Pair Formation[a]

	Ion-pair formation occurs in the	
Parameter	Mobile phase	Stationary phase
A	Charge and hydrophobic surface area	Charge and hydrophobic surface area
B	Charge and hydrophobic surface area	Charge and hydrophobic surface area
C	Charge ($C = K_2$)	—
$\eta = B/AC$	Charge	Charge and hydrophobic surface area

[a] Reprinted with permission from Horváth *et al.* (*34*), *Anal. Chem.* Copyright © 1977 by the American Chemical Society.

of the carbon number for a homologous series of hetaerons such as alkyl sulfates. Similar qualitative arguments were applied to the prediction of retention behavior from molecular properties of the sample components in ion-pair reversed-phase chromatography.

Logarithms of enhancement factors for the retention of four catechol-amines obtained with four alkyl sulfates having different chain lengths are plotted against the carbon number of the hetaerons in Fig. 49. The experimental data were obtained with plain aqueous eluents containing no organic solvent and with an octadecyl silica column of very low carbon load (5%). The observed linear dependence of the modulus on the carbon number of the hetaeron suggest that the retention process follows the ion-pair model (mechanism I), in view of the above analysis which does not predict such a behavior for the case of dynamic ion-exchange as the prevailing mechanism. Numerical values of the individual terms, A, B and C, in Eq. (93) have been found to be consistent with the ion-pairing mechanism rather than with dynamic ion-exchange. On the basis of this finding it was suggested that ion-pair formation in the plain aqueous eluent may be responsible for the increased retardation factors in the presence of C_4 to C_{10} alkyl sulfonates under the experimental conditions employed (*178*). It was noted, however, that with hetaerons that have larger nonpolar moieties than those investigated, retention would be expected to occur primarily through dynamic ion-exchange because the equilibrium constant for binding of long-chain hetaerons to the nonpolar stationary phase is greater than the ion-pair formation constant.

Other authors have also reported results consistent with the ion-pairing model. Riley *et al.* (*201*) examined the retention behavior of cromoglycate

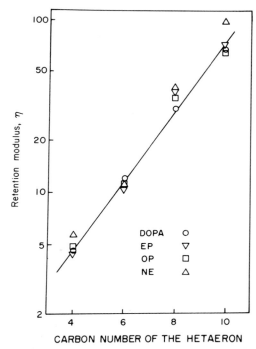

FIG. 49. The dependence of the enhancement factor of catecholamine derivatives on the carbon number of n-alkyl sulfates as the hetaerons. The straight line obtained by least-squares analysis fits the expression, $\log \eta = 0.225 \, (\pm 0.0317) N_0$ where η is the enhancement factor and N_0 is the carbon number. The intercept is zero within experimental error. Reprinted with permission from Horváth *et al.* (*34*), *Anal. Chem.* Copyright © 1977 by the American Chemical Society.

as well as several 8-azapuron-6-ones and 1,2,5-*s*-triazines in the presence of alkylbenzyldimethylammonium chlorides or sodium dodecyl sulfate in the eluent. They found that under their experimental conditions the ion-pair mechanism adequately described the dependence of retention factor on the chemical structure of the pairing ions as well as on the nature and concentration of the organic modifier and the ionic strength of the eluents.

On the other hand, the results of several systematic studies of the parameters which govern retention in this type of chromatography were consistent with the predictions of the model for dynamic ion-exchange. Knox and Laird (*53*) examined the effect of hetaeron concentration on the retention of sulfonic acids and related dyestuffs on a short alkyl silica (SAS) stationary phase. The hetaeron used was cetrimide (cetyltrimeth-

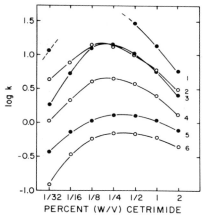

FIG. 50.   Dependence of the logarithm of the retention factor of six dyestuffs and inter-mediates upon cetrimide (cetyltrimethylammonium bromide). The data were obtained on 7 μm SAS silica (Wolfson Unit) with water–propranol (5:2, v/v) containing cetrimide at the indicated concentration as the mobile phase. The eluites are as follows: (1) Ponceau 4R; (2) Ponceau MX; (3) tartrazine; (4) Sunset Yellow; (5) Schaeffer's acid; (6) sulfanilic acid. Re-printed with permission from Knox and Laird (53).

ylammonium bromide) in water–propanol. The effect of cetrimide concentration on retention of sulfonic acids they observed is shown in Fig. 50. It is seen that at low cetrimide concentrations the addition of the hetaeron promotes retention whereas at higher concentrations further increase in the hetaeron concentration decreases binding to the stationary phase. They found that the hetaeron adsorbs to the stationary phase according to a Freundlich isotherm that can be described by the following equation:

$$[\mathcal{H}_s] = \alpha[\mathcal{H}_m]^{0.8} \tag{95}$$

where $[\mathcal{H}_s]$ and $[\mathcal{H}_m]$ are concentrations of hetaeron at the surface and in the mobile phase and α is the distribution coefficient of hetaeron between the mobile and stationary phases. In this model the distribution of eluite with negative charges between the mobile and stationary phases is measured by the coefficient β and the equilibrium constants K_2 and K_4 quantify complex formation in the mobile and stationary phases, respectively. The corresponding mass balances are given by

$$[E_s] = \beta[E_m] \tag{96}$$

$$[(\mathcal{H}_n E)_m] = K_2[\mathcal{H}_m]^n[E_m] \tag{97}$$

$$[(\mathcal{H}_n E)_s] = K_4[\mathcal{H}_s]^n[E_s] \tag{98}$$

This model also assumes that the binding of unpaired eluite to the stationary phase is negligible. Equations (96)–(98) can be combined to calculate the retention factor for the eluite as

$$k = \frac{\alpha^n \beta \, K_4 [\mathscr{H}_m]^{0.8n}}{1 + K_2 [\mathscr{H}_m]^n} \tag{99}$$

Reversed phase chromatography of sulfonated dyestuff intermediates with increasing concentrations of cetrimide was done by Knox and Laird (53) to test Eq. (99). In order to explain the rapid decrease in retention with increasing hetaeron concentration at high hetaeron concentrations (viz. the right-hand side of Fig. 50) the authors suggested that cetrimide could form micelles or agglomerates at sufficiently high hetaeron concentrations. The retention factor then takes the form

$$k = \frac{\alpha^n \beta \, K_4 [\mathscr{H}_m]^{0.8n}}{1 + K_2 [\mathscr{H}_m]^{n_1}} \tag{100}$$

where n_1 is the number of hetaerons per eluite in a cluster. It is necessary to assume that $n_1 > n$ in order to account for the data. They concluded that "it is probable that cetrimide–sulfonate ion-pairs are extracted from the water-rich eluent into an adsorbed layer rich in propanol and cetrimide." In a subsequent study (223) catecholamines were chromatographed in the presence of sodium decylsulfonate, sodium lauryl sulfate or sodium dodecylbenzenesulfonate. It was found that the retention factor decreases linearly with increasing acetonitrile concentration in the aqueous eluent, as shown in Fig. 51, and such results are in agreement with the generally observed regular behavior in RPC.

The effect of these "hydrophobic" anions on the retention factor of catecholamines was qualitatively similar to that observed with cetrimide in the separation of acids. In both cases the retention factor rose with increasing concentration of the hetaeron and either reached a limiting value or exhibited curvature suggestive of approaching an asymptotic value of retention factor at high hetaeron concentrations. The authors stated their belief that the mechanism of retention "operates rather like ion-exchange where amines (in the present instance) present in the eluent as cations exchange with counterions that are associated with detergent anions adsorbed onto or into the hydrocarbon surface of the packing."

Kissinger (218) argued that the changes in retention are not due to complex formation in the mobile phase but to coating of the stationary phase with the hetaeron so that it is in effect converted into an ion-exchanger. Terweij-Groen et al. (217) developed this concept further for the case of an acid, HE, the ionized form of which, E^{1-}, can combine with hetaeron

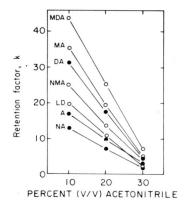

FIG. 51. Dependence of retention of catecholamines on volume percent acetonitrile in hetaeric chromatography. The eluent is water–acetonitrile at the volume percent indicated containing 0.2% (v/v) sulfuric acid and 0.1% (w/v) sodium dodecyl sulfate. The catecholamines separated are noradrenaline (NA), adrenaline (A), L-3,4-dihydroxyphenylalanine (LD), normetanephrine (NMA), dopamine (DA), metadrenaline (MA), and 3-methoxytyramine (MDA). Column 5-μm octadecyl silica treated with trimethylchlorosilane, 125 × 5 mm i.d. Reprinted with permission from Knox and Jurand (223).

HA to form a complex by displacement of counterion A^{1-} according to the following scheme

$$HE_m \xrightleftharpoons{K_0} HE_s \tag{101a}$$

$$H_m^+ + E_m^{1-} \xrightleftharpoons{K_{d1}} HE_m \tag{101b}$$

$$\mathscr{H}A_s + E_m^{1-} \xrightleftharpoons{K_s} \mathscr{H}E_s + A_m^{1-} \tag{101c}$$

$$E_m^{1-} \xrightleftharpoons{K_8} E_{mic}^{1-} \tag{101d}$$

The last reaction accounts for the partitioning of eluite into micelles of cetrimide in the example studied. The authors arrived at the following expression of the retention factor for the acid:

$$k = \frac{K_0 K_{d1}[H^+] + K_5[A_s]/[A_m^-]}{1 + K_{d1}[^{6+}] + K_8} \tag{102}$$

The above model to monobasic acids has been extended to develop an approximate expression for the retention factor of n-basic acids which bind only at "ion-exchange" sites and to a negligible extent elsewhere at the stationary phase surface (217). It is given by

$$k = \frac{K_5[(HA)_s]^n}{(1 + K_7)[A_m^-]^n} \tag{103}$$

Both expressions were found to conform to experimental results. Log–log plots of cetrimide concentration versus retention factor were linear with slope equal to the absolute value of the eluite charge as predicted by Eq. (99). Deviations were seen at low pH due to saturation effects. Decreases in retention factor were observed at high cetrimide concentrations and this was attributed to a partitioning of the sample components between the bulk mobile phase and a partitioning of the sample components between the bulk mobile phase and micelles of the hetaeron, the concentration of which increases with that of cetrimide. Added bromide ion caused decreased retention and this behavior as well as the observed pH dependence of the retention factor at fixed cetrimide concentration were in general conformity to Eq. (102). The volume fraction of propanol in the aqueous eluent was varied from 10 to 25% in these experiments (217). The retention factors decreased but the results cannot be satisfactorily explained on the basis of Eqs. (101a) to (103). As the authors' model predicts that the only effect of changing propanol concentration is a change in $[HA_s]$ so that plot of log $[HA_s]$ versus retention factor has a slope equal to the absolute value of charge on the anion. This was not the case, however, and the failure was attributed by the authors to changes in K_1 and K_8 with increasing propanol concentration. Nevertheless our estimate of the behavior of the ratio $K_1/(1 + K_8)$ from the data suggests that it decreases with increasing propanol concentration for monobasic acids although the ratio remains constant for the dibasic acid and even increases for the tribasic acid. Since it is not clear why the equilibrium constant for ion-exchange should decrease for monoprotic acids with increase in propanol concentration or why partitioning into micelles should be promoted by propanol, the explanation of the experimental data by the above model given in the literature (217) should be regarded with reservation.

3. Dynamic Complex Exchange

So far no single retention mechanism has accounted satisfactorily for all the observations made in ion-pair chromatography with detergents. Since objections can be raised to the mechanistic interpretations given by investigators to their experimental observations, it behooves us to propose a mechanism which appears to reconcile most of the data reported in the literature. In fact, a single mechanism is not likely to operate over the entire range of hetaeron concentrations and solvent compositions as was recognized by Horváth et al. (34). They have noted on the basis of data obtained with plain aqueous eluents only that whereas "chromatographic retention is enhanced by ion-pair formation in the mobile phase and subsequent binding of the complex to the nonpolar stationary phase, dynamic ion-exchange can be the underlying mechanism under some other conditions."

As we seen above, numerous studies have concluded that the dominant mechanism is dynamic ion-exchange which corresponds to mechanism II. The main arguments have been based on the findings that (i) the hetaeron binds strongly to the stationary phase, (ii) added salt decreases retention of eluite, (iii) added salt increases binding of hetaeron to the stationary phase, and (iv) organic solvents such as n-propanol decrease hetaeron binding and also decrease the retention of eluite. The latter effect is not due to decrease of hetaeron concentration at the surface; examination of the data of Terweij-Groen et al. (217) indicates the decrease in retention is greater than that expected on the basis of that effect alone. However, some of the observations mentioned above are not unique to dynamic ion-exchange and are rather consistent with the ion-pairing model, which corresponds to mechanism I.

A particularly compelling argument for dynamic ion-exchange has put forward the observation that retention of anionic and cationic sample components increases and decreases with increasing concentration of a cationic hetaeron, respectively. Whereas anionic hetaerons are expected to promote the elution of anionic eluites and to enhance the retention of cationic eluites, the quantitative data presented in this regard (226) are not wholly consistent with the model since the hetaeron concentration at which the effect is half-maximal is different for anionic and cationic eluites. If the observed phenomena were due to the presence of bound hetaeron in both cases, the two effects would have identical dependence on the hetaeron concentration in the mobile phase.

A chief argument for the dynamic ion-exchange mechanism is that detergents do bind "strongly" to the hydrocarbonaceous surface of the stationary phase. (209, 217, 223–227). The mechanistic implications of this observation favor of mechanism II, however, are greatly attenuated by the low surface coverage of the order of a few percent, which can be calculated from experimental data representing "strong" binding of hetaerons (201, 217–219, 223–225) and the existence of ion-pairs (634) under conditions usually employed in ion-pair RPC.

In contradistinction to these investigations the results by Horváth et al. (34) give sufficient support to the proposition that dynamic ion-exchange was not the dominant mechanism under their experimental conditions.

The experimental data conformed to Eq. (93) and therefore could be interpreted by either mechanism I or II; data analysis showed no linear dependence of the logarithm of parameter C in Eq. (93) on the carbon number of the alkyl sulfate hetaerons. However, in the case of dynamic ion exchange parameter C is the binding constant of the hetaeron to the stationary phase and its logarithm should be linearly dependent on the carbon number of the alkyl moiety. Even if the results of this study are not accepted as support for ion-pairing (mechanism I) uniquely, they cannot be used to validate dynamic ion-exchange (mechanism II) either.

In fact a great variety of mechanisms could be invoked in an effort to explain the data abundant in the literature. These include, in addition to dynamic ion-exchange and ion-pairing, complex exchange with displacement of hetaeron, mechanism IIIa [cf. Eq. (75)] and complex exchange with no displacement, mechanism IIIb [cf. Eq. (82)]. Of these, only the latter, ligand exchange without expulsion of bound hetaeron from the surface, appears to offer a single mechanistic framework for a consistent interpretation of all experimental data available so far. This model assumes that ion-pair formation occurs in the mobile phase and the ion-pair binds to the stationary phase. Thereafter the complex decomposes and the sample component combines with another hetaeron which is already bound to the surface. The details for this mechanism are given in Eqs. (76)–(82).

In cases when binding of hetaeron to the surface is found to be "strong," experiments usually show that the hetaeron concentration at which retention is half-maximal is significantly different from the concentration at which coverage of the stationary phase is half-maximal. This is contradictory to what would be expected if the mechanisms followed either dynamic-ion exchange [cf. Eq. (72)], or dynamic complex exchange with displacement [cf. Eq. (75)], or dynamic ion-exchange coupled with ion-pairing [cf. Eq. (83)]. These are mechanisms in which hetaeron is assumed to be strongly bound to the surface.

The above observation, however, can be reconciled by the predictions of mechanism IIIb: dynamic complex exchange without expulsion of hetaeron [cf. Eq. (82)]. According to this mechanism the retention factor rises with increasing hetaeron concentration to a constant plateau, the value of which is $K_6(1 + K_7)$. Plots of k versus $[\mathcal{H}_m]$ allow us to evaluate conveniently a "formation" constant which is defined as the reciprocal of the hetaeron concentration at which the retention factor is the mean of the values obtained in the absence of hetaeron and at saturation. If K_1, the binding constant of hetaeron to the stationary phase, is much greater than K_2, the formation constant of the ion-pair [see Eq. (66)], the above-defined parameter will be very nearly the ion-pair formation constant.

Thus, experimental data may yield a value for the parameter which is not the equilibrium constant for binding of the hetaeron to the surface even if the hetaeron was strongly bound to the surface and eluite interaction with the bound hetaeron played a significant role in determining the magnitude of retention.

According to our analysis most (635), if not all, literature data appear to be consistent with this model with an additional modification discussed below. It accounts for the observation that different hetaeron concentrations in the mobile phase are required to promote elution and to facilitate retention of sample components having the same and opposite charge, respectively. Whereas the promotion of elution would depend on the degree of surface coverage by hetaeron and would result in minimal retention in the limit of $[\mathcal{H}] \approx 10/K_1$, the retention factor of retained eluites would be maximal and reach its limiting value at $[\mathcal{H}] \approx 10/K_2$. Since, K_1 is expected to be much greater than K_2, the corresponding concentration of the hetaeron in the mobile phase would be much less for reducing than for increasing retention.

The model predicts decreased retention when at constant mobile phase hetaeron concentration the concentration of salt or organic solvent in the eluent is increased. It can also be easily reconciled with the main objection to the ion-pairing model, that significant hetaeron binding to the surface occurs. Indeed the complex-exchange model requires the hetaeron binding to occur to a significant extent over the hetaeron concentration range of interest, while it still predicts the observed concentration dependence will reflect ion-pair formation. This is hardly surprising because the comprehensive dynamic complex-exchange model, mechanism IIIb, requires that both participate in bringing about retention. It is also interesting to note that according to this model the dependence of the three parameters of Eq. (93) on eluite and hetaeron properties are the same as that for the ion-pair mechanism and shown in Tables XV and XVI. As a consequence, the arguments used by Horváth et al. (34) for ion-pairing in neat aqueous mobile phase can be used with equal force to argue not only against dynamic ion-exchange but also for the present dynamic complex-exchange model.

The chief defect of the proposed mechanism is its failure to predict decrease in retention at high hetaeron concentrations. This can be remedied by evoking micelle formation at relatively high detergent concentrations and substituting the factor $(1 + K_2[\mathcal{H}] + K_8[\mathcal{H}]^{n_1})$ for the factor $(1 + K_2[\mathcal{H}])$ in the denominator of Eq. (82). Thus, the term $K_8[\mathcal{H}]^{n_1}$, where n_1 is greater than one, is included to account for partitioning of eluite into hetaeron micelles in the mobile phase that do not bind to the stationary phase. There are ample literature data which suggest that detergent concentration in the eluent may exceed the critical micelle concentration when the alkyl chain is long and the limiting value of the retention factor is reached.

Of course sound scientific standards demand that the demonstration of a given mechanism is complemented by the demonstration that alternatives do not occur. The mechanism suggested here can be regarded as proven simply in the sense that all data reported on retention in ion-pair chromatography in the literature according to our best knowledge are consistent with it. It does have the attractive feature that it appears to account for all experimental data in one self-consistent mechanism. This unified mechanism may be misleading isofar as one could draw the conclusion that a single mechanism prevails for all hetaerons and solvent compositions. That assumption, however, is very likely not true and, therefore, investigations of mechanism when carried out under markedly different conditions may reveal the existence of a simpler mechanism which, of course, is a limiting case for the comprehensive model described above.

4. General Considerations

The results of the mechanistic studies, discussed in the previous section, offer little general guidance for the practicing chromatographer for optimal use of detergents in RPC. On the other hand, most observations in ion-pair chromatography are useful, even if the retention mechanism is not elucidated, for making predictions and optimizing separation. It has been observed that added salts usually decrease retention. Similarly, increased temperature also decreases retention but the selectivity of the chromatographic system is not affected strongly by changing temperature (*142*). The pH of the eluent determines the degree of ionization for both the component and hetaeron; consequently, retention is reduced if the pH change decreases the degree of ionization of either. The flexibility of the chromatographic system is maximized by the use of strongly acidic or basic hetaerons as in such cases only the degree of ionization of the eluite varies with changes in the pH of the eluent. This seems to be a safe rule although some data on pH effects in the literature cannot be strictly interpreted in this fashion.

The magnitude of retention can be affected by changing either the concentration or the chemical nature of the hetaeron. The dependence of retention factor on hetaeron concentration is schematically illustrated in Fig. 52. This result will be obtained whether dynamic ion-exchange, ion-pairing, or dynamic ligand exchange with or without displacement which respectively represent mechanisms I, II, IIIa, or IIIb occurs. If one is using a hetaeron at the concentration A on the figure, retention can be enhanced or reduced by increasing or decreasing the hetaeron concentra-

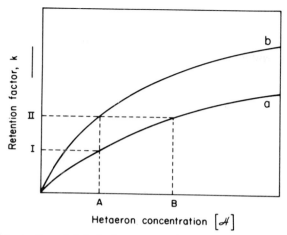

FIG. 52.   The retention of eluite by hetaeron can be modified by change in hetaeron concentration or changes in the chemical nature of the hetaeron. Retention factor obtained with hetaeron (a) at concentration A is I. It can be increased to II by increase of hetaeron concentration to B. Alternatively, another hetaeron that is more hydrophobic and therefore more retentive can be used at concentration A to obtain the same retention. This case is represented on curve (b).

tion, respectively. Alternatively similar effects can be achieved by using another hetaeron which is more or less hydrophobic in character than the reference hetaeron and therefore yields higher or lower values of the retention factor at the same hetaeron concentration as suggested by the schematic illustration.

Thus, independent of the retention mechanism, retention may be enhanced by either increasing the hetaeron concentration or using a more hydrophobic hetaeron at the same concentration. If, however, the hetaeron concentration is great enough to bring about close to maximum retention then modest changes in its concentration have little effect on retention, and retention can be further enhanced only by increasing the hydrophobicity of the hetaeron if the retention mechanism is not dynamic ion-exchange (mechanism II).

The effect of adding an organic solvent or increasing the concentration of the organic solvent in the mobile phase is determined by the balance of two opposing contributions to the overall retention. Generally, increasing the concentration of organic solvent such as CH_3OH and CH_3CN, in the aqueous mobile phase causes eluite retention to decrease in RPC in a fashion that the logarithm of the retention factor is usually linear in the volume fraction of organic solvent. Consequently increasing organic solvent concentration decreases the extent of surface coverage of hetaeron or decreases the binding of the complex in the ion-exchange or ion-pairing mechanism, respectively. An increase in the organic content of the eluent, however, is also accompanied by a decrease in the dielectric constant of the medium which will enhance coulombic interactions between eluite and hetaeron irrespective of the mechanism of the retention process.

The magnitude of the effect has been roughly estimated from the Born approximation of the energy of an ion in solution and the ratio of the ion-pair or ion-exchange equilibrium constant in a given eluent to that in a reference eluent, K/K_0, has been calculated (34, 129). The effect of increased organic solvent concentration is to increase this ratio and therefore the ion-pair or ion-exchange constant. It should be noted that the magnitude of the effect is highly sensitive to both the charges and the intermolecular separation distance of the hetaeron and eluite. Thus if either partner has a charge with absolute value greater than one, the ratio would increase very rapidly. On the other hand, if the distance between charges is increased, as would occur if a quaternary amine or an outer sphere complexing agent replaced a primary amine as the hetaeron, the rate of increase in the ion-pairing or ion-exchange constant with increase in organic solvent concentration would be reduced.

Figure 53 illustrates schematically how the dependence of the retention factor on the hetaeron concentration may be affected by increasing the concentration of the organic component in the eluent. On the basis of the previous discussion the predictions are applicable when the retention mechanism involves either ion-pairing or dynamic complex exchange without displacement. If retention occurs by ion-pairing, the addition of organic solvent causes the value of the limiting retention factor to decrease and the saturation curve to shift to the left due to an increase in the formation constant of the ion-pair complex. In the case of dynamic ion-exchange, however, the retention factor at saturating concentration of hetaeron would increase with the concentration of the organic component and the curve would shift to the right. In view of this behavior, the effect of added organic solvent is predictable only at the extreme of the retention factor versus hetaeron concentration plots and only if the mechanism is known.

As shown in Fig. 53 for the case of ion-pairing in the eluent and dynamic ligand exchange without expulsion, the addition of organic solvent will increase and decrease the retention factor at relatively low and high hetaeron concentration, and this is shown by points A and B, respectively. The opposite pattern obtains in the case of the dynamic ion-exchange mechanism, of course. At intermediate hetaeron concentrations

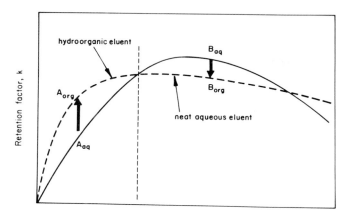

HETAERON CONCENTRATION

FIG. 53.   Schematic illustration of the effect obtained when a neat aqueous eluent is used instead of a hydroorganic mixture in ion-pair reversed-phase chromatography. Reprinted with permission from Horváth *et al.* (*34*), *Anal. Chem.* Copyright © 1977 by the American Chemical Society.

the results largely depend on the degree of saturation and could only be predicted if changes in the values of the pertinent equilibrium constant with changing solvent composition were known. The data of Terweij-Groen *et al.* (*217*), which show decrease in retention factor with increasing propanol concentration, can be interpreted in this light. It should be kept in mind that changes in the concentration of the organic solvent component in the eluent can bring about drastic changes in the magnitude of the various equilibrium constants with concomitant change in the retention mechanism proper.

This crude analysis is based on the behavior postulated by the Born equation. However, ion-pair formation equilibrium constants have been observed to deviate markedly from that behavior (*221–222*). Oakenful and Fenwick (*222*) found a maximum in the ion-pair formation constants of several alkylamines with carboxylic acids when determined at various methanol–water solvent compositions as shown by their data in Fig. 54. The results demonstrate that in this system the stability constant decreases with increasing organic solvent concentration above a critical value which yields maximum stability. The authors suggested that this was due to a weakening of hydrophobic interactions between the ion-pair forming species by increased alcohol concentrations. In practice the effect of added organic solvent has been either to decrease the retention factor or to have virtually no effect.

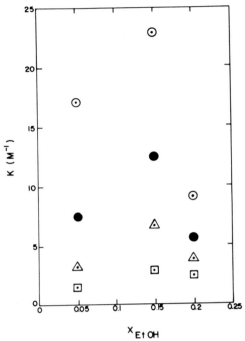

FIG. 54. Stability constants for ion-pair formation with decyltrimethylammonium car-
boxylates at 25°C in various mole fraction ethanol–water mixtures. The carboxylic acids
used are (○) decanoate, (●) nonanoate, (□) heptanoate, and (△) octanoate. The stability
constants were determined conductimetrically. The data are taken from Oakenful and Fenwick (222).

E. Analytical Applications

A variety of compounds have been separated by "soap chromatography," i.e., by using a hetaeron which is an ionic detergent or is closely
related to a detergent. Various dyestuffs and their intermediates have
been determined by this method (216, 228). Much attention has been
given to the analysis of catecholamines (207, 223, 229–231) as well as
other cations of pharmaceutical interest including niacin (232), opium alkaloids (233), and antihistamines (234). Tyramine, phenylethylamine, and
tryptamine in foods or leaves were determined in this way (235, 236). Separation of anions done with ease included that of 19 amino acids in 45 min
under acidic conditions (224), paracetamol and its derivatives (237), and
folic acid derivatives (238). In these studies the choice of organic modifier
in the solvent and stationary phase was observed to affect the chromatographic performance (224, 239). Added salts will reduce retention, pre-

sumably due to competition for the hetaeron (34,224). pH control appears to be important for amino acids where the cationic form binds better to anionic detergent than does the zwitterionic form (224, 235).

The use of nondetergent hetaerons is becoming more prevalent. Reverse argentation chromatography is observed with silver salts in the mobile phase which can form complexes with unsaturated molecules and thereby usually decrease retention. This has been observed with a variety of species (240–242) including vitamin D derivatives (241). Degree of unsaturation of an eluite could be determined from the magnitude of the change in retention factor. The effect of added Ag^+ on the retention and resolution of retinol and retinyl esters is shown in Fig. 55. The retention of the 18:1 and 18:2 esters can be changed at will by changing the concentration of Ag^+ in the mobile phase.

Amino acids, peptides, and proteins have been analyzed using hetaerons. Phosphoric acid or perchlorate salts in the mobile phase de-

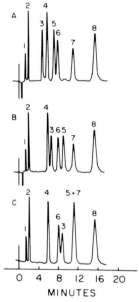

FIG. 55.   HPLC record of a standard mixture of retinol and retinyl esters. Conditions: column, 10 μm octadecyl silica; flowrate, 1 ml/min; mobile phase, (A) $CH_3OH/58.9 \times 10^{-3}$ M [Ag^+], (B) $CH_3OH/23.5 \times 10^{-3}$ M [Ag^+], (C) CH_3OH; detection 330 nm. Peak identity: (1) retinol; (2) retinyl propionate; (3) retinyl linoleate; (4) retinyl laurate; (5) retinyl oleate; (6) retinyl myristate; (7) retinyl palmitate; and (8) retinyl stearate. Reprinted with permission from DeRuyter and DeLeenheer (242), Anal. Chem. Copyright © 1979 by the American Chemical Society.

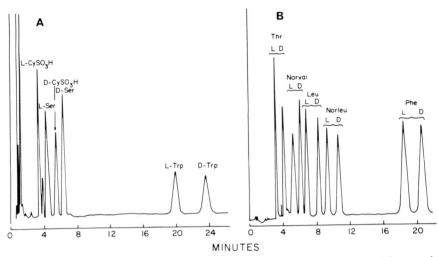

FIG. 56. Separation of D,L-dansyl amino acids. Conditions: 0.65 mM L-2-isopropyl-dien–Zn(II); 0.17 M NH$_4$Ac to pH 9.0 with aqueous NH$_3$; 35/65 CH$_3$CN/H$_2$O; T = 30°C; flowrate 2 ml/min; column: 15 cm by 4.6 mm i.d. 5 μm Hypersil C$_8$; solutes: CySO$_3$H = cysteic acid; Ser = serine; Trp = tryptophan; Thr = threonine; Norval = norvaline; Leu = leucine; Norleu = norleucine; Phe = phenylalanine. Detection at 254 nm. Reprinted with permission from LePage *et al.* (*246*), *Anal. Chem.* Copyright © 1979 by the American Chemical Society.

creased retention and improved reproducibility and dispersion in chromatography of insulin, glucagon and 1-24 ACTH pentaacetate (*170*) in addition to improving the chromatography of di- to decapeptides (*243*). Similar results have been obtained using trifluoroacetic acid (*244*) in the analysis of opioid peptides, e.g., pituitary endorphins. Cooke *et al.* (*245*) used a hydrophobic chelating agent, 4-dodecyldiethylenetriamine CH$_3$(CH$_2$)$_{11}$N(CH$_2$CH$_2$NH$_2$)$_2$, containing Zn^{2+} in acetonitrile–water as the mobile phase. This agent complexes with dansyl amino acids, dipeptides, or aromatic acids with enhanced retention. The peaks obtained are fairly narrow; this has been ascribed to the rapid kinetics of formation and decomposition associated with the outer sphere complexes that are known to exist in this system. In an extension of this work, asymmetric derivatives of this hetaeron have synthesized which permit the separation of dansyl amino acid racemates (*246*). Examples of such separations using L-2-isopropyl-dien–Zn(II) are shown in Fig. 56.

The scope of the technique can be estimated from Table XVII which is a representative, but not exhaustive, listing of the analyses in which hetaerons have been used. Ion-pair chromatography has been the subject of a

TABLE XVII

Partial Listing of Eluites Separated with the Aid of Hetaerons, Including the Mobile and Stationary Phases Used

Eluites	Hetaeron	Mobile phase	Ligate of siliceous stationary phase	Reference
Amino Acids and Peptides				
Amino acids	Anionic detergents	H_2O–MeOH	Octadecyl	237
Amino acids	H_3PO_4	H_2O	Octadecyl	248
Amino acid racemates	D- or L-proline + Cu(II)	H_2O	Octadecyl	249
Amino acid racemates	Asymmetric dien + Zn(II)	ACN–H_2O	Octadecyl	246
Dansyl amino acids	C_{12}-Dien–Zn(II)	ACN–H_2O (40:60)	Octadecyl	250
Thyroidal iodo-amino acids	H_3PO_4	H_2O–MeOH	Octadecyl	251
3,4-Dihydroxyphenylalanine derivatives	Heptanesulfonate	H_2O–MeOH–HOAc (95.5:05) or H_2O–HOAc (99:1)	Octadecyl	230
Tyramine, phenylethylamine, and tryptamine	Heptanesulfonate	MeOH–H_2O (35:65)	Octadecyl	235
Amino acids—dipeptides	C_{12}-dien + Zn(II)	CH_3CN–H_2O	Octyl	245
Peptides	R_4N RNH_2	MeOH–H_2O (50:50)	Octadecyl alkylphenyl	171
Opiates and opiate peptides	CF_3CO_2H	MeOH–H_2O formic acid	Octadecyl or alkylphenyl	252
Peptides	Dodecyl sulfate	2% PrOH	Octyl	217
Opiate peptides	CF_3CO_2H	H_2O–ACN–formic acid (10 mM)	Octadecyl	244
Peptides; LH-releasing factor, somatostatin, insulin, cytochrome c	Et_3N–PO_4	—	Octadecyl	253
Nucleotides				
Nucleotides	Mg(II)	H_2O	Octadecyl	116
Nucleotides	C_{12}–dien–Zn(II)	0.13 M NH_4OAc, pH 7.1 in ACN–H_2O (30:70)	Octyl	250
Mercaptopurine	Heptanesulfonate	MeOH–H_2O (10:90)	Octadecyl	254

(*Continued*)

TABLE XVII (*Continued*)

Eluites	Hetaeron	Mobile phase	Ligate of siliceous stationary phase	Reference
Catecholamines				
Catecholamines and phenylephrine	Heptanesulfonic acid	MeOH–H₂O (20:80)	Octadecyl	255
Catecholamines: epinephrine, norephrine, 3,4-dihydroxybenzylamine, dopamine	Heptanesulfonate	H₂O	Octadecyl	229
Catecholamines and derivatives	SDS, Na-1-dodecyl sulfonate, Na dodecyl benzenesulfonate	BuOH–HOAc–H₂O	SAS*	223
Catecholamines	Alkylsulfonates	H₂O	Octadecyl	207
Amines, catecholamines, tricyclic antidepressants, zimelidines	Nonyl(Me)₃NBr	Phosphate saturated with n-PrOH	Octyl	231
Pharmaceuticals				
Decongestants (phenylephrine-HCl, phenylpropanolamine-HCl and brompheniramine maleate)	Heptanesulfonate	CH₃CN–H₂O	Cyano	256
N,N-Trimethylene bis(pyridinium 4-aldoxine) dibromide and breakdown products	Heptanesulfonate	CH₃CN–H₂O	Octadecyl	257
Amylococaine, benzocaine, butacaine, cocaine, erythromycin	NH₄ HCO₂	MeOH–H₂O	Octadecyl	258
LSD, methamphetamine and amobarbital + secobarbital	Alkylsulfonate	MeOH–H₂O	Octadecyl	259
Cough syrup ingredients; guaiacolsulfonate and phenylephrine-HCl	Bu₄N⁺–PO₄	MeOH–H₂O	Octadecyl	260
Muscle relaxant—analgesic; methocarbamol and chlorzoxoneacetaminophen	Bu₄N⁺–PO₄	MeOH–H₂O (60:40)	Octadecyl	261
Tetrazolylchromone	Bu₄N⁺–PO₄	MeOH–H₂O (48:52)	Octadecyl	262
Antidiabetic agents: phenformin	Heptanesulfonate	MeOH–H₂O–HOAc	Octadecyl	263
Adrenaline	Hexanesulfonate	MeOH–H₂O (50:50)	Octyl	264
Paracetamol and (≥)3 metabolites	Dioctyl ammonium, Bu₄N	MeOH–H₂O–formic acid	Octadecyl	237
Sulfa drugs	Cetrimide	PrOH–H₂O (10:90)	Octyl	217
Quinine	Dodecyl sulfate	PrOH–H₂O (25:75)	Octyl	217

5,6,7,8-tetrahydrofolate, N-(p-aminobenzoyl)-L-glutamate

Compound	Reagent	Mobile phase	Column	Ref
Methotrexate and 7-hydroxymethotrexate	Bu_4N^+–PO_4	CH_3CN–H_2O (20:80)	Octadecyl	265
Vitamins				
Niacin, niacinamide, pyridoxine, thiamine, and riboflavin in multivitamin blend	Hexanesulfonate	MeOH–H_2O (25:75)	Octadecyl	266
B vitamins	Dodecyl sulfate	PrOH–H_2O (6:94)	Octyl	217
Unsaturates				
Vitamin D_2–D_3	$AgNO_3$	H_2O–MeOH (5:95)	Octadecyl	241
Estriol, equilin, estrone, and estradiol	$AgNO_3$	H_2O–MeOH (40:60)	Octadecyl	241
Unsaturated and heterocyclic compounds	$AgNO_3$	CH_3CN–H_2O MeOH–H_2O	Octadecyl	240
Dyes				
Tartrazine and intermediates	Bu_4N^+ Et_4N^+ Tridecylamine	H_2O–MeOH–CO_2H_2 (400:400:1)	Octadecyl	216
Dyestuff intermediates	Cetrimide	PrOH–H_2O	SAS*	53
Dyes	Cetrimide	PrOH–H_2O (25:75)	Octyl	217
Na cromoglycate	Alkylbenzyldimethyl-ammonium chlorides	MeOH–H_2O (50:50)	Octadecyl	267
Miscellaneous				
Aniline and metabolites	0.015 M NiOAc	MeOH–H_2O (15:85)	Octadecyl	268
Benzoic acids	C_{12}-dien–Ca(II)	CH_3CN–H_2O (30:70)	Octyl	246
Benzoic acid	Cetrimide	PrOH–H_2O (25:75)	Octyl	217
Amino benzoic acids	Zn(II)	MeOH–H_2O (4:96)	Octadecyl	269
Benzenesulfonic acids	Alkylammonium salt + Na_2SO_4	H_2O–MeOH	Octadecyl	270
Crown ethers	K^+, Na^+, Cs^+, Hg(II)	MeOH–H_2O	Octadecyl	271
Nitrophenol	Bu_4N–PO_4	MeOH–H_2O (45:55)	Octadecyl	272
α-Keto acid 2,4 dinitrophenylhydrazones	Bu_4–OAc	MeOH–H_2O	Octadecyl	273
Crown ethers	K^+, Na^+, Cs^+	MeOH–H_2O	Octyl, octadecyl, perfluoroheptyl	116

[a] "Short alkyl silica."

recent extensive review article (247). The subject of equilibria secondary to the actual chromatographic distribution equilibrium is extensively treated in LePage et al. (52).

VII. PHYSICOCHEMICAL MEASUREMENTS

A. Effects of Pressure on the Equilibria Involved in Chromatography

The magnitude of chromatographic retention is determined by the energetics of eluite interactions with both the mobile and stationary phases. Consequently retention data entail relevant thermodynamic information. According to Eq. (51) the retention factor, k, of a given eluite is proportional to the equilibrium constant, for the chromatographic distribution process. The proportionality constant is the so-called phase ratio, which is given by the relative magnitude of the stationary and mobile phases present in the particular column used. The close relationship between retention factor and equilibrium constant allows us to extract from the chromatogram thermodynamic information that may be useful in contexts other than chromatography.

Possible difficulties in obtaining accurate retention data for physicochemical measurements should be recognized, however. The evaluation of t_0, the elution time of an "unretained" peak (274) is often connected with systematic error and the measurement of the retention time of asymmetrical peaks may not be accurate. Moreover, no satisfactory methods are available for the precise evaluation of the phase ratio in the column. Consequently, the measurement of the equilibrium constant proper is beset with difficulties as discussed in Section VII,B.

The effect of pressure on chemical processes which occur in RPC has not been widely investigated. At first glance, the effect may appear to be significant as a consequence of the isothermal relationship between the equilibrium constant and pressure:

$$(\partial \ln K/\partial P)_T = -\Delta V/RT \tag{104}$$

where ΔV is the molar volume change for the process with equilibrium constant K. The change in the equilibrium constant upon changing pressure can easily be estimated from the integrated form of Eq. (104) which is given by

$$\ln(K_2/K_1) = \Delta V(P_1 - P_2)/RT \tag{105}$$

where K_1 and K_2 are the equilibrium constants at pressures P_1 and P_2, respectively. A simple calculation shows that the effect of pressure on retention can be neglected under conditions usually employed in HPLC. If we assume that the difference in the retention factors, i.e., equilibrium constants, is 5% in experiments conducted at average column pressures of 200 atm (~ 3000 psi) and 100 atm and at temperature of 300°K, respectively, the absolute molar volume change is given by

$$|\Delta V| \geq RT \ln(1.05)/ \Delta P \tag{106}$$

$$= 12 \text{ ml/mol}$$

This value is similar to that observed in many chemical processes, including ionization of weak acids in which case the volume change ranges in absolute value from 3 to 30 ml/mol (275, 276) with most values near 10 ml/mol. The above calculation shows that under such conditions the change in retention values due to variations in the column pressure would be relatively minor in chromatographic practice. For major molecular rearrangements, such as reversible denaturation of proteins, absolute values of volume change are much greater, yet less than 100 ml/mol (277–279). Such large molar volume differences may result in retention factor changes of 50% under the above conditions.

The above considerations apply to differences in the average column pressure. Large axial pressure gradients, when the flowrate through the column is high, may cause significant amounts of viscous dissipation with concomitant increase in average column temperature and in nonuniformity of radial temperature profiles. This effect has received a detailed investigation by Lin and Horváth (280).

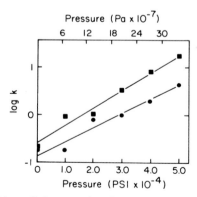

FIG. 57. The logarithm of the retention factor of methyl red (circles) and methyl orange (squares) in 50 vol % methanol (1 atm, pH 7) increases with inlet pressure. The data were obtained on Bondapak C_{18}/Corasil at room temperature. Reprinted from Prukop and Rogers (281), with permission from Marcel Dekker.

Prukof and Rogers (281) examined the effect of high pressure on the retention of methyl red and methyl orange by using an octadecyl silica column and 4-methoxy-2-nitroaniline in the presence and absence of ion-pairing reagents in 50% methanol. Data obtained in the absence of hetaeron are shown in Fig. 57. In the case of methyl red, increase of operating pressure from 1 to 3000 atm resulted in a retention factor nearly four times higher than that obtained at the lower pressure, whereas the corresponding increase for methyl orange was approximately sevenfold. Similar results were obtained in the presence of ion-pairing reagents. The numbers correspond to partial molar volume changes of -10.5 and -13.0 ml/mol, respectively. The volume changes need not be so large; that of 4-methyl-2-nitroaniline was about -4 ml/mol. As impressive as these effects are, it should be remembered that the operating pressure in HPLC is usually less than 300 atm and an increase from atmospheric pressure to 300 atm would result in a less than 20% change in the retention factor.

Another difference between the conditions of Prukof and Rogers (281) and those encountered in practice, is that they operated the column at a nearly uniform pressure throughout its length by using a suitable restrictor at the outlet. In chromatographic practice, however, the column outlet is at near atmospheric pressure so that there is an axial pressure gradient. As a consequence, an appropriate method for averaging the pressure effects must be used to estimate partial molar volume changes in ordinary practice. In any case the average pressure in the column is much lower than the inlet pressure so that the change in the retention factor, or in a physicochemical constant calculated from retention data, from that determined at atmospheric pressure is much less than the preceding calculations which used the inlet pressure would suggest.

B. Evaluation of Enthalpy and Entropy Changes

The entropy and enthalpy changes associated with the reversible binding of the eluite by the stationary phase are defined by applying the van't Hoff relationship to the chromatographic process as follows

$$\Delta S° = (\Delta G° - \Delta H°)/T = -R \ln k + R \ln \phi + \frac{R}{T}\frac{d(\ln k)}{d(1/T)} \quad (107a)$$

and

$$\Delta H° = -R\frac{d(\ln k)}{d(1/T)} \quad (107b)$$

As a result the dependence of the logarithm of the retention factor on the absolute temperature is given by

$$\ln k = -\frac{\Delta H^{\circ}}{RT} + \frac{\Delta S^{\circ}}{R} + \ln \phi \qquad (107c)$$

If the heat capacity change upon binding the eluite to the stationary phase is zero and the phase ratio ϕ is independent of temperature, then a plot of the logarithm of the retention factor against the reciprocal absolute temperature according to Eq. (107c), the so-called van't Hoff plot, is linear. In such cases the enthalpy change, ΔH°, can be obtained directly from the slope of the line obtained by plotting experimental data.

The evaluation of the corresponding entropy change, ΔS°, from the intercept is difficult because the phase ratio is usually not known. The necessity of determining the value of ϕ for the evaluation of Gibbs free energy and entropy are obvious from Eq. (107c). When a liquid stationary phase is used, the evaluation of ϕ is fairly straightforward since the value is given by the volume ratio of stationary and mobile phases in the column and both volumes are at least in principle measurable. With bonded phases, however, the "volume" or some other equivalent property of the stationary phase is not clearly defined.

Whereas Eq. (107c) and the methods adumbrated for the evaluation of binding enthalpy and entropy are valid when the binding heat capacity change is zero, that condition is not always satisfied. When the heat capacity change is nonzero, in which case the binding enthalpy is itself temperature-dependent, the van't Hoff plots will be curved. The retention enthalpy, which must be evaluated by use of the differential form of the van't Hoff relationship that is given by Eq. (107b) for chromatographic measurement, is obtained at each temperature from the slope of a line tangent $d \ln k/d(1/T)$ to the curve which describes the experimental data in the van't Hoff plot. The corresponding entropy change at each temperature is estimated by use of Eq. (107a). Curvature of van't Hoff plots is not commonly observed in RPC. It can be expected to occur whenever the concentrations of eluite bound to each of two or more different kinds of sites are similar or when the eluite can exist in two or more temperature-dependent conformations. A simple example of the latter effect is given in Section IV,C and an experimental example has been intensively examined (158). As shown in these, the observed retention enthalpy can be temperature-dependent although enthalpy of binding for each species and the enthalpy for the conformation change are each independent of temperature. This is a consequence of the nonlinear coupling of the enthalpies in the expressions for mass conservation [see Eq. (19)]. It should be noted, however, that apparently linear van't Hoff plots can be obtained even when multiple binding sites or eluite forms are present. Consequently, the retention enthalpy so obtained reflects the enthalpy

changes for the various phenomena which are involved in the chromatographic retention process. The enthalpy for conformation change can be
determined in an extrachromatographic experiment and, in principle, the
binding enthalpies of the individual forms can be determined from the limiting slopes at the extremes of temperature where only one form is present
to any sensible extent (158).

The evaluation of the phase ratio with bonded phases can involve the
measurement of mobile phase volume in the column and the determination of the surface area of the stationary phase that is accessible to the
eluite, together with the surface concentration of the covalently bound
ligates. The surface area of the stationary phase is usually estimated by
the BET method. The mobile phase volume can be estimated directly
from the elution volume and a "packing factor" characteristic for the stationary phase and the packing procedure. Alternatively, the volume of
bonded stationary phase can be estimated from the carbon content and
density as well as the mass of stationary phase in the column (178).

Neither of the above methods is immune to criticism. Area determination by the BET method measures the surface accessible to small molecules such as those of nitrogen. In porous silica, a significant fraction of
pores may be large enough to permit the passage of nitrogen molecules
but too small to accommodate larger molecules of chromatographic interest. As a consequence BET data are expected to overestimate the surface
area accessible to even relatively low molecular weight sample components in liquid chromatography. This might be acceptable if the ratio of
the chromatographically accessible surface area to the BET area is a constant. However this assurance does not exist and the constancy required
is unlikely to be found between silicas produced by different processes.
Consequently any estimation of "stationary phase volume" on the basis
of the BET surface area of the support is likely to be inaccurate. Development of novel methods for the determination of phase ratio with columns
packed with bonded stationary phases is clearly desirable and may be
imperative for accurate determination of changes in entropy and Gibbs
free energy upon reversible binding of eluites to bonded stationary
phases.

The uncertainty in the measurement of elution time t_0 or elution volume
of an unretained tracer is another potential source of error in the evaluation of thermodynamic quantities for the chromatographic process. It can
be shown that a small relative error in the determination of t_0, will give
rise to a commensurate relative error in both the retention factor and the
related Gibbs free energy. Thus, a 5% error in t_0 leads to errors of nearly
5% in both k and ΔG. An analysis of error propagation showed that if the

relative error in t remains constant with changing temperature, it results in a commensurable error in the enthalpy determination, i.e., 5% in the above example, where the error in elution time of an unretained eluite arises from systematic causes and not from temperature-dependent chemical effects. If, however, the "unretained eluite" interacts, albeit very weakly, with the stationary phase, and the enthalpy of binding comparable in absolute value to that of the eluite of interest, the resulting error in the binding enthalpy of the retained eluite can be as large or even larger than that postulated. For instance, when this type of error in t_0 is of the order of 5%, the error in the enthalpy can range from 5 to 15%, depending upon the respective signs of the enthalpies of retained and "unretained" eluites. Although estimates of error made in this fashion are conservative, they serve as caveat that great pains should be taken to reduce or control the uncertainty in data taken for physicochemical measurements and that the data should be interpreted with care and circumspection.

The determination of both the enthalpy and entropy change associated with a process is of value in determining whether a given chemical interaction, or a series of interactions, has a fundamental similarity to another set. For example, knowledge of the pH dependence of a reaction together with the enthalpy of ionization for each ionogenic group of potential kinetic significance allows one to determine by comparison with corresponding data for reference compounds the ionizations of which groups, e.g. a carboxyl or amine group, are kinetically important. This concept has been extended in the use of linear free energy relationships which postulate that the free energy change in one process is linearly related to that in another process if the underlying chemical features of the reactions are fundamentally related. Leffler and Grunwald (179) have discussed many examples for this phenomenon that is also responsible for the so-called enthalpy–entropy compensation. Compensation behavior is observed whenever similar transformations of various molecules, e.g. binding to a surface, differ in their respective enthalpies and entropies of reaction in such a way that a plot of the enthalpy against the entropy is linear.

Enthalpy–entropy compensation, is expected to occur only if unity of mechanism exists. From the slope of the enthalpy vs entropy plot the so-called compensation temperature, which is characteristic for the type of transformation under investigation, can be calculated.

Compensation temperatures have been found to be identical for the retention of a variety of eluites on three different types of reversed phase columns over a wide range of eluent composition (177). The authors noted that if compensation behavior occurs in chromatography, the retention

TABLE XVIII

Enthalpy–Entropy Compensation in RPC Using Octadecyl Silica as the Stationary Phase[a]

Eluite	Eluent	Column	T (°K)	Intercept	β (°K)	Range of β at 0.5 P (°K)
Aromatic carboxylic acids	50 mM aqueous phosphate buffer, pH 2.0, with and without 6% (v/v) acetonitrile	5 μm Spherisorb ODS	308	-2.97 ± 1.08	647	539–897
Substituted hydantoins, allantoin, and phenylacetic acid	50 mM aqueous phosphate buffer, pH 7.0, with and without 30% (v/v) acetonitrile	10 μm Partisil ODS-2	313	-2.17 ± 0.49	596	490–745
Substituted benzene derivatives (Data from Ref. 178)	Water–methanol 60:40 (v/v)	Permaphase ODS	313	-3.13 ± 0.56	639	554–755

[a] The intercepts of the compensation plots according to Eq. (108) and the compensation temperatures, β, calculated from the slope are indicated. Reprinted with permission from Melander et al. (177).

factor at temperature T is related to the enthalpy of binding to the stationary phase $\Delta H°$ by the relationship

$$\ln k = \frac{\Delta H°}{R}\left(\frac{1}{T} - \frac{1}{\beta}\right) - \frac{\Delta G_\beta°}{RT} + \ln \phi \tag{108}$$

where β is the compensation temperature and $\Delta G_\beta°$ is the free energy for the reversible binding at that temperature. From Eq. (108) it follows that a plot of the logarithm of the retention factors against the enthalpies obtained with a given chromatographic system and various eluites yields a straight line as long as the retention mechanism is invariant. The compensation temperature for the retention process can be readily evaluated from the slope. If the same value of the compensation temperature is obtained from measurements made with various chromatographic systems, i.e., under different conditions as far as the stationary and mobile phases are concerned, we can infer that major mechanistic features of the retention process are identical throughout.

The results shown in Table XVIII imply that with octadecyl silica as the stationary phase the same binding mechanism occurs in pure water as well as in water–methanol and water–acetonitrile mixtures up to at least 40% (v/v) of organic component. However, the results do not allow us to draw conclusions as to whether the binding occurs by adsorption or by partitioning into the octadecyl silica stationary phase under conditions of the study. Nevertheless from an analysis of the data of Frank and Evans (*182*) on the solubility of hydrocarbons in water we may infer that partitioning is unlikely. Their data reveal that the benzene–water partition coefficient for low molecular weight solutes does not show clear compensation behavior and the 450°K compensation temperature calculated from their data is significantly lower than that observed in RPC with octadecyl silica stationary phases (*177*).

C. Correlation between Retention Factor and Partition Coefficients Used in Quantitative Structure–Activity Relationships

In spite of the preceding observation that eluite retention in RPC with hydrocarbonaceous bonded phases may not occur by partitioning of the eluite between two liquid phases, theoretical considerations based on the solvophobic treatment of solvent effects shows that it might be possible to relate the observed retention factors to partition coefficients between water and an organic solvent. Such a relationship would be quite useful in light of the scale developed by Hansch and his co-workers (*282, 283*) to characterize hydrophobic properties of drugs and other biologically active

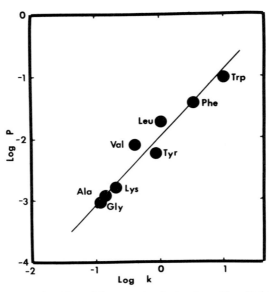

FIG. 58.   Plot of the algorithm of the retention factor, k, and log P, the water–n-octanol partition coefficient of eight amino acids. The chromatographic data were obtained on 5 μm LiChrosorb RP-8, 250 × 4.6 mm i.d.; eluent; 0.1 M aqueous phosphate buffer, pH 6.7, $T =$ 70°C. Eluites: Trp, tryptophan; Phe, phenylalanine; Leu, leucine; Val, valine; Tyr, tyrosine; Lys, lysine; Ala, alanine; Gly, glycine. Reprinted with permission from Molnar and Horváth (203).

substances on the basis of their partition coefficients evaluated in the system n-octanol–water. In this approach the logarithm of the partition coefficient, log P, is determined for molecules of interest and the useful-ness of the scale has been extended by establishing a substituent parame-ter, π, for each group in the solute molecule group. The latter is defined for a given substituent as the difference in log P values between a mole-cule containing that group and the parent compound. This approach, like enthalpy–entropy compensation, is based on linear free energy relation-ships (179). It has found use in understanding the interactions of small so-lutes with macromolecules and has been especially valuable in estab-lishing quantitative structure–activity relationships, QSAR, for drug de-sign. Rekker (284) has developed a hydrophobic index system similar to that of Hansch that is supposed to account for molecular folding.

There is experimental evidence that under certain conditions RPC may possibly be used as an alternative method to evaluate a parameter equiva-lent to log P. Linear relationships between logarithm of the retention factor and carbon number are anticipated in this optic and have been observed for alcohols (*159*), alkyl benzenes, alkyl bromides, alkyl disulfides, chlorbenzenes and methyl benzenes, alkanes, *n*-alcohols, ethers, esters, amides, and ketones (*143, 148, 160, 172*) *inter alia*. Differences in the chromatographic behavior of catecholamines and related species could be quantitatively interpreted in terms of contributions due to substituents (*200*). A simple relationship between the logarithm of the retention factor of an eluite and its corresponding log *P* value has been found by Molnár and Horváth (*203*) for amino acids as seen in Fig. 58. Tanaka *et al.* (*86*) furnished additional support for the existence of this relationship in RPC of various benzene derivatives with hydroorganic eluents containing methanol, acetonitrile, or tetrahydrofuran.

The evaluation of partition coefficients by HPLC has been expected to yield data for linear free energy relationships used in the establishment of structure–function or structure–physiological response relations. Carlson *et al.* (*215*) studied the retention of several anilines and phenols on reversed phase using acetone–water as the mobile phase and showed that the logarithm of the retention factor extrapolated to pure water could be directly related to log *P* for the particular group of substances investigated. It has also been suggested (284a) that logarithms of retention factors can be as useful as log *P* data in predicting biological effects. This was confirmed by the results of Yamana *et al.* (*285*) who examined the retention factors of penicillins and cephalosporins on octadecyl silica columns. The data were obtained with water–methanol and plots of log *k* vs eluent composition were extrapolated to obtain retention factors appropriate to pure water. These values, as well as those obtained in 30% methanol, correlated well with values of log *P* and this technique was used to determine log *P* for some highly ionizable β-lactams as these values could be obtained only with the greatest of difficulty with the classical partitioning method. Baker has also shown that the logarithm of the retention of several classes of drugs is related the log *P* factor values (*286*).

Retention in RPC with 34–60% (v/v) methanol–water as the mobile phase has been shown to correlate well to log *P* by Riley *et al.* (*192*) at each mobile phase composition. A mild dissent has been made by investigators (*289*) who found a correlation between RPC retention and log *P* but who claimed a closer relationship between the corresponding data obtained in thin-layer chromatography in the system oleyl alcohol–water and log *P* (*289*).

Direct chromatographic determination of octanol–water partition coefficients has been attempted using a Corasil I column impregnated with octanol (*287*) or a reversed-phase column which had been coated with octanol (*288*). In either case, the sample was injected into an aqueous mobile phase. When the octanol-coated octyldecyl silica support was used, the logarithm of the relative retention time was highly correlated with the logarithm of the octanol–water partition coefficients.

The exact nature of the relationship between chromatographic retention and the corresponding log P remains to be fully explicated and clarified. There is ample basis, however, to expect chromatography to develop into a powerful tool for extrabiological understanding and prediction of *in vivo* effects of pharmaceuticals and toxic agents and for measurement of biodynamic hydrophobicity. In a similar fashion, it may also serve as a tool for screening solvent extraction schemes of technological import. Of course, HPLC as an analytical technique *par excellence* can be conveniently used to measure concentration in the two phases when the conventional shake-flask method or its variant is used for the determination of log P values (*290*).

D. Measurement of Equilibrium Constants for Association Processes in Solution

Under certain conditions chromatographic retention measurements can be used to determine the magnitude of equilibrium constants for reversible associations between a sample component and a complexing agent present in the eluent. The use of migration rate data to obtain equilibrium constants for interactions between eluents of the solvent and the solute is not novel. It has been developed for paper electrophoresis (*291, 292*).

Chromatographic measurements discussed here are based on the evaluation of the effect of the concentration of the complexing agent (hetaeron) in the eluent on the retention factor of the eluite as treated in Section VI. If the stoichiometry of the hetaeron–solute association process is 1:1. then Eq. (89) of Section VI can be used to calculate the retention factor. A more general expression is available for the scheme presented in Fig. 59. In view of the equilibria involved mass balance yields an expression for the retention factor as follows

$$k = \frac{k_0 + \sum_{i=1}^{n} k_i [\mathscr{H}_m]^i \prod_{j=1}^{i} K_j}{1 + \sum_{i=1}^{n} [\mathscr{H}_m]^i \prod_{j=1}^{i} K_j} \tag{109}$$

where K_j is equilibrium constant for the jth binding, which forms $E\mathscr{H}_j$, $[\mathscr{H}_m]$ is the concentration of hetaeron in the mobile phase, k_i is the retention factor of the form of eluite combined with i hetaeron molecules per molecule and k_0 is the retention factor in absence of hetaeron (*204*).

In many cases it is convenient, or more conventional, to write Eq. (109) in terms of the corresponding dissociation constants. For the jth equilib-

$$E + L \xrightleftharpoons{K_1} EL$$

$$EL + L \xrightleftharpoons{K_2} EL_2$$

$$EL_2 + L \xrightleftharpoons{K_3} EL_3$$

.   .   .   .

.   .   .   .

.   .   .   .

$$EL_{n-1} + L \xrightleftharpoons{K_n} EL_n$$

FIG. 59.   A general representation of the binding of n molecules of hetaeron, L, to an eluite E. Each binding step has an associated equilibrium constant, K_i, for the formation of species containing i molecules of hetaeron from the reaction of a species containing $i - 1$ molecules of hetaeron with hetaeron. As a consequence, the retention factor can have a complex dependence on hetaeron concentration; the relationship is given by Eq. (109).

rium, the dissociation constant, $K_{k,j}$ is related to the association constant by

$$K_{d,j} = 1/K_j \tag{110}$$

so that the retention factor can be expressed by

$$k = \frac{k_0 + \sum_{i=1}^{n} k_i [\mathscr{H}_m]^i \prod_{j=1}^{i} K_{d,j}}{1 + \sum_{j=1}^{n} [\mathscr{H}_m]^i \prod_{j=1}^{i} K_{d,j}} \tag{111}$$

Similar expressions have been obtained for the particular cases of mono-protic acids and bases, diprotic acids and bases, and zwitterions (207, 208), and in each case the data conformed well to Eq. (111). It has also been shown (207) that the acid dissociation constants can be determined by using reversed phase chromatography. The pK_a values of 10 aromatic acids calculated from chromatographic data by employing Eq. (91) were

TABLE XIX

Comparison of the pK_a Values of Organic Acids as Obtained by Least-Squares Analysis of Chromatographic Data and by Potentiometric Titration[a]

| Acid | pK_{a_m} | | pK_a | |
	Chromatography	Titration	Literature	Ref.
Benzoic	3.93	3.78	4.19 $(I = 0)$[b]	293
3,4-Dihydroxyphenylacetic	4.20			
Homovanillic	4.29			
p-Hydroxyphenylacetic	4.30			
Mandelic	3.49	3.46	3.37	294
Phenylacetic	4.14	4.10	4.31	295
Salicylic	2.84	2.92	1.88 $(I = 3)$	296
Vanillylmandelic	3.25			
o-Phthalic	3.44	3.20	2.95 $(I = 0)$	295
	5.10	4.79	5.41 $(I = 0)$	295
Anthranilic	2.21		2.09 $(I = 0)$	297
	5.30	4.74	4.79 $(I = 0)$	297

[a] The chromatographic conditions are stated in Fig. 42. Available literature data are also listed. Reprinted with permission from Horváth *et al.* (*207*), *Anal. Chem.* Copyright © 1977 by the American Chemical Society.
[b] I = ionic strength.

in good agreement with those obtained by independent potentiometric determination of the equilibrium constants as shown in Table XIX.

Equation (111) has been used to evaluate pK_a values of some pteroyl-oligo-γ-L-glutamates. These species, which are usually not sufficiently pure or soluble for potentiometric titration, have several carboxylic groups with similar pK_a values in the molecule and therefore the determination of their acid dissociation constants is beset with great difficulties. The chromatographic approach outline here, however, yielded pK_a values which were in good agreement with those obtained for carboxylic groups in similar molecules (*204*). The agreement between experimental and calculated data is shown on the graph in Fig. 47.

The stability constant for the complex formed from ATP^{4-} with Mg^{2+} was determined from the dependence of the retention factor of ATP on the concentration of $MgCl_2$ in the eluent (*298*). The method has been extended to the measurement of similiar constants for complexes involving other nucleotides, crown ethers, and nitrosonaphtholsulfonic acids. The stability constants measured by the chromatographic approach were in good agreement with literature data obtained by other techniques. Some of the difficulties and pitfalls inherent in this method have been discussed (*116*). The results obtained so far strongly suggest that HPLC in the reversed phase mode can be used to obtain physicochemical data. Such data are significant, in contexts other than chromatography, particularly as far as the properties of biologically important complex molecules are concerned.

VIII. SELECTED ANALYTICAL APPLICATIONS

A. General Remarks

The use of nonpolar (reversed-phase) bonded phases has increased in a nearly explosive fashion since the introduction of HPLC as the most versatile chromatographic technique using sophisticated instrumentation. This is due in part to the high efficiency of microparticulate bonded phases which is superior to that of conventional column packings such as ion-exchanger resins. The use of secondary chemical equilibria, e.g. with detergents, has expanded the use of RPC to include the analysis of highly polar molecules which would otherwise be poorly retained and poorly resolved.

In this section a variety of analytical separations reported in the literature are reviewed to show the wide structural diversity of eluite which can be separated by RPC and to assist the reader in becoming similar with the use of this fluid chromatographic technique. The descriptions are arranged according to the matrix in which an analyte is found or the area of chemistry in which the samples are generally encountered. Thus theophylline, for example, is regarded as a nucleotide and, for the most part, its analysis in food samples is found with appropriate cross references. On the other hand, the separations of pharmaceuticals found in serum, urine, and pharmaceutical samples are cited separately. It is hoped that this method of classification may serve the purposes of those whose analytical interests are incidental to their primary research pursuits.

An indicator of the resolution inherent in RPC is given by the separation of various organic molecules from their perdeuterated congeners. The separation of deuterium or tritium-containing molecules from their protium-bearing analogs has been reported for perdeuterated benzene, toluene (205), and tritiated polynuclear aromatics (206). Tanaka and Thornton (159) have reported that a variety of perdeuterated fatty acids, alkanes, and cyclohexane all elute earlier than the corresponding protiated substance. This result is in agreement with those of the other workers cited. The resolution obtained demonstrates the sensitivity of RPC to nuance in molecular structure of the eluite.

Identification of unknown samples in gas chromatography can be facilitated by the use of Kovats indices (*299*) in which the retention observed is related to those of selected standard eluites. Attempts have been made to develop such a system for RPC in which *n*-2-ketones were used as standards. These solutes were assigned retention index values according to the number of methylene carbons (*202*). The system was applied to some rather polar species with poor results; the index of a single molecule changed by the equivalent of nearly two methylene groups as the solvent composition changed. A similar system was proposed earlier in which alkyl phenones were used as standards (*300*). These methods do not seem to consider adequately the effect of solvent change on polar groups in determining retention. A method for relating factors obtained under a wide variety of conditions including solvent composition and temperature to each other has been described for alkanes (*301*). A method for including the effect of polar groups has also been outlined (*302*). This last method is probably more likely to lead to a system analogous to the Kovats index system than all the earlier approaches described because it explicitly includes the effects of solvent composition and temperature on each functional group of a molecule. As a consequence, it may be possible to relate retention data obtained with hydroorganic mixtures to those obtained with water and thereby use a water-based retention index system for all RPC data. Of course, an alternative mobile phase may prove to be a better choice of an appropriate reference solvent.

The use of hetaerons such as those employed in ion-pair chromatography is not extensively discussed here; it is covered in Section VI. The reader is referred to that section and Table XVIII, in particular, for a partial listing of separations effected by exploiting secondary chemical equilibria.

The list of separations given here cannot be regarded as exhaustive. Many separations done in RPC are reported in such a fashion that it is difficult to discover them as RPC separations using standard methods of searching the literature. Furthermore, the reports of separations of certain chemical classes, e.g. amino acids or catecholamines, are so prevalent that only a fraction of the references were used in the interest of keeping the reference list to manageable size.

B. Analysis of Physiological Samples

The analysis of samples obtained from physiological matrices has been the object of many reports on the use of RPC. This use can be expected to expand in response to the need to understand the pharmacodynamics and pharmacokinetics of various pharmaceutical agents. In addition, the versatility of RPC will facilitate the analysis of various metabolites and may thus become an especially sensitive tool for the diagnosis of disease state.

1. Serum and Plasma Analyses of Drugs

A variety of drugs have been assayed in clinically interesting matrices including serum or plasma. A large literature exists on the determination of antibiotics. Various cephalosporins have been analyzed, including cephradine and cephalexin in blood, urine, and pharmaceutical preparations (303, 304), cephaloridine from serum and renal cortex homogenates (305), and cefazolin from serum (306) as well as cephalothin (307) using RPC with methanol–water mixtures containing ammonium salts or with methanol–acetic acid. Acidified aqueous ethanol was used as mobile phase for the determination of sulfisoxazole (308), whereas other sulfonamides were analyzed with acidic methanol in the mobile phase, in the case of sulfapyridine and its metabolites (309), and with acidic acetonitrile for sulfadiazine, sulfamerazine, and sulfamethazine (310). However no acidification was necessary in the determination of sulfamethoxazol and N^4-acetylsulfamethoxazole in aqueous methanol or acetonitrile (311). Salicylazosulfapyridine and sulfapyridine in plasma have also been analyzed (312). The retention and separation of diastereomers of penicillins (D- and L-ampicillin and D- and L-α-phenoxyethylpenicillin) and cephalosporins (D- and L-cephalexin) were found to depend upon the pH and methanol content of the mobile phase (313). LiChrosorb RP8 was used in the analysis of the antibiotics amoxycillin and ampicillin (313).

Other antibiotics studied using RPC include chloramphenicol which could be determined at serum concentrations of about 0.5 μg/ml using 0.1 ml samples (315) after extraction with ethyl acetate (316). A simple method for the analysis of chloramphenicol in serum and cerebrospinal fluid has been reported in which the analyte is extracted into methanol and the extract chromatographed with acetic acid–water–methanol (1:62:37) as the mobile phase (317).

The antifungal agent griseofulvin in plasma (318) in plasma concentrations of 50 ng/ml could be measured using 10-μl samples. The bacteriostat, 3,4,4'-trichlorocarbanilide, and its metabolic products have been determined subsequent to a single-step sample cleanup (319). The trichomonacide metronidazole, misonidazole, and their metabolites have been analyzed from serum and urine samples with similar sensitivity (320).

The presence of tetracyclines in urine and plasma can be readily and sensitively determined (321). Doxycycline, tetracycline, oxytetracycline, demethylchlortetracycline, and methacycline analyses were found to be highly sensitive to the pH of the medium (322). Determination of aminoglycoside antibiotics include gentamicin, sisomicin, and netilmicin utilizing ion-pair chromatography on reversed phase has been reported (323). Amphotericin B obtained from serum or cerebrospinal fluid has been analyzed subsequent to extraction into methanol (324). Binary solvents have been used effectively in antibiotic analyses (325, 326).

A large variety of agents used in cancer treatments has been examined on reversed phase. The alkylating agent melphalan was determined in acidic methanol–water (1:1) at the 50 ng/ml plasma level (327). ACNU, a pyrimidine-containing nitrosourea with antitumor activity was analyzed in methanol–water (1:1) containing ion-pairing agents (328). 5-Fluorouracil and 1-(2-tetrahydrofuryl)-5-fluorouracil in blood have been analyzed in 10% aqueous methanol subsequent to extraction from blood by acetonitrile (329). A great deal of attention has been paid to the analysis of methotrexate, MTX, a folic acid antagonist used as a cytostat. These include the study of Chatterji and Gallelli (330) in which analyses of MTX and also its degradation products were developed. A sensitive analysis with fluorometric detection can be done if MTX is oxidized prior to analysis (331). The use of ion-pairing agents and ultraviolet detection provides a simple but sensitive method for the determination of MTX impurities in preparations and its metabolites in urine (265).

RPC has found use in the analysis of barbiturates including the determination of drugs taken in an overdose (332). Thiopental was determined using a mobile phase comprised of methanol–0.1% sodium citrate buffer, pH 6.5 (45:55) (333). Hydantoins, along with other species which have anticonvulsant activity, have been determined with barbiturates. These include phenytoin in the presence of phenobarbital and primidone (334, 335) and the related anticonvulsants ethosuximide and carbamazepine (336).

Determination of various analgesic and antipyretic pharmaceuticals on reversed phase has included not only the analysis of serum levels of aspirin, salicylic acid and salicyluric acid using acidified acetonitrile (337), or methanol (338), but also sulfinpyrazone under isocratic conditions (339), and 6-chloro-α-methylcarbazole-2-acetic acid (340). The polar thiol metabolites of acetaminophen were analyzed by RPC and the method was found to be superior to other chromatographic techniques used in this analysis (341).

Diazepam, oxazepam, and N-desmethyldiazepam have been determined under isocratic conditions (342); a somewhat more sensitive assay for diazepam and N-methyldiazepam has been reported (343). The closely related compounds, chlordiazepoxide and its N-demethyl metabolite, have also been determined (344). Analysis of the antidepressant amitriptyline, its metabolites and related drugs have been carried out using a four-component solvent (345, 346) and also aqueous acetonitrile (347).

Tolbutamide and phenformin, oral antidiabetic agents, have been analyzed directly on μBondapak C_{18} using methanol–water which contained
1-heptanesulfonic acid in addition for the phenformin determination (263).

Analysis of vitamins in serum samples include that of retinoids (vitamin
A and its analogs) in studies on their metabolism in vitamin-A-deficient
hamsters (348) as well as the determination of vitamin A (349) with fluorometric detection and the determination of 13-cis retinoic acid and all-trans
retinoic acid (350). Reversed phase has been used to determine vitamins
D_2 and D_3 and also to separate 25-hydroxyvitamin D_2 from 25-
hydroxyvitamin D_3 after the prior separation in vitamin D and 25-
hydroxyvitamin D fractions (351). An octadecyl-coated column has also
been used for the determination of total 25-hydroxyvitamin D (352). Desmosterol, cholesterol, and other sterols have been separated in RPC in
~ 10 min (353). The conjugated and oxidized metabolites of diethylstilbesterol have been resolved with octyl and octadecyl silica as stationary
phases using water–methanol gradients (354).

Other analyses done on serum samples include the use of RPC for the
determination of cannabinoids (355) in post mortem analyses of fatally in-
jured drivers (356). Reserpine was analyzed by ion-pairing subsequent to
oxidation to a fluorophore (357).

2. Metabolic Studies

The determination of naturally occurring metabolites and those of drugs
in serum is frequently of interest in studies of the metabolic fate of pharmaceutical agents and in studies of their rates of interconversion and distribution to various tissues. Reversed-phase high-performance chromatography has been used to monitor the efficiency of hemodialysis by determining metabolites in the hemodialyzate as well as serum and urine (358).
Methods for the derivatization and determination of serum pyruvic and
α-ketoglutaric acids have been reported (359). A variety of glycosphingolipids and phospholipids have been analyzed as perbenzoyl or biphenyl-
carbonyl derivatives or in the free state (360). The analysis of ubiquinone,
coenzyme Q, on reverse-phase is much more rapid than the alternative
methods and the coenzyme can be detected by either ultraviolet absorption (361) or fluorescence (362).

Metabolism and pharmacokinetics of piprozolin were studied by RPC
without prior clean-up of the urine sample (363). Misonidazole, metronidazole, and their metabolites have been analyzed using a ternary solvent
system containing methanol–acetonitrile–5 mM KH_2PO_4, pH 4, at volume ratios 4:3:93. The results facilitated the determination of the respective serum levels in a particular regimen of chemotherapy (364). The sig-

nificance of the analysis of sulfisoxazole from plasma for clinical monitoring and pharmacokinetic studies has been noted (365).

Warfarin has been determined from plasma (366, 367). Studies of its metabolic fate due to the action of hepatic mixed function oxidases have profited from the use of reversed-phase columns for the determination of metabolites (368). Related studies on warfarin resistance have utilized vitamin K_1 2,3-epoxide to examine the rate of its reduction in rats to form vitamin K_1; the reaction was followed chromatographically (369).

The pharmokinetics of propranolol and that of its major active metabolite, 4-hydroxypropranol, have been studied in human plasma (370).

3. Biogenic Amines

Analysis of biogenic amines has received much attention of late. Molnar and Horváth (180) demonstrated that catecholamines can be separated on reversed phase columns using plain aqueous solvents containing only salts but no organic solvent. The retention factors and selectivity was modified by changes in the pH of the eluent. Separation schemes for vanillylmandelic acid and homovanillic acid (371), norepinephrine and dopamine (372, 373), and adrenaline and noradrenaline (374) have since been reported. The fluorometric determination of adrenaline can be used to advantage in the presence of large excess of noradrenaline. Ion-pair chromatography has been used by several groups for the determination of catecholamines (180, 207, 223, 229, 231, 255). Histamine has been determined as its o-phthalaldehyde derivative (375). Dopa, dopamine, noradrenaline, and cysteinyl dopas have been determined in sera of patients with malignant melanoma (209).

Much effort has been expended in the development of more sensitive methods for the analysis and detection of catecholamines. They have been analyzed as the dansyl derivatives (376) or after precolumn derivatization with o-phthalaldehyde (377, 378). Postcolumn derivatization followed by fluorometric analysis have been described in which the fluorophore was formed with o-phthalaldehyde (379) or with 9,10-dimethoxyanthracene-2-sulfonate as the ion-pair (380). Several laboratories have shown the sensitivity and specificity in electrochemical detection methods (381–383).

4. Urine Analysis

The analysis of many drugs and metabolites in urine samples has been reported also as the analysis of these substances in serum or plasma samples. Therefore, such works are cited in Section VIII,B,1 and will not be repeated here. Urinary components can be useful indicators of one's metabolic state and therefore their analysis is frequently helpful in estab-

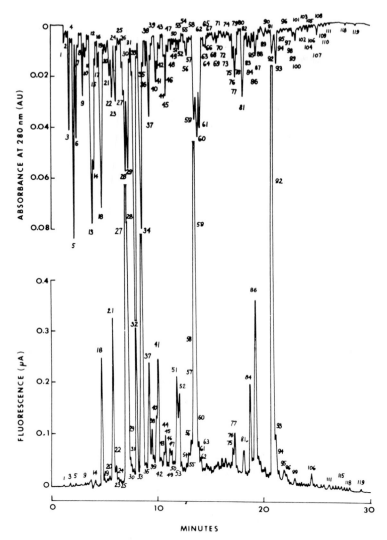

FIG. 60. Chromatogram of acidified urine extract. Column: 5-μm octadecyl silica, 25 cm × 4.6 mm i.d., temperature, 70°C; flowrate, 2.0 ml/min. Gradient elution from 0.1 *M* phosphate buffer, pH 2.1, with acetonitrile to about 40% (v/v) organic solvent concentration. Sample size 10 μl containing the extract of 100 μl of urine. Reprinted with permission from Horvath *et al*. (*384*).

lishing a diagnosis or confirming a clinical diagnosis. Thus the demonstration of the separation of over 100 urinary aromatic acids in less than 30 min by use of a gradient of acetonitrile in aqueous phosphoric acid suggests RPC may be especially useful for establishing metabolic profiles for clinical work (*384, 385*). A chromatogram showing this separation is reproduced in Fig. 60. Hippuric acid determination on reversed phase has been reported by several workers (*386, 387*). Pyruvic and α-ketoglutaric acids were converted to quinoxalones to enhance detectability prior to analysis using a linear gradient from aqueous ammonium acetate to methanol (*388*).

In addition to assaying the urine as an index of an individual's metabolic state, such assays can also be used to monitor the excretion of various drugs. Salicylic and salicyluric acids have been so monitored (*389*) as have the antimalarial agent, mefloquine (*390*), benzodiazepines (*391*), and the hypoglycemic agent metformin following derivatization (*392*). Other compounds analyzed in urine samples include 2-acetylaminofluorene and its metabolites (*393*), β-cetotetrine (*394*), 4-ethylsulfonylnaphthalene-1-sulfonamide and its metabolites (*395*), sulbenicillin and carbenicillin (*396*), nitrofurantoin and hydroxymethylnitrofurantoin (*397*), gentamicin (*398*), ketoprofen (*399*), and propranolol (*400*) and six of its metabolites (*401*), and benzoylecgonine, a polar metabolite of cocaine (*402*). Cyanophase with heptane–isopropanol (82:18) was used for the analysis of the antitumor agent *cis*-dichlorodiammineplatinum(II) and its degradation products in urine (*403*).

Porphyrias can be differentiated on the basis of chromatographic data taken on esterified preparations taken from fecal or urinary samples (*404*).

C. Biochemical Applications

The analysis of a variety of compounds of biochemical interest have been reported. Some of these were given in the preceding section. The versatility of RPC can be shown, however, by considering the large number of applications given to the analysis of purine and pyrimidine bases, their nucleosides and nucleotides, as well as the determination of amino acids and their corresponding peptides and proteins.

1. Purines, Pyrimidines, Nucleosides, and Nucleotides

Purines, pyrimidines, and the corresponding nucleosides have been separated in reversed phase systems. The five major nucleosides as well as inosine and xanthosine and the corresponding seven bases were separated by Hartwick and Brown (*405*). The selective determination of adenosine in the presence of other serum components has been reported (*406*).

Use of variable wavelength detectors and enzymic peak-shift, a technique in which the sample is incubated with an appropriate enzyme to produce a new chromatographic peak due to the product formed in the enzymic reaction, was found to be useful in peak identification (*406, 407*). Chromatographic identification of a number of minor nucleosides, including methylated nucleosides from urine and other biological matrices, has been reported (*408–412*). Some methylated purines in serum (*409*) and nucleosides in urine (*410, 411*) and 11 methylated bases from *Escherichia coli* tRNA have been separated (*412*). Purines from rabbit liver perfusates have been examined by RPC (*413*). A method for the prediction of retention factor of nucleosides and bases in gradient elution from isocratic data has been presented (*414*) and was found to predict retention times correctly to within 5%.

The technique has been used to separate breakdown products of reduced nicotinamide adenine dinucleotide (NADH) in acidic solution and to establish the reaction mechanism (*415*). It has also been used to monitor enzyme rates of reaction when at least one reactant is a nucleotide (*416–418*).

A variety of xanthines including caffeine, theobromine, and theophylline have been found from food materials including coffee, chocolate, and tea (*419–420*). Theophylline determination in sera has been much studied. The technique allows the determination of theophylline at serum levels of 1.5–20 mg/liter theophylline with sample sizes ranging from 50 to 10 μl (*421–425*). Hill (*426*) assayed theophylline using 50 μl of serum and an analysis time of 8 min with good interbatch precision and accuracy. Alternative methods which allow the determination of as little as 0.1 mg/ml (*427*) or 20 ng theophylline in 10 ml serum have been described (*428*).

In addition to theophylline, purine or pyrimidine derivatives of pharmaceutical interest which have been analyzed include ftorafur, a fluorouracil derivative (*429*), arabinosides of adenine and hypoxanthine (*408*), acephylline (*430*), and allopurinol and oxipurinol (*431*).

2. Nucleic Acids

At the time of this writing, the authors are unaware of any reports of separation of nucleic acids or large nucleic acid fragments by RPC with columns packed with hydrocarbonaceous bonded phases. On the other hand, a large literature exists on the fractionation of RNA, DNA, and their fragments by RPC on stationary phases that are prepared by coating a silanized diatomaceous earth or polychlorotrifluoroethylene resin support with a long-chain quaternary amine. First dimethyl dilauryl ammonium chloride on Chromosorb W was used as the stationary phase and subsequently the general system has gone through several evolutionary stages in which each new phase has been named RPC-1 through RPC-6. The properties of these phases in the separation of tRNAs has been the subject of an exhaustive review article (433). The phase of choice now seems to be RPC-5 which is prepared by coating small particles of Plaskon CTFE 2300 (Allied Chemical), a polychlorotrifluoroethylene resin, with a solution of Adogen 464 (Ashland Chemical), a trialkylmethyl ammonium chloride in which the predominant alkyl chain length is C_8-C_{10}, in either methylene chloride or chloroform. It has been used for the separation of ribo-oligomers (434, 435), tRNAs (431, 433, 436), and DNA restriction fragments (437, 438), and for the separation of strands of DNA-restriction fragments (439). Elution is usually achieved in a gradient of increasing electrolyte concentration. Control of column temperature is absolutely essential for reproducible results on RPC-5. The selectivity and, indeed, the elution order of tRNA's can be modified by control of the pH and Mg(II) concentration of the eluent (433).

DNA restriction fragments up to 440 base pairs long have been isolated on RPC-5 at neutral pH and rechromatographed under alkaline conditions to obtain the individual complementary strands comprising the duplex present in the neutral effluent (439). As a consequence, this method is of value as a preparative tool for the analysis and sequencing of DNA fragments (437–439). The elution order generally follows molecular size insofar as the smaller fragments are eluted before the larger species (437–440). However, fragments with "sticky ends," i.e., those prepared with an endonuclease which leaves unpaired bases at the end of the fragment, were eluted later than those with blunt ends, i.e., no unpaired bases (437, 440). Elution order also depends on the base composition or sequence of the fragments (441, 442). Adenine–thymine rich fragments or fragments which have regions of high A–T content bind more tightly to RPC-5 than is expected on the basis of size alone (442).

A systematic study has been conducted on the effect of temperature, pH, and salt used in the gradient on resolution (440). Best resolution was observed at 43°, pH 6.8 with KCl or NaCl as the gradient former for the fragments of pRZ2 DNA which is recombinant DNA containing 4600 base pairs.

Studies have been made to find an alternative to RPC-5 because it is no longer commercially available. "Homemade" RPC-5 or Kel-F beads coated with Adogen 464 as the stationary phase manifested resolution and elution order comparable to that obtained on RPC-5 under similar conditions (440). Recently, Usher (443) has shown that DNA restriction fragments, oligonucleotides, and tRNA can be fractionated by using only Kel-F powder without Adogen as the stationary phase. The mobile phase contained triethanolamine and gradient elution from water to 18% acetonitrile was employed.

3. Amino Acids, Peptides, and Proteins

Separation of amino acids, peptides, and proteins by RPC has received much attention recently. Amino acids have been analyzed as their phenylthiohydantoin derivatives in sequencing studies (444, 445). Furthermore, dansyl, dinitrophenyl, and dansyl(dimethylaminoazobenzenesulfonyl) (446–448) derivatives of amino acids as well as the amides formed with (−)-2-methoxyl-α-1-naphthaleneacetic acid (449) were separated. In all cases the analysis time was short. Zimmerman et al. (444) reported the separation of 20 amino acid phenylhydantoins in less than 20 min using an acetonitrile gradient in sodium acetate buffer. Molnár and Horváth (203) separated 11 underivatized amino acids on reversed phase in 8 min. Formation of derivatives is useful in enhancing the retention of amino acids which are too polar to be retained in free form in RPC even with plain aqueous eluents. Derivatization can also serve to incorporate a chromophore or fluorophore into the molecule and thus facilitates identification. Molnár and Horváth (203) monitored the absorbance of the effluent at 210 nm, which is a sensitive method of detection but has the disadvantage of being affected by changes in the pH and composition of the eluent. Hearn et al. (450) also used absorbance at 210 nm to detect iodoamino acids. Attention has been paid to the development of alternative methods of detection which include a sulfhydryl-group specific electrochemical detector (for cysteine) (451) and the use of postcolumn reactors for the formation of fluorescent derivatives (452, 453). Products of the fluorescamine–amino acid reaction were analyzed and two derivatives of each amino acid obtained. It was concluded that precolumn derivative formation with this reagent is to be avoided unless amino acid esters are used (454).

These techniques have been used successfully in the micro-Edman degradation of the enzyme mouse sarcoma dihydrofolate reductase to obtain the amino acid sequence of the first 25 amino acids (455). Similarly, RPC has been used in conjunction with the automated Edman technique for sequencing 32 residues of myoglobin (456). Methionine and its oxidation products, methionine sulfoxide and methionine sulfone, in methionine fortified foods have been analyzed as their dansyl derivatives (457). Lysine has been determined as its dansyl derivative in a study in which the stability of lysine in fortified wheat flour was evaluated (458).

The retention of polar amino acids can be enhanced by the inclusion of decyl sulfate (203) or other anionic surfactants (234) in the eluent. The use of such agents in the mobile phase modifies stationary phase interactions and leads to enhanced retention factors and improved resolution in the chromatography of amino acids and peptides. The use of sodium hexane-sulfonate or sodium dodecyl sulfate has been helpful in the analysis of peptides (459). On the other hand, tetrabutylammonium salts have found to increase substantially protein retention on μ-Bondapack alkylphenyl columns (460). Numerous other agents have also been found to improve resolution of such mixtures. Phosphoric acid is widely used in the separation of amino acids (461) and di- to decapeptides (462) as well as for the chromatography of insulin, glucagon, and 1-24 ACTH pentaacetate (463). A buffer comprising triethylammonium phosphate has been used with good effect in the analysis of peptides and proteins (464). Trifluoroacetic acid has been reported to decrease retention, sharpen peaks and eliminate tailing in the chromatography of pituitary endorphins (244).

More recently separations of D- and L-isomers have been reported using asymmetric diens (246) or D-L-proline in the presence of Cu(II) (249) in the mobile phase. These approaches to exploiting secondary equilibria are discussed in Section VI,E.

The separation of short peptides has been studied by many workers (203, 465–471). The retention is commonly observed to increase with increased hydrophobic character of the peptide sidechains. An especially intriguing observation is that whereas L,L and D,D dipeptide diastereomers have essentially identical retention factors on reversed phase, and the retention of the corresponding D,L and L,D diastereomers is also identical, the D,L and L,D peptides are retained longer than those containing only one type of optical isomer (469, 470). An explanation of this effect has been given by Kroeff and Pietrzyk (469) on the basis of the preferred configuration of such dipeptides. As shown in Fig. 61 the amino acid sidechains are trans in L-L and D-D enantiomers and cis in D-L and L-D enantiomers. The configurational differences, as noted by them, "should produce a higher overall hydrophobic surface area and increase the retention of the L-D and D-L enantiomers in comparison with the L-L and D-D enantiomers where the . . . groups are on opposite sides."

The analysis of larger peptides has been reported, including the eicosa-peptide ribonuclease-S peptide and the tripeptide α-melanotropin (203), oxytocin, lypressin and other nonapeptides (471), peptides from the tryptic digestion of the enzymes enolase, phosphoglycerate kinase, or glyceraldehyde-3-phosphate dehydrogenase (472). In addition, the separation of several proteins including neurohypophyseal proteins (473) TSH-releasing hormone, angiotensin II, LH-releasing hormone, somatostatin, leucine, and methionine encephalins have also been accomplished (474). The composition of acetonitrile–water mixture and the apparent pH were found to be important in the resolution of minor impurities of

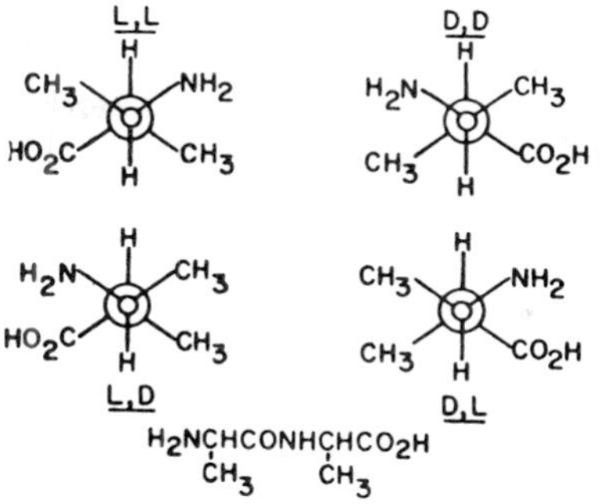

FIG. 61.   Structures of alanylalanine stereoisomers looking along the dipeptide bond. In the D,L- and L,D-isomers the methyl sidechains are cis whereas in the L,L- and D,D-isomers the methyls are trans to one another. As a consequence the D,D and L,L forms are expected to be retained less than the D,L- or L,D-isomers. Reprinted with permission from Kroeff and Pietrzyk (469), *Anal. Chem.* Copyright © 1978 by the American Chemical Society.

nonapeptides (475). Similar results were obtained with methanol–water eluents. Preparative RPC has been applied to various peptides, including the peptide opioid β-endorphin (476) and globin chains (477) among others (478).

4. Enzyme Systems

RPC has also been used successfully in the study of enzyme reactions. Ornithine aminotransferase has been assayed by chromatographic analysis of the reaction product, Δ'-pyrroline-5-carboxylic acid (479). Vasquez and Bieber (148) found the chromatographic assay of IMP-GMP:pyrophosphate phosphoriboxyltransferase to be superior to the spectral assay. Assays of acid and alkaline phosphatase (480), tryptophase (481) and S-methyl-L-homocysteine hydrolase (481a) by RPC have been reported. Adenosine deaminase has been assayed by several workers (416, 417, 482). Chromatographic analysis was found especially useful when assaying adenosine deaminase in the presence of other enzymes which can catalyze competing side reactions (482) such as those found in blood, or other physiological samples. Adenosine deaminase was also adsorbed in a thin layer at the top of a microparticulate octadecyl silica column which was thereby converted into a chromatographic enzyme reactor. Adenosine introduced was partially deaminated to inosine in the top layer and then the two substances were separated in the lower regions of the column (417). In a recent short review (483) the assays of enzymes which catalyze steroid conversions, prostaglandin synthetase and cytochrome P-450 were discussed.

The role of HPLC in the study of enzyme systems is expected to burgeon from these small beginnings. The technique can be used to analyze complex mixtures of reaction products with high speed and precision. As a consequence, the existence of competing reactions may be rapidly discovered and the kinetics of the reactions associated with the various pathways conveniently analyzed.

D. Analysis of Environmental Samples

If the size of the literature is a reliable indicator, the analysis of components found in environmental samples has not been developed to the same extent as clinical applications of reversed-phase chromatography. More attention has been paid to the analysis of volatile species by gas phase chromatography. This is due in part to the difficulty in identifying large molecular weight complex molecules which are present in water at trace levels. However, determination of a variety of analytes in water, soil, or other matrices has been reported and the wider use of RPC in the evaluation of water quality especially can be expected. The apolar phases used in RPC may be a boon in the determination of dilute analytes. Frei (484) has discussed how relatively unpolar compounds dissolved in water can be concentrated at the top of a reversed-phase column and then eluted as a narrow band with an appropriate solvent. This technique can be used for the analysis of environmental samples in which the analyte of interest is in exceedingly low concentration.

Methods for the analysis of 2,4-dichlorophenoxyacetic acid and associated impurities (485) and of its amino acid conjugates have been reported (486). The analysis of the former compound seems to be of more interest in the assay of formulations whereas the determination of the latter substances is of value in studying the persistance of the herbicide in food. The urea herbicide linuron and its metabolites were determined by using UV detection (487). A variety of herbicides (488) including two urea products chlortoluron degradation products have been chromatographed by the use of gradient elution (489). A triazine herbicide, atrazine, was determined in water samples after preconcentration on XAD-2 (490).

The pesticides methyl and ethyl parathion were determined in run-off water after preconcentration on XAD-2. This allowed analyses of these compounds at the parts per billion level (491). Parathion and paraoxon obtained from leaf extracts and orchard soil have also been determined (492). The separation of 30 carbamate pesticides by RPC has been described (493). Various modes of postcolumn fluorometric detection of carbamate insecticides have been reported including post-column reaction between o-phthalaldehyde and methylamine, a carbamate hydrolysis product (*494*), or inhibition of the reaction catalyzed by the enzyme cholestinase and employing a fluorogenic substrate (*495*).

Attention has been given to the analysis of polynuclear aromatic hydrocarbons. Christensen and May (*496*) have discussed the use of a variety of detectors with regard to selectivity and sensitivity in such analytical work. A reversed phase procedure for the analysis of benz[*a*]pyrene in coal tar pitch has been reported (*497*). Determination of fluoranthene, 3,4-benzofluoranthene, 11,12-benzofluoranthene, 1,12-benzoperylene, 3,4-benzopyrene and indeno(1,2,3-*cd*)pyrene from water have been described (*498*) as has been the analysis of chlorinated derivatives of polynuclear aromatics with a combined HPLC and gas chromatographic system (*499*). Aza-arenes have been determined in atmospheric samples (*500*). Acrylamide monomer is readily determined in environmental samples at less than 1 ppm levels (*501*).

RPC is also of value in monitoring the effects of various waste-water treatment systems. Felix *et al.* (*502*) report that columns packed with non-polar bonded phases can be used with gradient elution to obtain fingerprints from waste-water samples and comparison of such fingerprints sheds light on the effect of treatment on the quality of oil shale retort waters. On the other hand, benzyl alcohol extractable cobalamins in sewage sludge were analyzed by RPC and it was found that increased concentration of cobalamin correlates with increasing environmental temperature and depressed oxygen levels (*503*). From the results it follows that HPLC could possibly be used to monitor the state of activated sludge facilities.

E. Analysis of Natural Products and Foodstuffs

Although the analysis of plant materials and various foodstuffs using RPC has not developed a large literature, it is a field which can be expected to grow in importance. The technique affords an experimentally simple method which can yield a rich harvest of information about the composition of a given material or foodstuff to determine whether a foodstuff has been modified in a nontraditional manner, adulterated, or contains noxious substances, which may have been incorporated either before or during processing.

1. Natural Products

In addition to the substances cited elsewhere in this section, a variety of natural materials of vegetable origin have been analyzed using RPC. These include benzoic and cinnamic acids from plants (*504*) and a variety of phenolic compounds from tobacco (*505*). Alkaloids have been analyzed from a variety of sources including ergot (506, 507) and gum opium (508) among others (509). Silymarin, an extract from the fruit of *Silybum marianum* was shown to consist of at least four components on the basis of RPC (510, 511). The bitter substances found in gentian root extracts were found by RPC to consist mostly of gentiopicroside and additionally three minor components (512).

A variety of glycosides have been analyzed using RPLC including cardiac glycosides from milkweed and monarch butterflies (513), and glycoflavones (514, 515). A variety of anthroquinones, monoglycosides and diglycosides have been resolved using a gradient from water to 80% methanol–water (516). Various xanthone glycosides and the aglycones gentisin and isogentisin have also been separated (517).

The fatty acid composition of a bacterium, *Vibrio parahaemolyticus*, was obtained after saponification. The method gives results in good agreement with those of gas chromatography except that in RPC some previously unidentified constituents were resolved (518).

Norsesquiterpenes of defensive secretions of a water beetle were isolated and identified on silica and by RPC. Subsequently the method was used to obtain the norsesquiterpene titer of two species (519).

After preliminary work-up, cytokinins in plant extracts were resolved on RP-8 (520). Other investigators have reported a two-step separation exclusively done on reverse phase to isolate the two major cytokinins of sorghum leaf (521).

2. Foodstuff Analysis

The analysis of foodstuffs for their constituents and their nutritional value has been characterized as "an inviting field to high-pressure liquid chromatography" in a review with that title concerned with the determination of purines and vitamins in food and animal feeds (522). The invitation seems to have been accepted as evidence by the variety of analyses described at a recent symposium (523). The analysis of food quality is an important problem for the food technologist. The extent of decomposition in tuna, as well as other fish products, prior to canning can be determined from the histamine, putrescine, cadaverine, spermine, and spermidine found in the pack by RPC using gradient elution. The results agreed well with organoleptic data (524). In addition to freshness, the presence of toxic contaminants is an important index of food quality. A number of mycotoxins including rubratoxin, aflatoxin, and zearalenone have been determined chromatographically (525) from cereals (526–528) and wines (529). The use of laser fluorimetry allows the determination of as little as 0.75 pg of aflatoxin (530).

Characterization and quantitation of the various components of food-stuffs which are significant for the development of characteristic flavors is of interest. As a consequence analysis of xanthines found in tea and coffee, cited in nucleotide analyses in Section VIII,C,1, is of value as will be the analysis of other nucleosides and nucleotides including inosinic acid, a flavoring agent. Acesulfam, a sweetener, has been determined on reversed phase (531). Other flavor constituents which have been determined include vanillin and coumarin (532), 5-hydroxytryptamide in coffee (533), neohesperidin dihydrochalcone from grapefruit juice (534), and hesperidin in orange juice (535). RPC has been used in the analysis of organic acids of cranberry juice (536) and the hop bitter substances (537). HPLC has been used to analyze the methoxylated flavones found in orange and tangerine juice and thereby establish characteristic profiles for each (549). Saccharin determination in the analysis of an orange-based beverage has been reported (550).

Analysis of vitamin content of food materials appears to be a developing field. B vitamins in rice were analyzed using a mobile phase which contained pentanesulfonic acid and heptanesulfonic acid (538). Although the peaks were not sharp, the separation of the vitamins was satisfactory. Vitamin D in fortified milk has been analyzed after removal of cholesterol and carotenes in a preliminary cleanup (539, 540). Vitamin A has been analyzed in margarine, infant formula, and fortified milk (541, 542). Reports of the analysis of other vitamins in food are few to date but this mode of analysis can be expected to rapidly expand in the future in light of the variety of vitamin determinations in formulations which have been done (see Section VIII,F,1).

An important property in determining the acceptability of a food is its color. Modification of this property is achieved by the addition of synthetic food dyes, some of which have been analyzed using RPC including a large variety of red and orange dyes (543, 544). Alternately various plant pigments may be used as color supplements. Chlorophyll and carotenoid pigments have been determined with octadecyl silica column and aqueous methanol in the eluent (545). Red beet betacyanins and betaxanthins have been determined with the use of ion-pairing (546). Methyl anthranilate found in grape beverages could be analyzed with no prior work-up except dilution and detected at concentrations of less than 50 ng/ml (547). Of more interest (and possible distress) to the oenophile is perhaps the reported separation of anthocyanin pigments of Cabernet Sauvignon grapes into 20 discrete peaks, most of which were identified (548).

Analysis of the triglycerides present in soybean oil has been reported on reverse phase column using methanol–chloroform (9:1) as the mobile phase (551) whereas determination of the fatty acids formed upon oxidation of soybean oil could be separated using acetonitrile-water as the mobile phase (552).

The increased use of RPC in the analysis of pesticide residues in food material can be anticipated. Naphthaleneacetic acid residues were determined in apples (553). Methods for detection of imazalil, a fungicide, in citrus pulp, peel, and whole fruit samples has been described (554). Methods were developed for quantitative analysis of nitrofurazone in milk (555), benz[a]pyrene in paraffin oils used as food additive (556), and paraquat in sunflower seeds (557). Aflatoxins found in corn samples were obtained in quantities of the order of 40 mg utilizing a reversed phase column and acetonitrile–water as the mobile phase (558).

The separation of various organomercury complexes was accomplished in RPC by including 2-mercaptoethanol in the mobile phase. This approach has been used to evaluate mercury content of fish samples (559).

F. Analysis of Industrial Products

RPC has great promise in the analysis and quality control of industrial products. It can also be used to determine the quality or composition of the raw material or feedstock in a process and to monitor the product at each stage of processing. In view of the rapid growth of biotechnology and progress in fermentations, this use of RPC is likely to attain great significance in the future.

1. Pharmaceutical Preparations

The analysis of pharmaceutical agents in preparations has been one of the most important applications of modern RPC. A simple gradient maker to be used on the low pressure side of the pump for use in gradient elution of pharmaceuticals has been described (560). For the detection and identification of pharmaceuticals, micro internal reflection infrared spectroscopy (561) and ultraviolet scanning spectroscopy with stopped flow (562) were also employed.

Water-soluble vitamins in formulations have been determined by use of ion-pair chromatography. The vitamins include several B vitamins as well as niacin, folic acid, and ascorbic acid (563). Vitamins D_2 and D_3 were rapidly separated on reverse phase columns (241) as are vitamins A, D, and E in multivitamin tablets (564). Addition of silver ions to the mobile phase has been shown to increase the flexibility inherent in RPC by complexing with the unsaturated bonds and thereby decreasing the retention factor. This effect is also observed with other unsaturated drug molecules including steroids (241). Vitamin A and related compounds have been analyzed on reverse phase with separation of cis and trans isomers (565). A method for the determination of vitamins A acetate, D_2, and E acetate at the rate of 10 samples per hour has been described (566).

Methods for analysis of analgesics including salicylic acid and acetylsalicylic acid (567) in combination with a variety of other active ingredients have been reported (568). Antipyrine and benzocaine in ear drops have been determined by RPC and the method was found to be four to six times more rapid than the best alternative (569). Salicylic acid and benzoic acid have also been assayed in ointments (570).

RPC has been widely used in the analysis of antibiotic preparations containing trimethoprim and sulfonamide combinations (571), cycloheximide (572), bleomycins (573), and the macrolide antibiotic tylosin B (574). Penicillins, tetracyclines, and cephalosporins are readily analyzed on microparticulate reversed phases (575) as are tetracycline and its degradation products (576). A collaborative study indicated that ethion in formulations can be reliably determined (577). HPLC on reversed phase has been found to be sufficiently rapid to be used for monitoring the fermentative production of erythromycin and tetracycline (578).

Forensic analysis of street drugs include that of cocaine together with excipients frequently encountered (579), amphetamines (580), and dyes found in heroin samples (581). An on-line photochemical derivitization of cannabinoids has been described (582). Other pharmaceutical agents studied in formation include nortriptyline in tablets. (583), glycyrrhizic acid from licorice extract (584, 585), pirimiphos methyl (586), digitalis glycosides (587), pilocarpine (588), and its antagonist atropine (589).

A specific assay for nitroglycerin and its degradation products in dosage form has been described (590). In addition RPC has been used to determine methadone (591) and clindamycin and its phosphate, and palmitate derivatives (592). Mestranol and norethistereone found in oral contraceptive tablets have been analyzed using methanol–0.01 M phosphate buffer, pH 7, as the eluent (593); ethinylestriadiol has also been determined (594). The pilocarpine and its four degradation products were determined in opthalmic solutions by using isocratic elution with methanol–water (3:97) acidified with KH_2PO_4 (595). Idoxuridine in formulations has been determined (596).

2. Others

The pesticide rotenone and five other rotenoids were determined in formulations by using RPC (597, 598). The assay of warfarin in rodenticide preparations has been reported by Billings et al. (599). Methiocarb (3,5-dimethyl-4-(methylthio)phenylmethylcarbamate) has been determined in commercial preparations (600). The diastereomers of sumicidin, a pyrethroid insecticide, were separated on both reversed phase columns with acidic acetonitrile–water as the mobile phase and on silica gel column (601).

Two drugs found in animal feeds, sulfanitran and dinsed, have been separated and determined quantitatively by using RPC with acetonitrile–water as the mobile phase in ten minutes (602). An analysis of furazolidine and nitrofurazolidine in animal feeds has also been described (603). Various sulfobetaine amphoteric surfactants were analyzed using 80% methanol–water as the mobile phase (604). The quantitative separation of polyethylene glycols from ethoxylated fatty alcohols or p-alkylphenols in nonionic surfactants was accomplished by RPC with acetonitrile–water (65:35) as the eluent (605). Rapid analysis by RPC of amphoteric surfactants (606, 607) and various nonionic detergents (158, 607, 608) was also carried out by using RPC.

Among chromatographic separations of industrial interest is that of ether polycarboxylic acids (609). Triamcinolone acetonide in cream and suspension formulations was analyzed by using prednisolone as an internal standard (610). Various antioxidants and antiozonants used as stabilizers in elastomers have been determined by using a ternary eluent containing methanol–water–butanol (82:18:1) (611). The determination of oxidants in polyethylene and their various thermal photochemical degradation products was sought by Lichtenhaler and Ranfelt (612). Very low concentrations of rosin in shellac could be determined with good accuracy (613).

Many organic chemicals are analyzed by RPC. These include various arylhydroxylamines as the N-hydroxyurea derivative with methyl isocyanate (614) alkyl- and alkoxy-disubstituted azoxybenzenes (615), n-alkyl-4-nitrophenylcarbonate esters ranging in length from methyl to octyl (616), 4-nitrophenol in the presence of 4-nitrophenyl phosphate (617), benzilic acid, and benactyzine-HCl using ion-pair chromatography (618), as well as aniline and its various metabolites (619), stereoisomers of 4,4'-dihydroxyhydrobenzoin (620), and aldehydes and ketones as the 2,4-dinitrophenylhydrazones (621). The technique has also been used to analyze propellants and hydrazine and 1,1-dimethylhydrazine were quantitatively determined (622, 623).

G. Miscellaneous Separations

Many other analytical methods which are not readily included in the above categories use RPC and have been reported in the literature. For instance, various metal complexes can be conveniently analyzed by this technique. Transition metals have been separated on reverse phase as metal cluster complexes (624). Other complexes include the diethyldithio-carbamates of Se, Cr, Ni, Co, Pb, Cu, and Hg (625) as well as tris(2,2′-bipyridyl)ruthenium(II) derivatives. Crown ethers (626) have been analyzed in reverse phase as the Hg(II) halides (271) or in the presence of Cs^{2+}, K^+, or Na^+ (116). Phosphoric and silicic acids have been analyzed on reverse phase after reaction with ammonium molybdate to form the corresponding molybdoheteropoly acids (627).

Phthalate esters are readily separated by using hydrocarbonaceous bonded phases and a water–methanol gradient; the retention is directly related to the number of alkyl carbons (628). RPC has been used to separate o-, p-, and m-hydroxyphenylacetic acids (629). Separation of short- and long-chain fatty acids was carried out in the form of their p-bromo-phenacyl, p-nitrophenacyl, p-chlorophenacyl, and 2-naphthacyl esters in order to facilitate detection. The various fatty acids were well separated (630). Alkylphenol (631) and hydroxyphenol (632) isomers were also conveniently analyzed by using RPC. HPLC in the reversed-phase mode has been found to be twice as fast as gas chromatography in the analysis of nitrosoamines (633).

Acknowledgment

The authors with to thank Avi Nahum for assistance in preparation of some of the graphs, and Arlene Cashmore and Ursula Pedersen for the artwork. The authors gratefully acknowledge support of their research on the subject treated in this chapter by grants CA 21948 and GM 20993 from the National Institutes of Health, U.S. Public Health Service, Department of Health and Human Service.

Notation

Lower Case

a	distance of closest approach in Debye–Hückel theory, Eq. (43a)
a	constant in Lietzke–Stoughton–Fuoss theory, Eqs. (46), (47)
b	ionic radius in Debye–Hückel theory, Eq. (43a)
b	parameter used in theory of gradient elution; dimensionless rate of solvent strength increase, Eqs. (14), (15)
e	charge on electron (in electrostatic units)
f	volume fraction of salt solution which manifests fused-salt behavior in Lietzke–Stoughton–Fuoss theory, Eq. (46)
k	retention factor
k_{ij}	retention factor of species i in solvent j, Eq. (10).
k	Boltzmann constant, Eq. (48)
k_0	retention factor obtained in the absence of a gradient, Eq. (14)
k_0	retention factor of species with no hetaeron bound
k_0	retention factor of neutral species, Eqs. (89)–(111)

k_i	retention factor of species with i hetaerons bound, Eqs. (109)–(111)
k_{-1}	retention factor of monoanion, Eq. (89)
k_1	retention factor of monocation, Eq. (90)
k_i, k_j	retention factor of eluite i or j, Eq. (61)
k_w	retention factor in water, Eq. (13)
m	mobile phase (subscript)
n_B	solvent refractive index, Eq. (39b)
n	constant in van der Waals potential, Eq. (35), (39a)
n_A	refractive index of solute A
n	charge, number (superscript)
n^1	number of hetaerons in micelle cluster, Eq. (100)
r	distance between molecular centers, Eqs. (34), (54), (55)
r_i	molecular radius of species i, Eqs. (41a), (42)
s	surface, stationary phase (subscript)
t	parameter in integrated van der Waals potential, Eqs. (33), (37b)
t_e	retention time, Eq. (14)
t_0	retention time of an unretarded peak, Eqs. (14), (15)
v_i	molecular volume of species i
v_0	superficial velocity, cm/sec, Eq. (9)
w	mass of bound ligate per unit mass support, Eq. (2)
x	constant ($\sim$436) [Eq. (36b)]
x_w	mole fraction water, Eq. (11)
y	activity coefficient, Eqs. (44), (45), (47), (49)

Upper Case

A	retention factor in absence of hetaeron, Eqs. (93)–(94)
A	molecular surface area of eluite, Eqs. (20)–(22)
A_i	molecular surface area of species i, Eqs. (10), (23), (58b), (59)
A_m	ligate surface area requirement, Eq. (5)
A_m	mobile phase anion concentration, Eqs. (102)–(103)
A_s	molecular surface area of solvent, Eqs. (21)–(22), (58b)
ΔA	ligate eluite surface area change with binding, contact surface area, Eqs. (8), (59b), (62)
$B°$	specific permeability coefficient, Eq. (9)
B	coefficient of the Davies equation for ionic effects, Eq. (49)
B	parameter in phenomenal expression for hetaeric chromatography, Eqs. (93)–(94)
B	coefficient relating activity to ionic strength of fused salts, Eqs. (45), (47)
C	parameter in phenomenal expression for hetaeric chromatography. The reciprocal yields the hetaeron concentration at which the effect is half maximal, Eqs. (93)–(94)
C	coefficient relating activity to ionic strength, Eqs. (45), (47)
C_{AB}	London parameter in van der Waals potential, Eqs. (25)–(26), (36a), (53)–(55)
C_S	carbon load, Eq. (3)
D	solvent function, used to estimate reduction of van der Waals potential, Eq. (39a)–(39b), (55)–(56)
D	parameter in Onsager reaction field, Eqs. (41)–(42)
D_1	Clausius–Mosotti function for species i, Eq. (42)
ΔG_c	free energy of cavity formation, Eqs. (20), (22)

ΔG_{int}	free energy of interaction between solute and solvent, Eqs. (24), (58)
ΔG_r	Gibbs free energy of formation of complex in gas phase, Eqs. (57)–(58)
ΔG_{vdw}	free energy of interaction due to van der Waals interaction, Eqs. (24), (56), (58), (62)
ΔG_{es}	free energy of interaction due to van der Waals interaction, Eqs. (24), (41a), (43a), (58)
$\Delta G°$	standard Gibbs free energy of binding
$\Delta G_E°$	standard Gibbs free energy of solution of eluite E, Eq. (57)
$\Delta G_B°$	standard Gibbs free energy of solution of ligand B, Eq. (57)
$\Delta G_{EB}°$	standard Gibbs free energy of solution of complex EB, Eq. (57)
ΔG_B	standard Gibbs free energy at temperature B, Eq. (108)
$\Delta H_b°$	observed binding enthalpy, Eq. (19)
$\Delta H°$	standard isomerization enthalpy, Eqs. (18c)–(19)
$\Delta H_A°$	standard binding enthalpy of species A, Eqs. (18a)–(19)
$\Delta H_B°$	standard binding enthalpy of species B, Eqs. (18b)–(19)
HE	concentration of acidic form of eluite E
I_i	ionization potential of species i, Eq. (26)
I	ionic strength, Eqs. (43a), (45), (47), (49)
K	equilibrium constant in isomerization reaction, Eqs. (16)–(19)
K_0	equilibrium constant under reference conditions, Eq. (103)
K_0	distribution coefficient of eluite to surface, Eqs. (7), (51), (53), (65), (72), (75), (82), (88), (93)
K_1	Langmuir adsorption coefficient of hetaeron between surface and mobile phase, Eqs. (63), (64), (68), (69), (71)–(72), (74)–(75), (78), (80)–(84), (87)
K_2	ion-pair formation constant in mobile phase, Eqs. (66), (68), (69), (72), (74)–(75), (78), (80)–(84), (87)–(88), (93), (98), (100)–(101)
K_3	distribution coefficient between mobile and stationary phase for ion pairs, Eqs. (67)–(69), (83)–(84), (88)
K_4	stability constant for adsorption of eluite to surface bound hetaeron, Eqs. (70)–(72), (83)–(84), (97)–(100)
K_5	distribution coefficient for ligand exchange, Eqs. (73)–(75), (101c)–(103)
K_6	binding constant in ligand exchange, Eqs. (76), (78), (80)–(82), (84)
K_7	exchange constant in ligand exchange, Eqs. (77), (79)–(82), (84)
K_j	association constant for jth binding process, Eq. (109)
K_{d_j}	dissociation constant, reciprocal of K_j, Eqs. (89)–(92), (110)–(111)
L	column length, Eq. (9)
M	molarity, Eqs. (46), (47)
M	molecular weight, Eq. (2)
N	Avogadro's number (6.023×10^{23} mol^{-1})
P	water–octanol distribution coefficient
P	pressure, taken to be 1 atm
ΔP	pressure drop, Eqs. (9), (105)
Q'	dimensionless sum of gas phase van der Waals interactions, Eqs. (37a)–(37b), (40), (56)
Q''	dimensionless sum of solvent corrections to gas phase van der Waals interactions, Eqs. (38), (40)
R	gas constant (8.32 joules/deg-mol)
R_i	molecular diameter of species i, Eqs. (27)–(30)
S	coefficient relating logarithm of retention factor to solvent composition; solvent strength, Eqs. (13)–(15)

S_{Si}^*	BET surface area of silica, Eqs. (2), (3)
S_{Si}	BET surface area corrected for mass gain with silanization, Eq. (1)
$\Delta S°$	standard binding entropy
$\Delta S°$	entropy of isomerization, Eqs. (18c)–(19)
$\Delta S_A°$	binding entropy of species A, Eqs. (18a)–(19)
$\Delta S_B°$	binding entropy of species B, Eqs. (18b)–(19)
SCE	surface carbon equivalent, Eq. (3)
T	absolute temperature (degrees Kelvin)
V_{lig}	ligate volume
V	molecular volume of solvent
V_i	molar volume of species i, Eqs. (10), (12), (23)
V_{AB}	van der Waals potential between molecules A and B, Eqs. (25), (35), (39a)
V_{eff}	effective van der Waals potential, Eqs. (35), (39a), (54), (55)
V	molar volume of dry salt, Eqs. (46), (47)
ΔV	molar volume change in reaction, Eqs. (104)–(105)
X_{OH}	extent of surface silanization, Eq. (4)
Z	ionic charge, Eq. (43a)

Script

$\mathcal{D}$	Clausius–Mosotti function
$[\mathcal{H}_m]$	mobile phase concentration of hetaeron, Eq. (63) et seq.
$[\mathcal{H}_s]$	stationary phase concentration of hetaeron, Eq. (63) et seq.
$\mathcal{S}$	Debye–Hückel coefficient, Eqs. (47)–(49)
$[\mathcal{H}_s]^*$	maximum stationary phase concentration of hetaeron, Eq. (63) et seq.
ℓ_i	Kihara parameter of species i
$\overline{\ell}$	mean Kihara parameter, Eqs. (31), (34), (35), (54), (55)

Greek

$\overline{\alpha}_i$	molecular polarizability of species i, Eqs. (26), (41a)
α	coefficient in Freundlich equation, Eq. (95)
α_L	surface concentration of ligate (μmol/m²), Eqs. (2), (4), (6)
α_{OH}	surface concentration of silanol (μmol/m²), Eqs. (1), (4)
β	distribution coefficient of negatively charged solute between two phases, Eqs. (96), (99)
β	compensation temperature, Eq. (108)
γ	bulk surface tension
ϵ	dielectric constant of solvent, Eqs. (41b), (43)
ϵ_s	dielectric constant at surface, Eq. (60)
$\epsilon_j°$	eluotropic strength of solvent, Eq. (10)
η	viscosity coefficient, Eq. (9)
η	modulus, Eq. (94)
θ	surface coverage by a ligate, Eq. (6)
$\kappa^e(r)$	correction factor for surface tension for solute of radius r in solvent, Eqs. (20), (21)
κ^e	correction factor for surface tension for solute of radius r in solvent, Eqs. (21), (22), (58b), (59a)
κ	Debye–Hückel screening parameter, Eq. (43)
κ	common logarithm of the retention factor

μ_i	dipole moment of species i, Eq. (41a)
π	logarithm of substituent n-octanol–water partition coefficient
$\rho, \bar{\rho}$	parameter in Kihara formulation of van der Waals potential, Eqs. (25), (34), (35)
$\sigma, \bar{\sigma}$	parameter in Kihara formulation of van der Waals potential, Eqs. (25), (32), (35), (54), (55)
τ_{ji}	group selectivity coefficient, Eqs. (61)–(62)
ϕ	phase ratio
ϕ_0	volume fraction organic in mobile phase, Eq. (13)
$\dot{\Phi}$	rate of increase in composition less polar mobile phase component, Eq. (15)
ϕ_{AB}	dimensionless function relating the reduction of the potential between molecules A and B in solvent to the corresponding gas phase potential, Eqs. (35), (54)
ω_i	acentric factor of molecule i, Eq. (28)–(29)

REFERENCES

1. E. Lederer and M. Lederer, "Chromatography," 1st ed., p. 45. Elsevier, Amsterdam, 1952.
2. G. A. Howard and A. J. P. Martin, *Biochem J.* **46**, 532 (1950).
3. R. J. Boscott, *Nature (London)* **159**, 342 (1947).
4. J. Boldingh, *Experientia,* **4**, 270 (1948).
5. S. M. Partridge and T. Swain, *Nature (London)* **166**, 272 (1950).
6. J. F. Nye, D. M. Maron, J. B. Garst, and H. B. Friedgood, *Proc. Soc. Exp. Biol. Med.* **77**, 466 (1951).
7. L. Bosch, *Biochim. Biophys. Acta* **11**, 301 (1953).
8. M. H. Silk and H. H. Hahn, *Biochem. J.* **56**, 406 (1954).
9. F. P. W. Winteringham, A. Harrison, and R. G. Bridges, *Nature (London)* **166**, 999 (1950).
10. P. Savary and P. Desnuelle, *Bull. Soc. Chim. France* p. 939 (1953).
11. G. A. Howard and A. R. Tatohell, *Chem. Ind. (London)* p. 219 (1954).
12. W. M. L. Crombie, R. Comber, and S. G. Boatman, *Biochem. J.* **59**, 309 (1955).
13. E. Abderhalden and A. Fodor, *Fermentforschung* **2**, 74 and 151 (1910).
14. H. G. Cassidy, *J. Am. Chem. Soc.* **62**, 3073 and 3076 (1940).
15. H. G. Cassidy and S. E. Wood, *J. Am. Chem. Soc.* **63**, 2628 (1941).
16. V. H. Cheldin and R. J. Williams, *J. Am. Chem. Soc.* **64**, 1513 (1942).
17. R. T. Holman and W. T. Williams, *J. Am. Chem. Soc.* **73**, 5285 (1951).
18. A. Tiselius, ed., "The Svedberg," p. 370. Almquist & Wiksells, Uppsala, 1944.
19. A. Tiselius and L. Hagdahl, *Acta Chem. Scand.* **4**, 374 (1950).
20. L. Hagdahl, R. J. P. Williams, and A. Tiselius, *Arkiv Kemi* **4**, 193 (1952).
21. J. Porath and C. H. Li, *Biochim. Biophys. Acta* **13**, 268 (1954).
22. C. H. Li, A. Tiselius, K. O. Pedersen, L. Hagdahl, and H. Cartensen, *J. Biol. Chem.* **190**, 317 (1951).
23. C. Fromagest, M. Jutisz, and E. Lederer, *Biochim. Biophys. Acta* **2**, 487 (1948).
24. A. Tiselius, *Kolloid Z.* **105**, 101 (1943).
25. G. Schramm and J. Primosigh, *Chem. Ber.* **76**, 373 (1943).
26. T. Green, F. O. Howitt, and R. Preston, *Chem. Ind. (London)* p. 591 (1955).
27. T. W. Winchester, U.S. At. Energy Comm. Rep. CF-58-12-43 (1958).
28. E. Cerrai and G. Ghersini, *Adv. Chromatogr.* **9**, 3 (1970).
29. Cs. Horváth and S. R. Lipsky, *Nature (London)* **211**, 748 (1966).

304 Wayne R. Melander and Csaba Horváth

30. J. F. Weiss and A. D. Kelmers, *Biochemistry* **6**, 2507 (1967).
31. R. L. Pearson, J. F. Weiss, and A. D. Kelmers, *Biochim. Biophys. Acta* **228**, 770 (1975).
32. S. Eksborg, P. O. Lagerström, R. Modin, and G. Schill, *J. Chromatogr.* **83**, 99 (1973).
33. S. Eksborg and G. Schill, *Anal. Chem.* **45**, 2092 (1973).
34. Cs. Horváth, W. Melander, I. Molnár, and P. Molnár, *Anal. Chem.* **49**, 2295 (1977).
35. A. J. P. Martin and R. R. Porter, *Biochem. J.* **49**, 215 (1951).
36. E. W. Abel, F. H. Pollard, P. C. Uden, and G. Nickless, *J. Chromatogr.* **22**, 23 (1966).
37. H. N. M. Stewart and S. G. Perry, *J. Chromatogr.* **37**, 97 (1968).
38. J. J. Kirkland and J. J. DeStefano, *J. Chromatogr. Sci.* **8**, 309 (1970).
39. Cs. Horváth, B. A. Preiss, and S. R. Lipsky, *Anal. Chem.* **39**, 1422 (1967).
40. R. E. Majors, *Anal. Chem.* **44**, 1722 (1972).
41. R. K. Iler, "The Colloid Chemistry of Silica and Silicates." Cornell Univ. Press, Ithaca, New York, 1955.
42. C. Rossi, S. Munari, C. Cengari, and G. F. Tealdo, *Chim. Ind. (Milano)* **42**, 724 (1960).
43. I. Halász and J. Sebastian, *J. Chromatogr. Sci.* **12**, 161 (1974).
44. Cs. Horváth and S. R. Lipsky, *J. Chromatogr. Sci.* **7**, 109 (1969).
45. J. F. K. Huber and J. A. R. J. Hulsman, *Anal. Chim. Acta* **38**, 305 (1967).
46. J. J. Kirkland, *Anal. Chem.* **41**, 218 (1969).
47. A. Breyer and W. Rieman, *Anal. Chim. Acta* **18**, 204 (1958).
48. J. Sherma and W. Rieman, *Anal. Chim. Acta* **18**, 214 (1958).
49. B. Hofstee, *Anal. Biochem.* **52**, 430 (1973).
50. S. Shaltiel and Z. Er-El, *Proc. Natl. Acad. Sci. U.S.A.* **70**, 778 (1973).
51. N. A. Parris, *J. Chromatogr.* **149**, 615 (1978).
52. B. L. Karger, J. N. LePage, and N. Tanaka, *In* "HPLC: Advances and Perspectives" (Cs. Horváth, ed.), Vol. I, Ch. 3. Academic Press, New York, 1980.
53. J. H. Knox and G. R. Laird, *J. Chromatogr.* **122**, 17 (1976).
54. L. R. Snyder *In* "HPLC: Advances and Perspectives" (Cs. Horváth, ed.), Vol. I, Ch. 4. Academic Press, New York, 1980.
55. K. K. Unger, "Porous Silica" *J. Chromatogr.* Lib. Vol. **16**, Elsevier, New York, 1979.
56. C. Eaborn, "Organosilicon Compounds." Butterworths, London, 1960.
57. R. E. Majors *In* "HPLC: Advances and Perspectives" (Cs. Horváth, ed.), Vol. I, Ch. 2. Academic Press, New York, 1980.
57a. H. Englehardt and H. Elgass *In* "HPLC: Advances and Perspectives" (Cs. Horváth, ed.), Vol. II, pp. 57–111. Academic Press, New York, 1980.
58. S. Hjerten, *Arch. Biochem. Biophys.* **99**, 466 (1962).
59. J.-M. Egly and J. Porath *In* "Affinity Chromatography" (O. Hoffmann-Ostenhof, M. Breitenbach, F. Koller, D. Kraft and O. Scheiner, eds.), pp. 5–22 Pergamon, Oxford, 1978.
60. W. Melander and Cs. Horvath, *Arch. Biochem. Biophys.* **183**, 200 (1977).
61. R. K. Iler, "The Chemistry of Silica: Solubility, Polymerization, Colloid and Surface Properties, and Biochemistry." Wiley, New York, 1979.
62. R. G. Brownlee and J. W. Higgins, *Chromatographia* **11**, 567 (1978).
63. Cs. Horváth and H.-J. Lin, *J. Chromatogr.* **126**, 401 (1976).
64. L. Boksanyi, O. Liardon, and E. Kovats, *Adv. Colloid Interface Sci.* **6**, 95 (1976).
65. L. Boksanyi, O. Liardon, and E. Kovats, *Helv. Chim. Acta* **59**, 717 (1967).
66. F. Riedo, M. Czenca, O. Liardon, and E. Kovats, *Helv. Chim. Acta* **61**, 1912 (1978).
67. R. K. Gilpin and M. F. Burke, *Anal. Chem.* **45**, 1383 (1973).
68. R. E. Majors and H. J. Hopper, *J. Chromatogr. Sci.* **12**, 767 (1974).
69. B. B. Wheals, *J. Chromatogr.* **107**, 402 (1975).

70. J. H. Knox and A. Pryde, *J. Chromatogr.* **112,** 171 (1975).
71. J. J. Kirkland, *Chromatographia* **8,** 661 (1975).
72. D. G. I. Kingston and B. B. Gerhart, *J. Chromatogr.* **116,** 182 (1976).
73. K. Karch, I. Sebestian, and I. Halász, *J. Chromatogr.* **122,** 3 (1976).
74. K. K. Unger, N. Becker, and P. Roumeliotis, *J. Chromatogr.* **125,** 115 (1976).
75. E. J. Kikta, Jr., and E. Grushka, *Anal. Chem.* **48,** 1098 (1976).
76. M. C. Hennion, C. Picard, and M. Caude, *J. Chromatogr.* **166,** 21 (1978).
77. R. K. Gilpin, J. A. Korpi, and C. A. Janicki, *Anal. Chem.* **46,** 1314 (1974).
78. R. K. Gilpin, D. J. Camille, and C. A. Janicki, *J. Chromatogr.* **121,** 13 (1976).
79. H. Hemetsberger, W. Maasfeld, and H. Ricken, *Chromatographia* **9,** 303 (1976).
80. H. Hemetsberger, M. Kellermann, and H. Ricken, *Chromatographia* **10,** 726 (1977).
81. R. P. W. Scott, and P. Kucera, J. Chromatogr. **142,** 213 (1977).
82. R. E. Majors, *J. Chromatogr. Sci.* **15,** 334 (1977).
83. H. Hemetsberger, P. Behrensmeyer, J. Henning, and J. Ricken, *Chromatographia* **12,** 71 (1979).
84. P. Roumeliotis and K. K. Unger, *J. Chromatogr.* **149,** 211 (1978).
85. H. Colin, N. Ward, and G. Guiochon, *J. Chromatogr.* **149,** 169 (1978).
86. H. Colin and G. Guiochon, *J. Chromatogr.* **158,** 183 (1978).
87. J. H. Knox and J. Jurand, *J. Chromatogr.* **142,** 651 (1977).
88. J. D. Ramsay, *In* "Chromatography of Synthetic and Biological Polymers," Vol. I, "Column Packings, GPC, GF and Gradient Elution" (R. Epton, ed.), pp. 339–343. Horwood, Chichester, Sussex, 1978.
89. F. Eisenbeiss, *Merck Kontacte* (17) Feb. (1976).
90. J. M. Bather and R. A. C. Gray, *J. Chromatogr.* **122,** 159 (1976).
91. J. Shapiro and J. M. Kolthoff, *J. Am. Chem. Soc.* **72,** 776 (1950).
92. N. Tanaka, H. Goodell, and B. L. Karger, *J. Chromatogr.* **158,** 233 (1978).
93a. A. Nahum and Cs. Horváth, *J. Chromatogr.* in press.
93b. K. E. Bij, Cs. Horváth, W. R. Melander, and A. Nahum, *J. Chromatogr.* in press.
94. R. E. Majors, *Anal. Chem.* **44,** 1722 (1972).
95. J. J. Kirkland, *J. Chromatogr. Sci.* **10,** 593 (1972).
96. W. Strubert, *Chromatographia* **6,** 50 (1973).
97. J. C. Kraak, H. Poppe, and F. Smedes, *J. Chromatogr.* **122,** 340 (1974).
98. R. M. Cassidy, D. S. Legay, and R. W. Frei, *Anal. Chem.* **46,** 340 (1974).
99. H. R. Linder, H. P. Keller, and R. W. Frei, *J. Chromatogr. Sci.* **14,** 234 (1976).
100. M. Broquaire, *J. Chromatogr.* **170,** 43 (1979).
101. J. Asshauer and I. Halász, *J. Chromatogr. Sci.* **12,** 139 (1974).
102. C. J. Little, A. D. Dale, D. A. Ord, and T. R. Marten, *Anal. Chem.* **49,** 1311 (1977).
103. B. Coq, C. Gonnet, and J.-L. Rocca, *J. Chromatogr.* **106,** 249 (1975).
104. W. E. F. Gerbacia, J. Chromatogr. **166,** 261 (1978).
105. S. Bakalyar, J. Yuen, and D. Henry, Chromatography Technical Bulletin, TB 114-76, Spectra-Physics, Santa Clara, California.
106. P. A. Bristow, P. N. Brittain, C. M. Riley, and B. F. Williamson, *J. Chromatogr.* **131,** 57 (1977).
107. Cs. Horváth, W. Melander, and I. Molnár, *J. Chromatogr.* **125,** 129 (1976).
108. Cs. Horváth and W. Melander, *Am. Lab.* **10,** (10), 17 (1978).
109. Cs. Horváth and H.-J. Lin, *J. Chromatogr.* **149,** 43 (1978).
110. J. C. Giddings, *In* "Advances in Chromatography 1967" (A. Zlatkis, ed.), pp. 202–208. Preston Technical Abstracts 6, Evanston, Illinois 1967.
111. G. L. Gaines, Jr., *In* "Insoluble Monolayers at Liquid-Gas Interface" (I. Prigogine, ed.). Interscience, New York, 1966.

112. W. Melander, J. Stoveken, and Cs. Horváth, *J. Chromatogr.* in press.
113. J. G. Atwood, G. J. Schmidt, and W. Slavin, *J. Chromatogr.* **171**, 109 (1974).
114. A. Wehrli, J. C. Hildenbrand, H. P. Keller, R. Stanpfli, and R. W. Frei, *J. Chromatogr.* **149**, 199 (1978).
115. V. Rehak and E. Smolkova, *Chromatographia* **9**, 219 (1976).
116. Cs. Horváth, W. Melander, and A. Nahum, *J. Chromatogr.* **186**, 371 (1979).
117. I. Halász and Cs. Horváth, *Anal. Chem.* **36**, 1178 (1964).
118. Cs. Horváth, *In* "75 Years of Chromatography" (L. S. Ettre and A. Zlatkis, eds.), p. 8. Elsevier, Amsterdam.
119. H. Colin and G. Guiochon, *J. Chromatogr.* **126**, 43 (1976).
120. M. J. Telepchak, *Chromatographia* **6**, 234 (1973).
121. N. K. Bebris, A. V. Kiseley, Yu S. Nikitin, I. I. Frolov, L. V. Tarasova, and Ya. I. Yashin, *Chromatographia* **11**, 206 (1978).
122. R. Leboda and A. Waksmundzki, *Chromatographia* **12**, 207 (1979).
123. D. J. Pietrzyk and C. H. Chu, *Anal. Chem.* **49**, 757 (1977).
124. D. J. Pietrzyk and C. H. Chu, *Anal. Chem.* **49**, 860 (1977).
125. R. E. Kunin, E. Meitzner, and N. Bortnick, *J. Am. Chem. Soc.* **84**, 305 (1962).
126. R. G. Baum, R. Saetre, and F. F. Cantwell, *Anal. Chem.* **52**, 15 (1980).
127. F. F. Cantwell and S. Puon, *Anal. Chem.* **51**, 623 (1979).
128. S. R. Bakalyar, *Am. Lab.* **10**(6), 43, 52 (1978).
129. Cs. Horváth and W. Melander, *J. Chromatogr. Sci.* **15**, 393 (1977).
130. L. R. Snyder, "Principles of Adsorption Chromatography," pp. 194–195. Dekker, New York, 1968.
131. C. Eon, *Anal. Chem.* **47**, 1871 (1975).
132. J. C. Giddings, "Dynamics of Chromatography, Part I." Dekker, New York, 1965.
133. B. L. Karger, L. R. Snyder, and Cs. Horváth, "An Introduction to Separation Science," p. 78. Wiley (Interscience), New York, 1973.
134. J. Timmermans, "The Physico-Chemical Constants of Binary Systems in Binary Solution," Vol. 4. Wiley (Interscience), New York, 1960.
135. C. Carr and J. A. Riddick, *Ind. Eng. Chem.* **43**, 692 (1951).
136. G. Akerlof, *J. Am. Chem. Soc.* **54**, 4125 (1932).
137. G. Douheret and M. Morenas, *C.R. Acad. Sci. Paris* **264**, 4125 (1967).
138. W. Hayduk, H. Laudie, and O. H. Smith, *J. Chem. Eng. Data* **18**, 373 (1973).
139. F. E. Critchfield, J. A. Gibson, Jr., and T. L. Hall, *J. Am. Chem. Soc.* **75**, 6044 (1953).
140. C. Salceanu, *C.R. Acad. Sci. Paris* **261**, 4403 (1965).
141. H. Colin, C. Eon, and G. Guiochon, *J. Chromatogr.* **122**, 223 (1976).
142. H. Colin, J. C. Diez-Masa, G. Guiochon, T. Czajkowska, and I. Miedziak, *J. Chromatogr.* **167**, 41 (1978).
143. K. Karch, I. Sebastian, I. Halász, and H. Englehardt, *J. Chromatogr.* **122**, 171 (1976).
144. T. A. Schmidt, R. A. Henry, R. C. Williams, and T. F. Dieckman, *J. Chromatogr. Sci.* **9**, 645 (1971).
145. M. Riedmann, *Z. Anal. Chem.* **279**, 154 (1976).
146. L. R. Snyder, J. W. Dolan, and J. R. Gant, *J. Chromatogr.* **165**, 3 (1979).
147. F. Erni and R. W. Frei, *J. Chromatogr.* **130**, 169 (1977).
148. B. L. Karger, J. R. Gant, A. Hartkopf, and P. H. Weiner, *J. Chromatogr.* **128**, 65 (1976).
149. R. E. Majors, *In* "Bonded Stationary Phases in Chromatography," (E. Grushka ed.), Ch. 8. Ann Arbor Sci. Publ., Ann Arbor, Michigan, 1974.
150. H. Hemets, W. Maasfeld, and H. Richer, *Chromatographia* **9**, 303 (1976).

151. M. LaFosse, G. Keravis, and M. H. Durand, *J. Chromatogr.* **118**, 283 (1976).
152. S. R. Abbott, J. R. Berg, P. Achener, and R. L. Stevenson, *J. Chromatogr.* **126**, 421 (1976).
153. A. P. Grafeo and B. L. Karger, *Clin. Chem.* **22**, 184 (1976).
154. D. Westerlund and A. Theodorsen, *J. Chromatogr.* **144**, 29 (1977).
155. J. W. Dolan, T. R. Gant, and L. R. Snyder, *J. Chromatogr.* **165**, 31 (1979).
156. P. J. Schoenmakers, H. A. H. Billiet, R. Tijssen, and L. de Galan, *J. Chromatogr.* **149**, 519 (1978).
157. S. R. Bakalyar, R. McIlwrick, and E. Roggendorf, *J. Chromatogr.* **142**, 353 (1977).
158. W. R. Melander, A. Nahum, and Cs. Horváth, *J. Chromatogr.* **185**, 129 (1979).
159. N. Tanaka and E. R. Thornton, *J. Am. Chem. Soc.* **99**, 7300 (1977).
160. N. E. Hoffman and J. C. Liao, *Anal Lett.* **A11**, 287 (1978).
161. H. Englehardt and H. Elgass, *J. Chromatogr.* **158**, 249 (1978).
162. P. J. Schoenmakers, H. A. H. Billiet, and L. de Galan, *J. Chromatogr.* **185**, 179 (1979).
163. M. Martin, G. Blu, C. Eon, and G. Guiochon, *J. Chromatogr.* **112**, 399 (1975).
164. M. Martin, G. Guiochon, G. Blu, and C. Eon, *J. Chromatogr.* **130**, 458 (1977).
165. S. R. Bakalyar and R. A. Henry, *J. Chromatogr.* **126**, 327 (1976).
166. H. Engelhardt and H. Elgass, *J. Chromatogr.* **112**, 415 (1975).
167. R. Majors, *Varian Instrum. Appl.* **10**(3), 8 (1976).
168. R. L. Sampson, *Am. Lab.* **9**(5), 109, 112 (1977).
169. S. R. Bakalyar, M. P. T. Bradley, and R. Honganen, *J. Chromatogr.* **158**, 277 (1978).
170. W. S. Hancock, C. A. Bishop, R. L. Prestidge, D. R. K. Harding, and M. T. W. Hearn, *J. Chromatogr.* **153**, 391 (1978).
171. W. S. Hancock, C. A. Bishop, J. E. Battersby, D. R. K. Harding, and M. T. W. Hearn, *J. Chromatogr.* **168**, 377 (1979).
172. H. J. Moeckel and B. Masloch, *Fresenius' Z. Anal. Chem.* **290**, 305 (1978).
173. W. R. Melander, J. Stoveken, and C. Horvath, in preparation.
174. W. R. Melander, J. Stoveken, and Cs. Horvath, *J. Chromatogr.* **185**, 111 (1979).
175. J. R. Gant, J. M. Dolan, and L. R. Snyder, *J. Chromatogr.* **185**, 153 (1979).
176. R. J. Perchalski and B. J. Wilder, *Anal. Chem.* **51**, 775 (1979).
177. W. Melander, D. E. Campbell, and Cs. Horváth, *J. Chromatogr.* **158**, 215 (1976).
178. J. H. Knox and G. Vasvari, *J. Chromatogr.* **83**, 181 (1973).
179. J. Leffler and E. Grunwald, "Rates and Equilibria of Organic Reactions." Wiley, New York, 1963.
180. I. Molnar and Cs. Horváth, *Clin. Chem.* **22**, 1497 (1976).
181. W. Kauzmann, *Adv. Protein Chem.* **14**, 1 (1959).
182. H. S. Frank and M. W. Evans, *J. Chem. Phys.* **13**, 507 (1945).
183. G. Nemethy and H. A. Scheraga, *J. Chem. Phys.* **36**, 3382 (1962).
184. H. Eyring and M. S. Jhon, "Significant Liquid Structures." Wiley, New York, 1969.
185. R. B. Hermann, *J. Phys. Chem.* **76**, 2754 (1972).
186. M. J. Harris, T. Higuchi, and J. H. Rytting, *J. Phys. Chem.* **77**, 2694 (1973).
187. R. B. Amidon, S. H. Yalkowsky, S. T. Anik, and S. C. Valvani, *J. Phys. Chem.* **79**, 2239 (1975).
188. O. Sinanoğlu and S. Abdulnur, *Photochem. Photobiol.* **3**, 333 (1964).
189. O. Sinanoğlu and S. Abdulnur, *Fed. Proc.* (Pt. III), S-12 (1965).
190. O. Sinanoğlu, *Chem. Phys. Lett.* **1**, 340 (1967).
191. O. Sinanoğlu, *In* "Advances in Chemical Physics" (J. O. Hirschfelder, ed.), Vol. XII, pp. 283–326. Wiley, New York, 1967.

308   Wayne R. Melander and Csaba Horváth

192. T. Halicioğlu, Ph.D. Thesis, Yale University, New Haven, Connecticut, 1968.
193. O. Sinanoğlu, In "Molecular Associations in Biology" (B. Pullman, ed.), pp. 427–445. Academic Press, New York, 1968.
194. O. Sinanoğlu, personal communication (1979).
195. W. Melander and Cs. Horváth, In "Activated Carbon Adsorption from the Aqueous Phase" (M. J. McGuire and I. H. Suffet, eds.). Ann Arbor Science Publ., Ann Arbor, Michigan, 1980.
196. L. Onsager, J. Am. Chem. Soc. 58, 1486 (1936).
197. M. H. Lietzke, R. W. Stoughton, and R. M. Fuoss, Proc. Natl. Acad. Sci. U.S.A. 50, 39 (1968).
198. J. G. Kirkwood, In "Proteins, Amino Acids and Peptides" (E. J. Cohn and J. T. Edsall, eds.). Reinhold, New York, 1943.
199. K. Linderstrøm-Lang, C.R. Trav. Lab. Carlsberg Ser. Chim. 28, 281 (1953).
200. I. Molnár and Cs. Horváth, J. Chromatogr. 145, 371 (1978).
201. C. M. Riley, E. Tomlinson, and T. M. Jefferies, J. Chromatogr. 185, 197 (1979).
202. J. K. Baker and C.-Y. Ma, J. Chromatogr. 169, 107 (1979).
203. I. Molnár and Cs. Horváth, J. Chromatogr. 142, 623 (1977).
204. B. T. Bush, Jr., J. H. Frenz, W. Melander, Cs. Horváth, A. R. Cashmore, R. N. Dreyer, J. K. Coward, J. O. Knipe, and J. R. Bertino, J. Chromatogr. 168, 343 (1979).
205. G. P. Cartoni and I. Ferretti, J. Chromatogr. 122, 287 (1976).
206. A. M. Jeffrey and P. P. Fu, Anal Biochem. 77, 298 (1976).
207. Cs. Horváth, W. Melander, and I. Molnár, Anal. Chem. 49, 142 (1977).
208. E. P. Kroeff and D. J. Pietrzyk, Anal. Chem. 50, 502 (1978).
209. C. Hansson, G. Agrup, H. Rorsmann, A.-M. Rosengren, E. Rosengren, and L.-E. Edholm, J. Chromatogr. 162, 7 (1979).
210. R. P. Scott and P. Kucera, J. Chromatogr. 142, 213 (1977).
211. E. H. Slaats, J. C. Kraak, W. J. T. Brugman, and H. Poppe, J. Chromatogr. 149, 255 (1978).
212. D. C. Locke, J. Chromatogr. Sci. 12, 433 (1974).
213. R. Defay and I. Prigogine, "Tension Superficielle et Adsorption." Desoer, Liege, 1951.
214. F. F. Cantwell and S. Puon, Anal. Chem. 51, 623 (1979).
215. R. M. Carlson, R. E. Carlson, and H. L. Kopperman, J. Chromatogr. 107, 219 (1975).
216. D. P. Wittmer, N. O. Nuessle, and W. G. Haney, Jr., Anal. Chem. 47, 1422 (1975).
217. C. P. Terweij-Groen, S. Heemstra, and J. C. Kraak, J. Chromatogr. 161, 69 (1978).
218. P. T. Kissinger, Anal. Chem. 49, 883 (1977).
219. J. L. M. van de Venne, J. L. H. M. Hendrikx, and R. S. Deelder, J. Chromatogr. 167, 1 (1978).
220. E. Tomlinson, S. S. Davis, and G. I. Mukhayer, In "Solution Chemistry of Surfactants" (K. L. Mittal, ed.), Vol. I, pp. 3–43. Plenum, New York, 1979.
221. A. Packter and M. Dowbrow, Proc. Chem. Soc. p. 220 (1962).
222. D. G. Oakenful and D. E. Fenwick, J. Phys. Chem. 78, 1759 (1974).
223. J. H. Knox and J. Jurand, J. Chromatogr. 125, 89 (1976).
224. J. C. Kraak, K. M. Jonker, and J. F. K. Huber, J. Chromatogr. 142, 671 (1977).
225. E. Tomlinson, C. M. Riley, and T. M. Jefferies, J. Chromatogr. 173, 89 (1979).
226. B. A. Bidlingmeyer, S. N. Deming, W. P. Price, Jr., B. Sachok, and M. Petrusek "Advances in Chromatography 1979" (A. Zlatkis, ed.), Chromatography Symposium, Houston, Texas, p. 435.
227. A. P. Konijnendijk and J. L. M. van de Venne, In "Advances in Chromatography 1979" (A. Zlatkis, ed.), p. 451. Chromatography Symposium, Houston, Texas.

228. E. Tomlinson and S. S. Davis, *J. Colloid Interfac. Sci.* **66**, 335 (1978).
229. T. P. Moyer and N.-S. Jiang, *J. Chromatogr.* **153**, 365 (1978).
230. J. Mitchell and C. J. Coscia, *J. Chromatogr.* **145**, 295 (1978).
231. I. M. Johansson, K.-G. Wahlund, and G. Schill, *J. Chromatogr.* **149**, 281 (1978).
232. S. P. Sood, D. P. Wittmer, S. A. Ismaiel, and W. G. Haney, *J. Pharm. Sci.* **66**, 40 (1977).
233. C. Olieman, L. Maat, K. Waliszewski, and H. C. Beyerman, *J. Chromatogr.* **133**, 382 (1977).
234. A. G. Ghanekar and V. A. Gupta, *J. Pharm. Sci.* **67**, 873 (1978).
235. P. E. Koehler and R. R. Eitenmiller, *J. Food Sci.* **43**, 1245 (1978).
236. M. F. Balandrin, A. D. Kinghorn, S. J. Smolenski, and R. H. Dobberstein, *J. Chromatogr.* **157**, 365 (1978).
237. J. H. Knox and J. Jurand, *J. Chromatogr.* **149**, 297 (1978).
238. A. R. Branfman and M. McComish, *J. Chromatogr.* **151**, 87 (1978).
239. C. P. Terweij-Groen and J. C. Kraak, *J. Chromatogr.* **138**, 245 (1977).
240. B. Vonach and G. Schomburg, *J. Chromatogr.* **149**, 417 (1978).
241. R. J. Tscherne and G. Capitano, *J. Chromatogr.* **136**, 337 (1977).
242. M. G. M. De Ruyter and A. P. De Leenheer, *Anal. Chem.* **51**, 43 (1979).
243. W. S. Hancock, C. A. Bishop, R. L. Prestidge, D. R. Harding, and M. T. W. Hearn, *Science* **200**, 1168 (1978).
244. C. E. Dunlap, III, S. Gentleman, and L. I. Lowney, *J. Chromatogr.* **160**, 191 (1978).
245. N. H. C. Cooke, R. L. Viavattene, R. Eksteen, W. S. Wong, G. Davies, and B. L. Karger, *J. Chromatogr.* **149**, 391 (1978).
246. J. N. LePage, W. Lindner, G. Davies, D. E. Seitz, and B. L. Karger, *Anal. Chem.* **51**, 433 (1979).
247. E. Tomlinson, T. N. Jeffries, and C. M. Riley, *J. Chromatogr.* **159**, 315 (1978).
248. W. S. Hancock, C. A. Bishop, and M. T. W. Hearn, *Anal. Biochem.* **92**, 170 (1979).
249. P. E. Hare and E. Gil-Av, *Science* **204**, 1226 (1979).
250. B. L. Karger, W. S. Wong, R. L. Viavattene, J. N. LePage, and G. Davies, *J. Chromatogr.* **167**, 253 (1978).
251. M. T. W. Hearn, W. S. Hancock, and C. A. Bishop, *J. Chromatogr.* **157**, 337 (1978).
252. C. E. Dunlap, III, S. Gentleman, and L. I. Lowney, *J. Chromatogr.* **160**, 191 (1978).
253. J. E. Rivier, *J. Liq. Chromatogr.* **1**, 343 (1978).
254. J. L. Day, L. Tterlikkis, R. Niemann, A. Mobley, and C. Spikes, *J. Pharm. Sci.* **67**, 1027 (1978).
255. A. G. Ghanekar and V. Das Gupta, *J. Pharm. Sci.* **67**, 1247 (1978).
256. A. G. Ghanekar and V. Das Gupta, *J. Pharm. Sci.* **67**, 873 (1978).
257. N. D. Brown and H. K. Sleeman, *J. Chromatogr.* **138**, 449 (1977).
258. K. Sugden, D. B. Cox, and C. R. Loscombe, *J. Chromatogr.* **149**, 377 (1978).
259. I. S. Lurie and J. M. Weber, *J. Liq. Chromatogr.* **1**, 587 (1978).
260. V. Das Gupta, *J. Pharm. Sci.* **68**, 118 (1979).
261. J. T. Stewart, I. L. Honigberg, and J. W. Coldren, *J. Pharm. Sci.* **68**, 32 (1979).
262. R. J. Helms, I. A. Muni, and J. L. Leeling, *J. Pharm. Sci.* **67**, 1185 (1978).
263. H. M. Hill and J. Chamberlain, *J. Chromatogr.* **149**, 349 (1978).
264. M. Wermeille and G. Huber, *J. Chromatogr.* **160**, 297 (1978).
265. J. L. Wisnicki, W. P. Tong, and D. B. Ludlum, *Cancer Treat. Rep.* **62**, 529 (1978).
266. R. L. Kirchmeier and R. P. Upton, *J. Pharm. Sci.* **67**, 1444 (1978).
267. E. Tomlinson, C. M. Riley, and T. M. Jefferies, *Methodol. Surv. Biochem.* **7**, 333 (1978).
268. L. A. Sternson and W. J. DeWitte, *J. Chromatogr.* **137**, 305 (1977).

269. V. Walters and N. V. Rahavan, *J. Chromatogr.* **176**, 470 (1979).
270. H. U. Ehmcke, H. Kelker, K. H. Koenig, and H. Ullner, *Fresenius' Z. Anal. Chem.* **294**(4), 251 (1979).
271. A. Mangia, G. Parolari, E. Gaetani, and C. F. Laureri, *Anal. Chim. Acta* **92**, 111 (1977).
272. P. H. Culbreth, I. W. Duncan, and C. A. Burtis, *Clin. Chem.* **23**, 2288 (1977).
273. B. C. Hemming and C. J. Gubler, *Anal. Biochem.* **92**, 31 (1979).
274. W. J. Al-Thamir, J. H. Purnell, C. A. Wellington, and R. J. Laub, *J. Chromatogr.* **173**, 388 (1979).
275. N. A. North, *J. Phys. Chem.* **77**, 931 (1973).
276. R. C. Neuman, Jr., W. Kauzmann, and A. Zipp, *J. Phys. Chem.* **77**, 2687 (1973).
277. J. F. Brandts, R. J. Oliviera, and C. Westort, *Biochemistry* **9**, 1038 (1970).
278. S. A. Hawley, *Biochemistry* **10**, 2436 (1971).
279. A. Zipp and W. Kauzmann, *Biochemistry* **12**, 4217 (1973).
280. H.-J. Lin and Cs. Horváth, *Chem. Eng. Sci.* (1980) in press.
281. G. Prukof and L. B. Rogers, *Sep. Sci.* **13**, 59 (1978).
282. C. Hansch, *Acc. Chem. Res.* **2**, 232 (1969).
283. A. Leo, C. Hansch, and D. Elkins, *Chem. Rev.* **71**, 525 (1971).
284. R. Rekker, "The Hydrophobic Fragmental Constant." Elsevier, Amsterdam, 1977.
284a. E. Tomlinson, *J. Chromatogr.* **113**, 1 (1975).
285. T. Yamana, A. Tsuji, E. Miyamoto, and O. Kubo, *J. Pharm. Sci.* **66**, 747 (1977).
286. J. K. Baker, *Anal. Chem.* **51**, 1693 (1979).
287. K. Miyake and H. Terada, *J. Chromatogr.* **157**, 387 (1978).
288. S. H. Unger, J. R. Cook, and J. S. Hollenberg, *J. Pharm. Sci.* **67**, 1364 (1978).
289. A. Hulshoff and J. H. Perrin, *J. Chromatogr.* **129**, 263 (1976).
290. A. Nahum and Cs. Horváth, *J. Chromatogr.* **192**, 315 (1980).
291. V. Jokl, *J. Chromatogr.* **71**, 523 (1972).
292. V. Jokl and I. Valaskova, *J. Chromatogr.* **72**, 373 (1972).
293. "Handbook of Chemistry and Physics," p. D-120. The Chemical Rubber Co., Cleveland, Ohio, 1972.
294. A. W. Walde, *J. Phys. Chem.* **43**, 431 (1939).
295. H. C. Brown, D. H. McDaniel, and O. Häfligu, *In* "Determination of Organic Structures by Physical Methods," p. 567. Academic Press, New York, 1955.
296. V. P. Vasilév and L. A. Kochergina, *Russ. J. Phys. Chem.* **41**, 1149 (1967).
297. J. J. Christensen, D. P. Wrathall, R. M. Izatt, and D. O. Tolman, *J. Phys. Chem.* **71**, 3001 (1967).
298. Cs. Horváth, W. Melander, and I. Molnár, *Abstr. Nat. Meet. Am. Chem. Soc., 173rd* Anal 031 (1977).
299. E. sz. Kovats, *In* "Advances in Chromatography" (J. C. Giddings and R. A. Keller, eds.), Vol. I. Dekker, New York, 1965.
300. E. J. Kikta, Jr. and A. E. Stange, *J. Chromatogr.* **138**, 41 (1977).
301. W. R. Melander, B.-K. Chen, and Cs. Horváth, *J. Chromatogr.* **185**, 99 (1979).
302. W. R. Melander, B.-K. Chen, and Cs. Horváth, in preparation.
303. M. A. Carroll, E. R. White, Z. Jáncsik, and J. E. Zarembo, *J. Antibiot* **30**, 397 (1977).
304. T. Nakagawa, J. Haginaka, K. Yamaoka, and T. Uno, *J. Antibiot.* **31**, 769 (1978).
305. J. S. Wold and S. A. Turnipseed, *J. Chromatogr.* **136**, 170 (1977).
306. J. S. Wold and S. A. Turnipseed, *Clin. Chim. Acta* **78**, 203 (1977).
307. I. Nilsson-Ehle, T. T. Yoshikawa, M. C. Schotz, and L. B. Guze, *Antimicrob. Agents Chemother.* **13**, 221 (1978).

308. W. G. Peng, M. A. F. Gadalla, and W. L. Chiou, *Res. Commun. Chem. Pathol. Pharmacol.* **18**, 233 (1977).
309. C. Fischer and U. Klotz, *J. Chromatogr.* **146**, 157 (1978).
310. T. J. Goehl, L. K. Mathur, J. D. Strum, J. M. Jaffe, W. H. Pitlick, V. P. Shah, R. I. Poust, and J. L. Colaizzi, *J. Pharm. Sci.* **67**, 404 (1978).
311. K. Harzer, *J. Chromatogr.* **155**, 399 (1978).
312. K. Lanbeck and B. Lindstrom, *J. Chromatogr.* **154**, 321 (1978).
313. F. Salto, *J. Chromatogr.* **161**, 379 (1978).
314. T. B. Vree, Y. A. Hekster, A. M. Baars, and E. van der Kleijn, *J. Chromatogr.* **145**, 496 (1978).
315. J. R. Koup, B. Brodsky, A. Lau, and T. R. Beam, Jr., *Antimicrob. Agents Chemother.* **14**, 439 (1978).
316. J. M. Wal, J. C. Peleran, and G. Bories, *J. Chromatogr.* **145**, 502 (1978).
317. I. Nilsson-Ehle and P. Nilsson-Ehle, *In Curr. Chemother., Proc. Int. Cong. Chemother.*, (W. Siegenthaler and R. Luethy, eds.), Vol. I, pp. 527–9. *Am. Soc. Microbiol. Washington, D.C.*
318. R. L. Nation, G. W. Peng, V. Smith, and W. L. Chiou, *J. Pharm. Sci.* **67**, 805 (1978).
319. T. A. Taulli, J. T. Hill, and G. W. Pounds, *J. Chromatogr. Sci.* **15**, 111 (1977).
320. R. A. Marques, B. Stafford, N. Flynn, and W. Sadee, *J. Chromatogr.* **146**, 163 (1978).
321. J. P. Sharma, E. G. Perkins, and R. F. Bevill, *J. Chromatogr.* **134**, 441 (1977).
322. A. P. de Leenheer and H. J. C. F. Nelis, *J. Chromatogr.* **140**, 293 (1977).
323. J. P. Anhalt, *Antimicrob. Agents Chemother.* **11**, 651 (1977).
324. I. Nilsson-Ehle, T. T. Yoshikawa, J. E. Edwards, C. M. Schotz, and L. B. Guze, *J. Infect. Dis.* **135**, 414 (1977).
325. A. Gulaid, G. W. Houghton, O. R. W. Lewellen, J. Smith, and P. S. Thorne, *Br. J. Clin. Pharmacol.* **6**, 430 (1978).
326. K. Lanbeck and B. Lindstrom, *J. Chromatogr.* **162**, 117 (1979).
327. S. Y. Chang, D. S. Albert, L. R. Melnick, P. D. Walson, and S. E. Salmon, *J. Pharm. Sci.* **67**, 679 (1978).
328. K. Nakamura, M. Asami, K. Kawada, and K. Sashara, *Sankyo Kenkyusho Nempo* **29**, 66 (1977).
329. N. Hobara and A. Watanabe, *J. Chromatogr.* **146**, 518 (1978).
330. D. C. Chatterji and J. F. Gallelli, *J. Pharm. Sci.* **66**, 1219 (1977).
331. J. A. Nelson, B. A. Harris, W. J. Decker, and D. Farquhar, *Cancer Res.* **37**, 3970 (1977).
332. P. F. Dixon and M. S. Stoll, *In* "High Pressure Liquid Chromatography in Clinical Chemistry" (P. F. Dixon, C. H. Gray, C. K. Lim, and M. S. Stolle, eds.), pp. 165–74. Academic Press, London, 1976.
333. G. L. Blackman, G. J. Jordan, and J. D. Paull, *J. Chromatogr.* **145**, 492 (1978).
334. R. W. Dykeman and D. J. Ecobichon, *J. Chromatogr.* **162**, 104 (1979).
335. P. M. Kabra, D. M. McDonald, and L. J. Marton, *J. Anal. Toxicol.* **2**, 127 (1978).
336. P. M. Kabra, B. E. Stafford, and L. J. Marton, *Clin. Chem.* **23**, 1284 (1977).
337. G. W. Peng, M. A. F. Gadalla, V. Smith, A. Peng, and W. L. Chiou, *J. Pharm. Sci.* **67**, 710 (1978).
338. D. Blair, B. H. Rumack, and R. G. Peterson, *Clin. Chem.* **24**, 1543 (1978).
339. L. T. Wong, G. Solomonraj, and B. H. Thomas, *J. Chromatogr.* **150**, 521 (1978).
340. G. Palmskog and E. Hultman, *J. Chromatogr.* **140**, 310 (1977).
341. A. R. Buckpitt, D. E. Rollins, S. D. Nelson, R. B. Franklin, and J. R. Mitchell, *Anal. Biochem.* **83**, 168 (1977).

342. P. M. Kabra, G. L. Stevens, and L. J. Marton, *J. Chromatogr.* **150**, 355 (1978).
343. R. R. Brodie, L. F. Chasseaud, and T. Taylor, *J. Chromatogr.* **150**, 361 (1978).
344. H. B. Greizerstein and C. Wojtowicz, *Anal. Chem.* **49**, 2235 (1977).
345. J. C. Kraak and P. Bijster, *J. Chromatogr.* **143**, 459 (1977).
346. H. F. Proelss, H. J. Lohmann, and D. G. Miles, *Clin. Chem.* **24**, 1948 (1978).
347. S. R. Biggs, R. R. Brodie, D. R. Hawkins, and I. Midgley, *Proc. Eur. Soc. Toxicol.* **18**, 174 (1977).
348. C. A. Frolik, T. E. Tavela, and M. B. Sporn, *J. Lipid Res.* **19**, 32 (1978).
349. K. Abe, K. Ishibashi, M. Ohmae, K. Kawabe, and G. Katsui, *Vitamins* **51**, 275 (1977).
350. C. A. Frolik, T. E. Tavela, G. L. Peck, and M. B. Sporn, *Anal. Biochem.* **86**, 743 (1978).
351. G. Jones, *In* "Vitamin D: Biochemical, Chemical and Clinical Aspects Related to Calcium Metabolism: Proceedings of the Third Workshop on Vitamin D" (A. W. Norman, K. Schaefer and J. W. Coburn, eds.), pp. 491–500. de Gruyter, Berlin, 1977.
352. K. T. Koshy and A. L. VanderSlik, *Anal. Lett.* **10**, 523 (1977).
353. E. Hansbury and T. J. Scallen, *J. Lipid Res.* **19**, 742 (1978).
354. R. Gottschlich and M. Metzler, *Anal. Biochem.* **92**, 199 (1979).
355. P. L. Williams, A. C. Moffat, and L. J. King, *J. Chromatogr.* **155**, 273 (1978).
356. J. D. Teale, J. M. Clough, L. J. King, V. Marks, P. L. Williams, and A. C. Moffat, *J. Forensic Sci. Soc.* **17**, 177 (1977).
357. R. Sams, *Anal. Lett.* **B11**, 697 (1978).
358. E. J. Knudson, Y. C. Lau, H. Veening, and D. A. Dayton, *Clin. Chem.* **24**, 686 (1978).
359. H. Terada, T. Hayashi, S. Kawai, and T. Ohno, *J. Chromatogr.* **130**, 281 (1977).
360. R. H. McCluer and F. B. Jungalwala, *Adv. Exp. Med. Biol.* **68**, 533 (1976).
361. K. Abe, K. Ishibashi, M. Ohmae, K. Kawabe, and G. Katsui, *Vitamins* **51**, 111 (1977).
362. K. Abe, K. Ishibashi, M. Ohmae, K. Kawabe, and G. Katsui, *Vitamins* **51**, 119 (1977).
363. A. von Hodenberg, W. Klemisch, and K. O. Vollmer, *Arznei-Forsch.* **27**, 508 (1977).
364. R. A. Marques, B. Stafford, N. Flynn, and W. Sadée, *J. Chromatogr.* **146**, 163 (1978).
365. G. W. Peng, M. A. F. Gadalla, and W. L. Chiou, *Res. Commun. Chem. Pathol. Pharmacol.* **18**, 233 (1977).
366. L. T. Wong, G. Solomonraj, and B. H. Thomas, *J. Chromatogr.* **135**, 149 (1977).
367. W. B. Forman and J. Shlaes, *J. Chromatogr.* **146**, 522 (1978).
368. M. J. Fasco, L. J. Piper, and L. S. Kaminsky, *J. Chromatogr.* **131**, 365 (1977).
369. G. R. Elliott, E. M. Odam, and M. G. Townsend, *Biochem. Soc. Trans.* **4**, 615 (1976).
370. A. M. Taburet, A. A. Taylor, J. R. Mitchell, D. E. Rollins, and J. L. Pool, *Life Sci.* **24**, 209 (1979).
371. A. Yoshida, M. Yoshioka, T. Yamazaki, T. Sakai, and Z. Tamura, *Clin. Chim. Acta* **73**, 315 (1976).
372. L. D. Mell and A. B. Gustafson, *Clin. Chem.* **23**, 473 (1977).
373. G. Schwedt, *Chromatographia* **10**, 92 (1977).
374. G. Schwedt, *J. Chromatogr.* **143**, 463 (1977).
375. Y. Tsuruta, K. Kohashi, and Y. Ohkura, *J. Chromatogr.* **146**, 490 (1978).
376. G. Schwedt and H. H. Bussemas, *Fresenius' Z. Anal. Chem.* **283**, 23 (1977).
377. L. D. Mell, Jr., A. R. Dasler, and A. B. Gustasfson, *J. Liq. Chromatogr.* **1**, 261 (1978).
378. T. P. Davis, C. W. Gehrke, C. W. Gehrke, Jr., K. O. Gerhardt, H. D. Johnson, and C. H. Williams, *Clin. Chem.* **24**, 1317 (1978).
379. G. Schwedt, *Anal. Chim. Acta* **92**, 337 (1977).
380. J. C. Gfeller, G. Frey, J. M. Huen, and J. P. Thevenin, *J. Chromatogr.* **172**, 141 (1979).
381. P. T. Kissinger, C. S. Bruntlett, G. C. Davis, L. J. Felice, R. M. Riggin, and R. E. Shoup, *Clin. Chem.* **23**, 1449 (1977).

382. H. Hashimoto and Y. Maruyama, *J. Chromatogr.* **152**, 387 (1978).
383. R. J. Fenn, S. Siggia, and D. J. Curran, *Anal. Chem.* **50**, 1067 (1978).
384. I. Molnár, Cs. Horváth, and P. Jatlow, *Chromatographia* **11**, 260 (1978).
385. I. Molnár and Cs. Horváth, *J. Chromatogr.* **143**, 391 (1977).
386. M. Ogata, R. Sugihara, and S. Kira, *Sangyo Igaku* **19**, 46 (1977).
387. H. Matsui, M. Kasao, and S. Imamura, *J. Chromatogr.* **145**, 231 (1978).
388. J. C. Liao, N. E. Hoffman, J. J. Barboriak, and D. A. Roth, *Clin. Chem.* **23**, 802 (1977).
389. I. Bekersky, H. G. Boxenbaum, M. H. Whitson, C. V. Puglisi, R. Pocelinko, and S. A. Kaplan, *Anal. Lett.* **10**, 539 (1977).
390. J. M. Grindel, P. F. Tilton, and R. D. Shaffer, *J. Pharm. Sci.* **66**, 834 (1977).
391. K. Harzer and R. Barchet, *J. Chromatogr.* **132**, 83 (1977).
392. M. S. F. Ross, *J. Chromatogr.* **133**, 408 (1977).
393. F. R. Fullerton and C. D. Jackson, *Biochem. Med.* **16**, 95 (1976).
394. S. E. Magic, *J. Chromatogr.* **129**, 73 (1976).
395. J. W. Stanley and T. E. Shellenberger, *Proc. Eur. Soc. Toxicol.* **18**, 177 (1977).
396. K. Yamaoka, S. Narita, T. Nakagawa, and T. Uno, *J. Chromatogr.* **168**, 187 (1979).
397. C. B. Vree, Y. A. Hekster, A. M. Baars, J. E. Damsma, E. van der Kleijn, and J. Bron, *J. Chromatogr.* **162**, 110 (1979).
398. J. P. Anhalt, *Antimicrob. Ag. Chromatogr.* **11**, 651 (1977).
399. T. M. Jefferies, W. O. A. Thomas, and R. T. Parfitt, *J. Chromatogr.* **162**, 122 (1979).
400. B. L. Nation, G. W. Peng, and W. L. Chiou, *J. Chromatogr.* **145**, 429 (1978).
401. J. F. Pritchard, D. W. Schneck, and A. H. Hayes, Jr., *J. Chromatogr.* **162**, 47 (1979).
402. A. P. Graffeo, D. C. K. Lin, and R. L. Foltz, *J. Chromatogr.* **126**, 717 (1976).
403. S. J. Bannister, L. A. Sternson, and A. J. Repta, *J. Chromatogr.* **173**, 333 (1979).
404. C. H. Gray, C. K. Lim, and D. C. Nicholson, *Clin. Chim. Acta* **77**, 167 (1977).
405. R. A. Hartwick and P. R. Brown, *J. Chromatogr.* **126**, 383 (1977).
406. A. M. Krstulovic, P. R. Brown, and D. M. Rosie, *Anal. Chem.* **49**, 2237 (1977).
407. A. M. Krstulovic, R. M. Harwick, P. R. Brown, and K. Lohse, *J. Chromatogr.* **158**, 365 (1978).
408. H. G. Schneider and A. J. Glazko, *J. Chromatogr.* **139**, 370 (1977).
409. P. R. Brown, A. M. Krstulovic, and R. A. Hartwick, *Monogr. Hum. Genet.* **10**, 135 (1978).
410. G. E. Davis, R. D. Suits, K. C. Kuo, C. W. Gehrke, T. P. Waalkes, and E. Borek, *Clin. Chem.* **23**, 1427 (1977).
411. C. W. Gehrke, C. K. Kuo, G. E. Davis, R. D. Suits, T. P. Waalkes, and E. Borek, *J. Chromatogr.* **150**, 455 (1978).
412. C. E. Salas and O. Z. Sellinger, *J. Chromatogr.* **133**, 231 (1977).
413. K. M. Taylor, L. Chase, and M. Bewick, *J. Liq. Chromatogr.* **1**, 849 (1978).
414. R. A. Hartwick, C. M. Grill, and P. R. Brown, *Anal. Chem.* **51**, 34 (1979).
415. J. R. Miksic and P. R. Brown, *Biochemistry* **17**, 2234 (1978).
416. J. Uberti, J. J. Lightbody, and R. M. Johnson, *Anal. Biochem.* **80**, 1 (1977).
417. W. Melander and Cs. Horváth, *In* "Enzyme Engineering" (G. B. Broun, G. Manecke and L. B. Wingard, Jr., eds.), Vol. 4, pp. 355–364. Plenum, New York, 1978.
418. B. Vasquez and A. L. Bieber, *Anal. Biochem.* **79**, 52 (1977).
419. R. F. Adams, G. Schmidt, and W. Slavin, *Chromatogr. Newsl.* **4**(1), 10 (1976).
420. A. C. Hoefler and P. Coggon, *J. Chromatorg.* **129**, 460 (1976).
421. J. J. Orcutt, P. P. Kozak, S. A. Gillman, and L. H. Cummins, *Clin. Chem.* **23**, 599 (1977).
422. G. W. Peng, V. Smith, A. Peng, and W. L. Chiou, *Res. Commun. Chem. Pathol. Pharmacol.* **15**, 341 (1976).

423. G. W. Peng, M. A. F. Gadalla, and W. L. Chiou, *Clin. Chem.* **24**, 357 (1978).
424. M. J. Cooper, B. L. Mirkin, and M. W. Anders, *J. Chromatogr.* **143**, 324 (1977).
425. S. J. Soldin and J. G. Hill, *Clin. Biochem.* **10**, 74 (1977).
426. R. E. Hill, *J. Chromatogr.* **135**, 419 (1977).
427. J. G. Kelly and W. J. Leahey, *Br. J. Clin. Pharmacol.* **3**, 947 (1976).
428. D. J. Popovich, E. T. Butts, and C. J. Lancaster, *J. Liq. Chromatogr.* **1**, 469 (1978).
429. J. A. Benvenuto, K. Lu, and T. L. Loo, *J. Chromatogr.* **134**, 219 (1977).
430. J. Zuidema and F. W. H. M. Merkus, *J. Chromatogr.* **145**, 489 (1978).
431. W. G. Kramer and S. Feldman, *J. Chromatogr.* **162**, 94 (1979).
432. H. Yanagawa, *Nucleic Acids Res., Spec. Publ.* **5**, 461 (1978).
433. R. M. Kothari and M. W. Taylor, *J. Chromatogr.* **86**, 289 (1973).
434. G. C. Walker, O. C. Uhlenbeck, E. Bedows, and R. I. Gumport, *Proc. Natl. Acad. Sci. U.S.A.* **72**, 122 (1975).
435. B. Z. Egan, *Biochim. Biophys. Acta* **299**, 245 (1973).
436. R. Dion and R. J. Cedergren, *J. Chromatogr.* **152**, 131 (1978).
437. S. C. Hardies and R. D. Wells, *Proc. Natl. Acad. Sci. U.S.A.* **73**, 3117 (1976).
438. A. Landy, C. Foeller, R. Reszelbach, and B. Dudock, *Nucleic Acids Res.* **3**, 2575 (1976).
439. H. Esaghour and D. M. Crothers, *Nucleic Acids Res.* **5**, 13 (1978).
440. J. E. Larson, S. C. Hardies, R. K. Patient, and R. D. Wells, *J. Biol. Chem.* **254**, 5535 (1979).
441. R. K. Patient, S. C. Hardies, and R. D. Wells, *J. Biol. Chem.* **254**, 5542 (1979).
442. R. K. Patient, S. C. Hardies, J. E. Larson, R. B. Inman, L. E. Maquat, and R. D. Wells, *J. Biol.. Chem.* **254**, 5548 (1979).
443. D. A. Usher, *Nucleic Acids Res.* **6**, 2289 (1979).
444. C. L. Zimmerman, E. Appella, and J. J. Pisano, *Anal. Biochem.* **77**, 569 (1977).
445. J. van Beeumen, J. van Damme, and J. de Ley, *FEBS Lett.* **93**, 373 (1978).
446. P. Kusch, *Wollforschungsinst. Tech. Hochsch. Aachen* **25**, 119 (1977).
447. J. M. Wilkinson, *J. Chromatogr. Sci.* **16**, 547 (1978).
448. J. Lammens and M. Verzele, *Chromatographia* **11**, 376 (1978).
449. J. Goto, M. Hasegawa, S. Nakamura, K. Shimada, and T. Nambara, *J. Chromatogr.* **152**, 413 (1978).
450. M. T. W. Hearn, W. S. Hancock, and C. A. Bishop, *J. Chromatogr.* **157**, 337 (1978).
451. D. L. Rabenstein and R. Saetre, *Anal. Chem.* **49**, 1036 (1977).
452. R. W. Frei, L. Michel, and W. Santi, *J. Chromatogr.* **126**, 665 (1976).
453. R. W. Frei, L. Michel, and W. Santi, *J. Chromatogr.* **142**, 261 (1977).
454. W. McHugh, R. A. Sandmann, W. G. Haney, S. P. Sood, and D. P. Wittmer, *J. Chromatogr.* **124**, 376 (1976).
455. J. A. Rodkey and C. D. Bennett, *Biochem. Biophys. Res. Commun.* **72**, 1407 (1976).
456. M. N. Margolies and A. Brauer, *J. Chromatogr.* **148**, 429 (1978).
457. L. S. O'Keefe and J. J. Warthesen, *J. Food Sci.* **43**, 1297 (1978).
458. J. J. Warthesen and P. L. Kramer, *Cereal Chem.* **55**, 481 (1978).
459. W. S. Hancock, C. A. Bishop, L. J. Meyer, D. R. K. Harding, and M. T. W. Hearn, *J. Chromatogr.* **161**, 291 (1978).
460. W. S. Hancock, C. A. Bishop, J. E. Battersby, D. R. K. Harding, and M. T. W. Hearn, *J. Chromatogr.* **168**, 377 (1979).
461. W. S. Hancock, C. A. Bishop, and M. T. W. Hearn, *Anal. Biochem.* **92**, 170 (1979).
462. W. S. Hancock, C. A. Bishop, R. L. Prestidge, D. R. K. Harding, and M. T. W. Hearn, *Science* **200**, 1168 (1978).

463. E. C. Nice and M. J. O'Hare, *J. Chromatogr.* **162**, 401 (1979).
464. J. E. Rivier, *J. Liq. Chromatogr.* **1**, 343 (1978).
465. R. Burgus and J. Rivier, *In* "Peptides 1976: Proceedings European Peptide Symposium, 14th" A. Loffet, ed., Univ. Bruxelles, Brussels, Belg., 1976, pp. 85–94.
466. W. Moench and W. Dehnen, *J. Chromatogr.* **140**, 260 (1977).
467. J. Rivier and R. Burgus, *Chromatogr. Sci.* **10**, 147 (1979).
468. M. T. Hearn and W. S. Hancock, *Trends Biochem. Sci.* **4**, N-58 (1979).
469. E. P. Kroeff and D. J. Pietrzyk, *Anal. Chem.* **50**, 1353 (1978).
470. E. Lundanes and T. Greibrokk, *J. Chromatogr.* **149**, 241 (1978).
471. K. Krummen and R. W. Frei, *J. Chromatogr.* **132**, 429 (1977).
472. M. J. Holland and J. P. Holland, *Biochemistry* **17**, 4900 (1978).
473. J. A. Glasel, *J. Chromatogr.* **145**, 469 (1978).
474. J. A. Feldman, M. L. Cohn, and D. Blair, *J. Liq. Chromatogr.* **1**, 833 (1978).
475. J. T. Stoklosa, B. K. Ayi, C. M. Shearer, and N. J. DeAngelis, *Anal. Lett.* **B11**, 889 (1978).
476. S. Gentleman, L. I. Lowney, B. M. Cox, and A. Goldstein, *J. Chromatogr.* **153**, 274 (1978).
477. L. F. Congote, H. P. J. Bennet, and S. Solomon, *Biochem. Biophys. Res. Commun.* **89**, 851 (1979).
478. M. Rubinstein, *Anal. Biochem.* **98**, 1 (1979).
479. J. J. O'Donnell, R. P. Sandman, and S. R. Martin, *Anal. Biochem.* **90**, 41 (1978).
480. A. M. Krstulovic, R. A. Hartwick, and P. R. Brown, *J. Chromatogr.* **163**, 19 (1979).
481. A. M. Krstulovic and C. Matzura, *J. Chromatogr.* **176**, 217 (1979).
481a. B. E. Chabannes, J. N. Bidard, N. N. Sarda, and L. A. Cronenberger, *J. Chromatogr.* **170**, 430 (1979).
482. R. Hartwick, A. Jeffries, A. Krstulovic, and P. R. Brown, *J. Chromatogr. Sci.* **16**, 427 (1978).
483. J. Studebaker, *Chromatogr. Sci.* **10**, 261 (1979).
484. R. W. Frei, *Int. J. Environ. Anal. Chem.* **5**, 143 (1978).
485. N. E. Skelly, T. S. Stevens, and D. A. Mapes, *J. Assoc. Off. Anal. Chem.* **60**, 868 (1977).
486. M. Arjmand, R. H. Hamilton, and R. O. Mumma, *J. Agric. Food Chem.* **26**, 971 (1978).
487. G. Glad, T. Popoff, and O. Theander, *J. Chromatogr. Sci.* **16**, 118 (1978).
488. R. A. Hoodless, J. A. Sidwell, J. C. Skinner, and R. D. Treble, *J. Chromatogr.* **166**, 279 (1978).
489. T. H. Byast, *J. Chromatogr.* **134**, 216 (1977).
490. D. Paschal, R. Bicknell, and K. Siebenmann, *J. Environ. Sci. Health* **B13**, 105 (1978).
491. D. C. Paschal, R. Bicknell, and D. Dresbach, *Anal. Chem.* **49**, 1551 (1977).
492. J. Kvalvag, D. L. Elliott, Y. Iwata, and F. A. Gunther, *Bull. Environ. Contam. Toxicol.* **17**, 253 (1977).
493. C. M. Sparacino and J. W. Hines, *J. Chromatogr. Sci.* **14**, 549 (1976).
494. R. T. Krause, *J. Chromatogr. Sci.* **16**, 281 (1978).
495. H. A. Moye and T. E. Wade, *Anal. Lett.* **9**, 891 (1976).
496. R. G. Christensen and W. E. May, *J. Liq. Chromatogr.* **1**, 385 (1978).
497. H. Boden, *J. Chromatogr. Sci.* **14**, 391 (1976).
498. F. Eisenbeiss, H. Hein, R. Joester, and G. Naundorf, *Chem. Tech.* **6**, 227 (1977).
499. A. R. Oyler, D. L. Bodenner, K. J. Welch, R. J. Liukkonen, R. M. Carlson, H. L. Kopperman, and R. Caple, *Anal. Chem.* **50**, 837 (1978).
500. M. Dong, D. C. Locke, D. Hoffmann, and N. Dana, *J. Chromatogr. Sci.* **15**, 32 (1977).

501. N. E. Skelly and E. R. Husser, *Anal. Chem.* **50,** 1959 (1978).
502. W. D. Felix, D. S. Farrier, and D. E. Poulson, *Energy Res. Abstr.* **2,** 50981 (1977).
503. R. A. Beck and J. J. Brink, *Environ. Sci. Technol.* **12,** 435 (1978).
504. J. B. Murphy and C. A. Stutte, *Anal. Biochem.* **86,** 220 (1978).
505. W. A. Court, *J. Chromatogr.* **130,** 287 (1977).
506. L. Szepesy, I. Fehér, G. Szepesi, and M. Gazdag, *J. Chromatogr.* **149,** 271 (1978).
507. J. Dolinar, *Chromatographia* **10,** 364 (1977).
508. C. Y. Wu and J. H. Wittick, *Anal. Chem.* **49,** 359 (1977).
509. S. Gorog, B. Herenyi, and K. Jovanovics, *J. Chromatogr.* **139,** 203 (1977).
510. G. Tittel and H. Wagner, *J. Chromatogr.* **153,** 227 (1978).
511. G. Tittel and H. Wagner, *J. Chromatogr.* **135,** 499 (1977).
512. O. Sticher and B. Meier, *Pharm. Acta Helv.* **53,** 40 (1978).
513. J. M. Enson and J. N. Seiber, *J. Chromatogr.* **148,** 521 (1978).
514. G. J. Niemann and J. van Brederode, *J. Chromatogr.* **152,** 523 (1978).
515. H. Becker, G. Wilking, and K. Hostettmann, *J. Chromatogr.* **136,** 174 (1977).
516. V. Quercia, *Boll. Chim. Farm.* **115,** 309 (1976).
517. M. J. Pettei and K. Hostettmann, *J. Chromatogr.* **154,** 106 (1978).
518. L. D. Mell, Jr., S. W. Joseph, and N. E. Bussell, *Gov. Rep. Anno. Index* **78**(23), 60 (1978).
519. A. T. Newhart and R. O. Mumma, *J. Chem. Ecol.* **4,** 503 (1978).
520. J. A. Holland, E. H. McKerrell, K. J. Fuell, and W. J. Burrows, *J. Chromatogr.* **166,** 545 (1978).
521. T. Kannangara, R. C. Durley, and G. M. Simpson, *Physiol. Plant.* **44,** 295 (1978).
522. A. J. Clifford, *Adv. Chromatogr.* **14,** 1 (1976).
523. G. Charalambous, "Liquid Chromatographic Analysis of Food and Beverages," Vol. I and II. Academic Press, New York, 1979.
524. J. L. Mietz and E. Karmas, *J. Food Sci.* **42,** 155 (1977).
525. G. W. Engstrom, J. L. Richard, and S. J. Cysewski, *J. Agric. Food Chem.* **25,** 833 (1977).
526. T. E. Moller and E. Josefsson, *J. Assoc. Off. Anal. Chem.* **61,** 789 (1978).
527. G. M. Ware and C. W. Thorpe, *J. Assoc. Off. Anal. Chem.* **61,** 1058 (1978).
528. E. Josefsson and T. Moeller, *J. Assoc. Off. Anal. Chem.* **62,** 1165 (1979).
529. D. M. Takahashi, *J. Chromatogr.* **131,** 147 (1977).
530. G. J. Diebold and R. N. Zare, *Science* **196,** 1439 (1977).
531. M. Uihlein, *Dtsch. Lebensm-Rundsch.* **73,** 157 (1977).
532. K. Kamata, T. Kan, H. Yamanobe, K. Mizuishi, R. Yamazoe, and T. Totani, *Tokyotoritsu Eisei Kenkyusho Kenkyu Nempo* **27,** 112 (1976).
533. H. R. Hunziker and A. Miserez, *Mitt. Geb. Lebensmittelunters. Hyg.* **68,** 267 (1977).
534. J. F. Fisher, *J. Agric. Food Chem.* **25,** 682 (1977).
535. J. F. Fisher, *J. Agric. Food Chem.* **26,** 1459 (1978).
536. E. D. Coppola, E. C. Conrad, and R. Cotter, *J. Assoc. Off. Anal. Chem.* 71, 1490 (1978).
537. M. Verzele and D. de Potter, *J. Chromatogr.* **166,** 320 (1978).
538. R. B. Toma and M. M. Tabekhia *J. Food. Sci.* **44,** 263 (1979).
539. S. K. Henderson and A. F. Wickroski, *In* "Application of High Pressure Liquid Chromatographic Methods for Determination of Fat Soluble Vitamins A, D, E, and K in Foods and Pharmaceuticals, Symposium Proceedings" Association of Vitamin Chemists Chicago, Ill. 1978, pp. 48–61.
540. S. K. Henderson and A. F. Wickroski, *J. Assoc. Off. Anal. Chem.* **61,** 1130 (1978).

541. J. N. Thompson and W. B. Maxwell, *J. Assoc. Off. Anal. Chem.* **60**, 766 (1977).
542. W. O. Landen, Jr. and R. R. Eitenmiller, *J. Assoc. Off. Anal. Chem.* **62**, 283 (1979).
543. W. Frede, *Dtsch. Lebensm. Rundsch.* **74**, 263 (1978).
544. J. Chudy, N. T. Crosby, and I. Patel, *J. Chromatogr.* **154**, 306 (1978).
545. K. Eskins, C. R. Scholfield, and H. J. Dutton, *J. Chromatogr.* **135**, 217 (1977).
546. K. R. Vincent and R. G. Scholz, *J. Agric. Food Chem.* **26**, 812 (1978).
547. A. T. R. Williams and W. Slavin, *J. Agric. Food Chem.* **25**, 756 (1977).
548. L. W. Wulf and C. W. Nagel, *Am. J. Enol. Vitic.* **29**, 42 (1978).
549. S. V. Ting, R. L. Rouseff, M. H. Dougherty, and J. A. Attaway, *J. Food Sci.* **44**, 69 (1979).
550. L. Ciraolo, R. Calapaj, and M. T. Clasadonte, *Rass. Chim.* **29**, 181 (1977).
551. S. Wada, C. Koizumi, and J. Nonaka, *Yukagaku* **26**, 95 (1977).
552. J. Graille, P. Ottaviani, and M. Naudet, *Actes Congr. Mond. Soc. Int. Etude Corps Gras, 13th, Sect. D* (M. Naudet, E. Ucciani and A. Uzzan, eds.), pp. 81–7. ITERG Paris, 1976.
553. W. P. Cochrane and M. Lanoutette, *J. Assoc. Off. Anal. Chem.* **62**, 100 (1979).
554. S. M. Norman and D. C. Fouse, *J. Assoc. Off. Anal. Chem.* **61**, 1469 (1978).
555. A. B. Vilim and A. I. MacIntosh, *J. Assoc. Off. Anal. Chem.* **62**, 19 (1979).
556. T. Nakagawa, Y. Sato, A. Watabe, T. Kawamura, and M. Morita, *Bull. Environ. Contam. Toxicol.* **19**, 703 (1978).
557. D. C. Paschal, L. L. Needham, Z. J. Rollen, and J. A. Liddle, *J. Chromatogr.* **177**, 85 (1979).
558. R. D. Stubblefield and O. L. Shotwell, *J. Assoc. Off. Anal. Chem.* **60**, 784 (1977).
559. W. A. MacCrehan and R. A. Durst, *Anal. Chem.* **50**, 2108 (1978).
560. F. Erni, R W. Frei, and W. Lindner, *J. Chromatogr.* **125**, 265 (1976).
561. S. C. Pattacini and J. Stoveken, *Chromatogr. Newsl.* **4**, 5 (1976).
562. R. W. Yost, *Chromatogr. Newsl.* **5**, 44 (1977).
563. R. B. H. Wills, C. G. Shaw, and W. R. Day, *J. Chromatogr. Sci.* **15**, 262 (1977).
564. M. Eriksson, T. Eriksson, and B. Sorenson, *Acta Pharm. Suec.* **15**, 274 (1978).
565. A. M. McCormick, J. L. Napoli, and H. F. Deluca, *Anal. Biochem.* **86**, 25 (1978).
566. J. W. Dolan, J. R. Gant, N. Tanaka, R. W. Giese, and B. L. Karger, *J. Chromatogr. Sci.* **16**, 616 (1978).
567. S. L. Ali, *J. Chromatogr.* **126**, 651 (1976).
568. K. Harzer and R. Barchet, *Dtsch. Apoth-Z.* **116**, 1229 (1976).
569. V. Das Gupta and S. Sachanandani, *J. Pharm. Sci.* **66**, 897 (1977).
570. V. Das Gupta, *J. Pharm. Sci.* **66**, 110 (1977).
571. P. Helboe and M. Thomsen, *Arch. Pharm. Chemi. Sci. Ed.* **6**, 453 (1978).
572. L. W. Brown, *J. Pharm. Sci.* **67**, 669 (1978).
573. C. R. Williams, *Proc. Anal. Div. Chem. Soc.* **14**, 242 (1977).
574. S. Bhuwapathanapun and P. Gray, *J. Antibiot.* **30**, 673 (1977).
575. E. R. White, M. A. Carrol, and J. E. Zarembo, *J. Antibiot.* **30**, 811 (1977).
576. S. L. Ali and T. Strittmatter, *Int. J. Pharm.* **1**, 185 (1978).
577. J. E. Launer, *J. Assoc. Off. Anal. Chem.* **62**, 11 (1979).
578. K. Tsuji and J. F. Goetz, *J. Antibiot.* **31**, 302 (1978).
579. W. A. Trinler and D. J. Reuland, *J. Forensic Sci.* **23**, 37 (1978).
580. W. A. Trinler, D. J. Reuland, and T. B. Hiatt, *J. Forensic Sci. Soc.* **16**, 133 (1976).
581. A. B. Clark and M. D. Miller, *J. Forensic Sci.* **23**, 21 (1978).
582. P. J. Twitchett, P. L. Williams, and A. C. Moffat, *J. Chromatogr.* **149**, 683 (1978).
583. J. R. Salmon and P. R. Wood, *Analyst* **101**, 611 (1976).

584. J. Killacky, M. S. F. Ross, and T. D. Turner, *Planta Med.* **30**, 210 (1976).
585. O. Sticher and F. Soldati, *Pharm. Acta Helv.* **53**, 46 (1978).
586. J. M. Zehner and R. A. Simonaitis, *J. Assoc. Off. Anal. Chem.* **60**, 14 (1977).
587. F. Erni and R. W. Frei, *J. Chromatogr.* **130**, 169 (1977).
588. S. Khalil and K. Wahba, *J. Pharm. Sci.* **66**, 1625 (1977).
589. B. Kreilgard, *Arch. Pharm. Chemi. Sci. Ed.* **6**, 109 (1978).
590. W. G. Crouthamel and B. Dorsch, *J. Pharm. Sci.* **68**, 237 (1979).
591. J.-W. Hsieh, J. K. H. Ma, J. P. O'Donnell, and N. H. Choulis, *J. Chromatogr.* **161**, 366 (1978).
592. L. W. Brown, *J. Pharm. Sci.* **67**, 1254 (1978).
593. P. Helboe and M. Thomsen, *Arch. Pharm. Chemi. Sci. Ed.* **6**, 397 (1978).
594. K. R. Bagon and E. W. Hammond, *Analyst* **103**, 156 (1978).
595. A. Noordam, K. Waliszewski, C. Olieman, L. Maat, and H. C. Beyerman, *J. Chromatogr.* **153**, 271 (1978).
596. G. P. R. Carr, *J. Chromatogr.* **157**, 171 (1978).
597. R. Bushway and A. Hanks, *J. Chromatogr.* **134**, 210 (1977).
598. R. I. Freudenthal, D. C. Emmerling, and R. L. Baron, *J. Chromatogr.* **134**, 207 (1977).
599. T. J. Billings, A. R. Hanks, and B. M. Colvin, *J. Assoc. Off. Anal. Chem.* **59**, 1104 (1976).
600. M. Fuzesi, *J. Assoc. Off. Anal. Chem.* **62**, 5 (1979).
601. D. Mourot, B. Delepine, J. Boisseau, and G. Gayot, *J. Chromatogr.* **168**, 277 (1979).
602. K. L. Eaves, B. M. Colvin, A. R. Hanks, and R. J. Bushway, *J. Assoc. Off. Anal. Chem.* **60**, 1064 (1977).
603. U. R. Cieri, *J. Assoc. Off. Anal. Chem.* **62**, 168 (1979).
604. N. Parris, W. M. Linfield, and R. A. Barford, *Anal. Chem.* **49**, 2228 (1977).
605. L. P. Turner, D. McCullough, and A. Jackewitz, *J. Am. Oil Chem. Soc.* **53**, 691 (1976).
606. N. Parris, *J. Am. Oil Chem. Soc.* **55**, 675 (1978).
607. D. Thomas and J. L. Rocca, *Analysis* **7**, 386 (1979).
608. A. C. Hayman and N. A. Parris, *Conf. Anal. Chem. Appl. Spectrosc. Pittsburg Abstr.* **294**, 1979.
609. W. Winkle, *Chromatographia* **10**, 13 (1977).
610. G. Gordon and P. R. Wood, *Analyst* **101**, 876 (1976).
611. D. Gross and K. Strauss, *Kautsch. Gummi. Kunstst.* **29**, 741 (1976).
612. R. G. Lichtenthaler and F. Ranfelt, *J. Chromatogr.* **149**, 553 (1978).
613. A. F. Cunningham, G. D. Furneaux, and D. E. Hillman, *Anal. Chem.* **48**, 2192 (1976).
614. L. A. Sternson, W. J. DeWitte, and J. G. Stevens, *J. Chromatogr.* **153**, 481 (1978).
615. S. Shiono, T. Miyakura, J. Enomoto, and T. Imamura, *Anal. Chem.* **49**, 1963 (1977).
616. C. R. Clark and J.-L. Chan, *J. Chromatogr.* **150**, 273 (1978).
617. P. H. Culbreth, I. W. Duncan, and C. A. Burtis, *Clin. Chem.* **23**, 2288 (1977).
618. N. D. Brown and H. K. Sleeman, *J. Chromatogr.* **140**, 300 (1977).
619. L. A. Sternson and W. J. DeWitte, *J. Chromatogr.* **137**, 305 (1977).
620. D. F. Tomkins and J. H. Wagenknecht, *J. Electrochem. Soc.* **125**, 372 (1958).
621. S. Selim, *J. Chromatogr.* **136**, 271 (1977).
622. J. C. Paterson-Jones and C. D. McAuley, *In* "Internationale Jahrestagung-Institut für Chemie der Treib-und Explosivstoff" Fraunhofer-Ges. 1977, pp. 63–78.
623. H. M. Abdou, T. Medwick, and L. C. Bailey, *Anal. Chim. Acta* **93**, 221 (1977).
624. C. T. Enos, G. L. Geoffroy, and T. H. Risby, *J. Chromatogr. Sci.* **15**, 83 (1977).
625. G. Schwedt, *Chromatographia* **11**, 145 (1978).
626. S. Valenty and P. E. Behnken, *Anal. Chem.* **50**, 834 (1978).
627. N. Ichinoise, *Bunseki Kagaku* **27**, 671 (1978).

628. A. Otsuki, *J. Chromatogr.* **133**, 402 (1977).
629. J. D. Warthen, Jr., and N. Mandava, *J. Chromatogr.* **144**, 263 (1977).
630. H. C. Jordi, *J. Liq. Chromatogr.* **1**, 215 (1978).
631. J. F. Schabron, R. J. Hurtubise, and H. F. Silver, *Anal. Chem.* **50**, 1911 (1978).
632. N. V. Raghavan, *J. Chromatogr.* **168**, 523 (1979).
633. H. Roeper, *J. Chromatogr.* **166**, 305 (1978).
634. W. R. Melander, K. Kalghati, and Cs. Horváth, *J. Chromatogr.* in press.
635. W. R. Melander and Cs. Horváth, *J. Chromatogr.* in press.

INDEX

A

Acenaphthene, 105
Acentric factor, 206
Acephyline, 287
Acesulfam, 295
Acetaminophen, 282
Acetic acid, 72, 174, 188
Acetonaphthalene, 76
Acetone, 166, 174, 176, 180
Acetonitrile, 72, 75, 166, 167, 173, 174, 176, 180, 190, 275
Acetonitrile–water, 182, 184, 225, 262
Acetonitrile–water mixtures, dependence of viscosity on composition, 194
  dependence of viscosity on temperature, 194
  properties of, 168,
  thermodynamic properties as a function of composition, 216
Acetophenone, 42, 74, 86, 182, 183
2-Acetylaminofluorene, 286
Acetylsalicylic acid, 297
N^4-Acetylsulfamethoxazole, 281
Acid dissociation constants, 190
  measurement of by RPC, 239, 278
  weak acids, 238–239
  weak bases, 239
Acid phosphatase, 291
ACNU, 282
Acrylamide, 293
1-24 ACTH pentaacetate, 262
Activated sludge, 293
Activation energies for binding and dissociation in RPC, 227
Activity coefficient, 209, 210
Adamantylethylmethyldichlorosilane, 135
Adamantylethyltrichlorosilane, 135
Adamantyl silica, 158
Adenine, arabinosides of, 287
Adenosine, 286, 291
Adenosine deaminase, 291

Adjusted retention time, 3
Adrenaline, 223, 243, 251, 264, 284
Adsorbed eluent monolayer, 102
Adsorption, 57
  comparison to partition, 225
  polar interactions in, 60
Adsorption chromatography, 58
  competition model, 101–103
  solvent interaction model, 101, 103, 104
Adsorption–desorption kinetics, effect on chromatographic efficiency in RPC, 227
Adsorption of detergents, 237
Adsorption isotherm
  Langmuir type, 58
  linear type, 58
Adsorption milieu, 58
Adsorption model in RPC, 224
Aflatoxins, analysis by RPC, 294, 296
  on polar sorbents, 105
Agarose, 120, 129
  hydrocarbonaceous, 129, 130
DL-Alanine, 222
Alanylalanine isomers, 291
Alcohols, 150
Aldehydes, 298
Alkaline phosphatase, 291
Alkaloids, analysis by RPC, 293
  on polar sorbents, 105
Alkanes, 275, 279, 280
Alkyl agarose, 129, 130
Alkyl ammonium hetaerons, 240
Alkylbenzyldimethylammonium chlorides, 248
Alkyl benzenes, 275
Alkyl bromides, 275
Alkyldimethylchlorosilanes, 136
Alkyl disulfides, 275
Alkyl-4-nitrophenylcarbonate esters, 298
Alkylphenols, 299
Alkylpolysiloxanes, 138, 139

Alkyl silica bonded phases, 136, 137, 143, 149–151
Alkyl sulfates, 244, 248
Alkylsulfonate hetaerons, 240
Alkyltrichlorosilanes, 138
Allantoin, 198, 272
Allopurinol, 287
Alumina, 60, 68
  acidic, 69
  basic, 69
  catalytic activity of, 69
  commercial stationary phases, 62–64
  comparison of retention behavior to that on silica, 69
  forms of, 68
  heat treatment of, 69–70
  Lewis acid sites at the surface of, 69
  mechanism of adsorption, 69
  neutral, 69
  pore size of, 68
  specific surface area of, 68
  surface properties of, 69
Amides, 275
Amine phosphate buffers, 190
Amines, 275
Amino acids, 211, 260, 262, 286, 289
  dansylated, 262
  separation of optical isomers by hetaeric chromatography, 290
Amino acid racemates, resolution of, 263
Aminobenzoic acid, 265
Amitriptyline, 282
Amobarbital, 264
Amoxycillin, 281
Amphetamines, 297
Amphiphilic ions, 243
Ampholytes, retention factor as a function of pH, 239
Amphoteric surfactants, 298
Amphotericin B, 282
Ampicillin, 281
Amylococaine, 264
Analgesics, 297
Analytical applications of RPC, classification of, 279
Angiotensin II, 290
Aniline, 265
Anisocratic elution, 92
Anisole, 182, 183
Anisopycnic slurries, 147

Anthanthrene, 105
Anthocyanin pigments, 295
Anthracene, 69, 74, 83, 94, 105, 107, 154, 179
Anthranilic acid, 278
Anthroquinones, 294
Antibiotics, 281
Antidiabetic drugs, 264
Antihistamines, 260
Antipyrine, 297
Argentation, 68
Argentation chromatography, 261
Aromatic acids in human urine, 285
Aromatic hydrocarbons, 69
Arylhydroxylamines, 298
Ascorbic acid, 296
Aspirin, 282
Asymmetric diens, 290
Asymmetrical peaks, 58, 82, 160
ATP, stability constants of metal complexes, 278
Atrazine, 292
Atropine, 297
Axial diffusion
  mobile phase, 8
  stationary phase, 8, 9
Aza-arenes, 293
Azoxybenzenes, 298

B

B vitamins, 265
Background electrolyte, 122, 161, 188
Band splitting, 98
Band spreading, 167
Band spreading, due to slow detector response, 44–46
  effect of sample introduction, 48
  extracolumn, 4
  in extracolumn dead space, 46
Barbital, 76
Barbiturates, 76, 77, 282
Bayerite, 68
Benactyzine-HCl, 298
Benzaldehyde, 182, 183
Benzamide, 182, 183
Benzanthracene, 83, 105, 179
Benzene, 66, 72, 83, 105, 107, 173, 179, 182, 183, 279
Benzenesulfonic acids, 265

Benzfluoranthene, 105, 293
Benzilic acid, 298
Benzocaine, 264, 297
Benzodiazepines, 286
Benzoic acid, 42, 221, 265, 278, 293, 297
Benzonitrile, 74, 182, 183
Benzoperylene, 105, 293
Benzophenone, 74
Benzopyrene, 105, 107, 179, 293, 296
Benzoyl peroxide, 42
Benzoylecgonine, 286
Benzyl alcohol, 86, 182, 183, 293
Benzyl silica, 158
Benzyldimethylchlorosilane, 135
Benzyltrichlorosilane, 135
BET method for surface area measurement, 270
Betacyanins, 295
Betaxanthins, 295
Bicycloheptyl-2-trichlorosilane, 135
Biodynamic hydrophobicity, 276
Biogenic amines, 284
Biopolymers, 123, 151
Biotechnology, 296
Biphenyl, 66, 83, 105, 107
Bis(cyanopropyl)dichlorosilane, 162
Bleomycins, 297
Bonded phases, 95
    advantages of, 106
    brush-type, 120, 136
    estersils, 120
    hydrocarbonaceous, 122, 139
    with hydrocarbonaceous ligates, 122, 126
    microparticulate, 119
    molecular fur, 136
    monomeric, 136, 137, 151
    pellicular, 119
    polymeric, 138, 151
    preparation of, 140
    rejuvenation, 140
    selectivity, 107, 108
    from silica, 68
    stability, 120, 126
Bonded stationary phases, 162
Born equation, 259
Brasan, 94
Bromide ion, 252
Bromobenzene, 94
3-Bromo-N-methyl cinnamamide, 195
Brompheniramine, 264

Bubble formation, 96
Buffer, 67, 122, 161, 188–190
Buffering capacity, 190
Butacaine, 264
Butanol, 177
Butyl chloride, 173
t-Butyldimethylchlorosilane, 135
n-Butyldimethyl silica, 143
t-Butyl perbenzoate, 42
n-Butyltrichlorosilane, 135

C

Cadaverine, 294
Caffeine, 287
Cannabinoids, 283, 297
Capacity factor, 3
Capacity ratio, 3, 58; see also Retention factor
Capping off, 139, 140
Carbamate
    insecticides, 292
    pesticides, 292
Carbamazipine, 282
Carbenicillin, 286
Carbon
    in stationary phase in RPC, 163
    surface properties, 163
Carbon black, 163
Carbon load, 140, 141, 153, 154, 157, 160
Carbon number, 153, 246, 248, 274
Carbon tetrachloride, 72
Cardiac glycosides, 105, 294
Carotenes, 295
Carotenoid, 295
Catalytic activity
    of alumina, 69
    of silica, 67
    of stationary phase, 118
Catecholamines, 244, 260, 264, 275, 284
Cavity formation, 218
Cefazolin, 281
DL-Cephalexin, 281
Cephaloridine, 281
Cephalosporins, 275, 281, 297
Cephalothin, 281
Cephradine, 281
Cerebrospinal fluid, 281
β-Cetotetrine, 286
Cetrimide, 250, 251, 252; see also Cetyltrimethylammonium bromide

Cetyltrimethylammonium bromide, 248–249
Charcoal, 116, 118, 163
Charge location in eluite, effect on binding in RPC, 224
Chemical reaction during chromatography, 67, 69
Chemisorption, 60
Chloramphenicol, 281
Chlorbenzenes, 275
Chlordiazepoxide, 282
Chlorobenzene, 182, 183
m-Chlorobenzoic acid, 230
Chloroform, 72, 173
6-Chloro-α-methylcarbazole-2-acetic acid, 282
Chlorophyll, 295
Chlorosilanes, 135, 140
m-Chlorotoluene, 182, 183
Chlortoluron, 292
Chlorzoxoneacetaminophen, 264
Δ^4-Cholestenone, 74
Cholestanyl silica, 158
Cholesterol, 283, 295
Cholinesterase, 293
Chromatographic enzyme reactor, 291
Chrysene, 69, 105, 107
Cinnamic acid, 293
Cinnamyl acetate, 198
Cinnamyl alcohol, 198
Cinnamyl aldehyde, 198
Clausius–Mosotti function, 208
Clindamycin, 297
Clonazepam, 195
Clusters of hetaeron, 250
Coating of polar adsorbent by polar solvent, 86
Cobalamins, 293
Cocaine, 264, 286, 297
Coenzyme Q, 283
Columns, heavily loaded, 68
Column chromatography
   comparison to TLC, 88–90
   stages of development, 1, 2
Column, conditioning of, 82
   inlet pressure, 167
   role of, 123–124
Column diameter, effect on observed efficiency, 43
Column efficiency, 5, 159
   as a function of carbon load, 160

Column-end frits, 148
Column length, 6, 16, 18, 23
   calculation of, 27
   in practice, 40
Column life, 161
Column overloading, 52
Column packing procedure, 125, 145, 146
   apparatus, 146–148
   preparation of, 146
Column packing structure, 125
Column packing, effect on efficiency, 11
   microparticulate, 132, 133
   particle size distribution of, 6, 7
   porosity of, 6
   structure of, 11
Column permeability, 7, 132, 161
Column pressure
   effect on chromatographic equilibrium constant, 266
   effect on retention factor, 267–268
Column regeneration, see Reequilibration of column
Column stability, 161
Column temperature, 267
   effect on retention, 192, 200
   effect on viscosity, 192
   nonuniform temperature profile, 194
   optimization of, 193
Column tubing, 145
   cleaning of, 145
   inner surface of, 145
Compensation behavior, see Enthalpy–entropy compensation
Compensation temperature, 199, 271, 272
   for reversed-phase chromatography, 273
   for solubility of hydrocabons in water, 273
Complex exchange with displacement of hetaeron, 254; see also Dynamic complex exchange
Complex exchange with no displacement, 233, 254, 258
Complex formation, 122, 229, 231, 249; see also Secondary equilibria
   with amphiphilic hetaerons, 244
   chromatographic measurement of stability constants, 276
   multiple, 277
Complexes in the mobile phase, stability constant of, 232
Complexes, outer sphere, 262
Complexing agent, 229; see also Hetaeron

Connecting tubes, contribution to extracolumn band spreading, 47–48
Coronene, 105
Coumarin, 295
Coupled columns, 98
Cresol, 182, 183
Cristobalite, 65
Critical micelle concentration, 256
Cromoglycate, sodium, 265
Crown ethers, 265, 299
  stability constants of metal complexes, 278
Cyano bonded phases, 162
β-Cyanoethylmethyldichlorosilane, 162
β-Cyanoethyltrichlorosilane, 162
Cyanogen bromide, 129
Cyanopropyldimethylchlorosilane, 162
Cyclohexane, 72, 279
2-(4-Cyclohexenyl)ethyldimethylchlorosilane, 135
2-(4-Cyclohexenyl)ethylmethyldichlorosilane, 135
2-(4-Cyclohexenyl)ethyltrichlorosilane, 135
Cycloheximide, 297
Cyclohexyltrichlorosilane, 135
DL-Cysteic acid, resolution of, 262
Cysteinyl dopa, 284
Cytochrome c, 263
Cytochrome P-450, 291
Cytokinins, 294

D

Darcy's law, 167, 192
Data transfer from TLC to column chromatography, 80–90
Davies equation, 210
Debye–Hückel theory, 209, 222
Decanoic acid, 260
Decyl sulfate, 289
Decyl trimethylammonium ion, 260
Degree of ionization, 224
Dehydration of silica, 61
Density, 169
N-Desmethyldiazepam, 282
Desmosterol, 283
Detection limit, 51
Detector cell volume, maximum allowable, 46
Detergent-based cation-exchange chromatography, 242

Detergents, 279
  effect on column stability, 161
  in hetaeric chromatography, 229–230, 236–238
Deuterated eluites, 219, 279
Diastereomers, chromatography on polar sorbents, 105
Diazepam, 282
Dibenzanthracene, 94, 105
1,4-Dibromobenzene, 153
4,4'-Dibromobiphenyl, 153
Dibutyldichlorosilane, 135
p-Dichlorobenzene, 182, 183
cis-Dichlorodiammineplatinum(II), 286
Dichloroethane, 72
Dichloromethane, 105
2,4-Dichlorophenoxyacetic acid, 292
Dielectric constant, 71, 165, 169, 208, 209, 217, 258
  effects on retention in RPC, 217
  mean value, 215
Diethyl ether, 72
1,1-Diethyldithiocarbamates, metal complexes of, 299
Diethylstilbesterol, 283
Diffusion, intraparticulate, 149
Diffusivity of eluite, dependence on molecular weight, 26
  effect on chromatographic parameters, 25
  effect on efficiency, 24–26
Diffusivity
  in liquids, calculation of, 29
  temperature dependence of, 192
Digitalis glycosides, 297
Diglycosides, 294
Dihexyldichlorosilane, 135
Dihydrofolate reductase, from mouse sarcoma, 289
Dihydroxyhydrobenzoin, 298
3,4-Dihydroxybenzylamine, 264
3,4-Dihydroxymandelic acid, 197
3,4-Dihydroxyphenylacetic acid, 197, 221, 223, 278
3,4-Dihydroxyphenylalanine, 251
3,4-Dihydroxyphenylalanine derivatives, 263
Diisopropyl ether, 72
Dilution in elution, 51
9,10-Dimethoxyanthracene-2-sulfonate, 284
Dimethyl sulfoxide, *see* DMSO

1,4-Dimethyl-9,10-anthraquinone, 175
Dimethyldichlorosilane, 117, 135
Dimethylformamide, 174
1,1-Dimethylhydrazine, 298
3,5-Dimethyl-4-(methylthio)phenylmethyl-
     carbamate, 297
Dimethylphthalate, 42, 86
Dinitrobenzenes, 182–184
Dinitronaphthalene, 76
Dinitrophenyl groups, 76
Dinitrophenylhydrazones, 298
Dinsed, 298
Dioxane, 72, 166, 174–176
Dipeptides, 263
Diphenyldichlorosilane, 135
Dipole moment, 208, 217
Diprotic acids, equilibrium constants and
     retention factors in hetaeric chroma-
     tography, 241
  retention factor as a function of pH, 239
Dissociation constants of complexes, mea-
     surement by chromatography, 276
DMSO, 175
DNA fragments, 288
Dodecyldiethylenetriamine, 262
Dodecyldimethyl silica, 143
Dodecyltrichlorosilane, 135
Dopa, 284
Dopamine, 223, 251, 264, 284
Drugs, in serum and plasma, 281
Dual site adsorption, 201
Dyes, in foods, 295
Dyestuffs, analysis, 260, 265
Dynamic complex exchange, 233, 252, 255
Dynamic ion exchange, 233, 244–248, 250,
     253, 254, 257, 258
Dynamically coated ion exchangers, 231,
     243

E

Efficiency, of commercial columns, 40
Eicosyltrichlorosilane, 135
Electrophoresis, comparison to RPC, 224
Electrostatic interactions, 120, 208, 213
  calculation of, 208–211
  dipoles, 208, 209
  at high ionic strength, 209–210
  in ion-pair chromatography, 258
  at low ionic strength, 209
  monopoles, 209–211

Eluent, in adsorption chromatography, 70,
     71
  dielectric constant of, 71
  effect of dissolved gases, 188
  elution strength of, 71
  helium sparge of, 188
  isohydric, 80
  mixed, 73, 88
  modulated with water, 80
  moisture content, 80
  polarity of, 71
  selection of, 87–89
     by using TLC data, 88–90
  ternary, 122
  viscosity, 6, 23–25, 71
Eluent composition, 190
  in chromatography on polar sorbent, 74
  effect on eluent strength, 73–75
  effect on selectivity, 77
  effect on the reproducibility of retention
     values, 77
  estimation, 190
  optimization of, 189–191
  on polar sorbent, 74
  rate of change in gradient elution, 186–
     187
  scouting by gradient elution, 190
  viscosity, 186, 187
Eluent mixtures, 88
Eluent properties, effect on retention in
     RPC, 215
Eluent strength, 4, 70, 169, 175, 181
  adjustment of, 73–77
  calculation of, 87
  of commonly used solvents, 72
  dependence on composition, 73–77
  effect on retention, 102
  modulation of, 59
  of mixed eluents, 73–77
  of neat eluents, 72
  optimization, 87
Eluite, chemical structure, effect on reten-
     tion in chromatography with polar
     sorbents, 90–92
  binding to stationary phase, 211–212
  concentration of, 16
  effect of molecular size on retention in
     RPC, 217–219
Eluotropic series, 71
  of commonly used solvents, 72

Eluotropic strength, 165, 166, 170–172, 174, 181
  definition of, 170
Elution order, as a function of chemical structure in chromatography with polar eluents, 91, 92
Elution time, 186
  error associated with the measurement of, 270, 271
  in gradient elution, 186
  of inert tracers, 266
Endorphins, 262, 290, 291
Energetics of adsorption, 58
Enhancement factor, 245, 248
  dependence on the chain length of alkyl sulfates, 246–247
Enolase, 290
Enthalpy for conformation change, 270
Enthalpy of retention, 192, 268, 269
  compensation, 197–199
  dependence on eluent composition, 200
  dual site adsorption, 201
  evaluation of, 196, 269
  heat capacity effects, 195
  magnitude in RPC, 198, 199
  with secondary equilibria, 195, 200
Enthalpy–entropy compensation, 197, 271–273
Entropic contributions, to free energy of solute binding, 212
Entropy change, 269
Entropy of mixing, 202
Environmental analysis, 292
Enzyme assay, 291
Enzyme reactions, kinetic study by RPC, 292
Enzymic peak-shift, 287
Epinephrine, 264
Equilibration of the chromatographic system, batch method, 78
  interaction of water with polar sorbents, 78
  perfusion method, 78
Equilibrium constant, 149
  measurement by chromatography, pressure dependence, 266
  for secondary equilibria with diprotic acids and zwitterions, 241
  for solute distribution between mobile and stationary phase, 211, 228, 266

Equilibrium constant for the chromatographic distribution process, 266
Equilibrium constant for distribution between two phases
  measured by chromatography, 266
  pressure dependence, 266
Equilibrium constant for retention in RPC, estimation of, 228
Equilin, 265
Ergot, 294
Erythromycin, 264, 297
Esters, 275
Estersils, 120
Estradiol, 265
Estragole, 198
Estriol, 265
Estrone, 265
Ethanol, 72, 166, 173–176
Ethanol–water mixtures, properties of, 171
Ethinylestriadiol, 297
Ethosuximide, 282
Ethylbenzene, 74, 182, 183
Ethyl benzoate, 182, 183
Ethyl bromide, 72
Ethylene glycol, 72, 175
Ethyl parathion, 292
4-Ethylsulfonylnaphthalene-1-sulfonamide, 286
Exclusion effects, 150
Extrachromatographic experiments, 243, 270
Extracolumn band spreading, 4, 7, 46–51
Extracolumn contributions to band broadening, 54
Extraction chromatography, 118–119

F

Fatty acids, 294, 296, 299
  derivatives for analysis by RPC, 299
  perdeuterated, 279
Fermentation, 296
Filtration packing, 145, 146
Fines, 133
  effect on column stability, 161
  removal of, 132, 133
Flavones, 295
Flow maldistribution, 8
Flow programming, 93–95
Flow velocity, optimum, 14, 19
Flowrate, optimum, 17

Fluoranthene, 83, 105, 293
Fluorescamine, reaction with amino acids, 289
Fluorocarbonaceous bonded phases, 162
Fluorouracil, 282, 287
Folic acid, 282, 296
  derivatives, 260, 265
Food dyes, 295
Foodstuffs, analysis by RPC, 293
Formation constant, apparent, 254
  of complexes in the mobile phase, 232
Free energy change, 204
  for cavity formation in liquid, 204–205
  for electrostatic interactions, 208–211
  entropic contributions, 212
  for retention, 211
  for solute–solvent interactions, 206
  for transfer from gas into liquid, 204
Free energy change and resolution, 4
Free volume, 206
Freundlich isotherm, 249
Ftorafur, 287
Fur configuration of ligates in bonded phases, 157
Furazolidine, 298

                  G

Gaussian peaks, 3, 8
Gentamicin, 282, 286
Gentian root, 294
Gentiopicroside, 294
Gentisin, 294
Ghost peaks, 188
Gibbs free energy for retention, 270
Globins, 291
Glucagon, 262, 290
Glucose-1-phosphate, 162
Glyceraldehyde-3-phosphate dehydrogenase, 290
Glycoflavones, 294
Glycosides, 294
Glycosphingolipids, 283
Glycyrrhizic acid, 297
Gradient elution, 80, 93, 98–101, 123, 169, 184
  column reequilibration, 84
  control of the modulator concentration in the eluent by using a precolumn and temperature programming, 85

  elution time, 186
  estimation of retention, 185, 186
  formation of solvent gradient, 100
  gradient slope, 186
  importance of solvent purity, 100
  for optimization of eluent composition in isocratic elution, 100
  prediction of retention, 185–186
  by using syringe pumps, 187
  viscosity changes, 187
Gradient shape, 184
Gradient volume, 186
Griseofulvin, 281
Group selectivity, 180, 216
Guaiacolsulfonate, 264
Gum opium, 294

                  H

Halides, detrimental effect on stainless steel, 189
Heat capacity change, 269
Height equivalent to a theoretical plate, 8
Helium sparge, 188
Hemodialysis, 283
Heptane, 72, 76, 173
Heptanoic acid, 260
1-Heptanol, 182, 183
Herbicides, analysis by RPC, 292
Herbicides, chromatography on polar sorbents, 105
Hesperidin, 295
Hetaeric chromatography, 230, 251
  effect of charge on hetaeron, 253
  retention model of, 231–238
Hetaeron, 191, 230, 231, 240, 243, 249, 280; see also Complexing agent
  adsorption on the stationary phase, 231, 249, 250
  amphiphilic, 243
  cetrimide, 248
  decylsulfonate, 250
  dodecylbenzenesulfonate, 250
  formation constant of complexes, 276
  lauryl sulfate, 250
  metal chelating, 262
  micelle formation, 250
  optically active, 262
  surface concentration of, 232

Hetaeron binding to the stationary phase, 249, 253
Hetaeron concentration, effect on adrenaline retention, 243
HETP, *see* Height equivalent to a theoretical plate
Hexadecyldimethyl silica, 143
Hexadecyltrichlorosilane, 119
Hexamethyldisilazane, 139
Hexane, 72, 173
1-Hexanol, 178, 182, 183
Hexylmethyldichlorosilane, 135
Hexyltrichlorosilane, 135
High-performance thin-layer chromatography, 89
High pressure, role in high performance in HPLC, 2
High pressure liquid chromatography, 2
Hippuric acid, 286
Histamine, 284, 294
Homologous series, 219
Homovanillic acid, 221, 223, 278, 284
HPLC, 2
HPTLC, *see* High-performance thin-layer chromatography
Hydantoins, 198, 272, 282
Hydraulic radius, 7
Hydrazine, 298
Hydrocarbonaceous agarose, 130
Hydrocarbonaceous bonded phases, 115, 123, 128, 131, 135, 142, 148, 151, 211
    with aromatic ligates, 158
    carbon load of, 140, 142
    characterization of, 141–145
    hydrolysis of, 161
    morphology of, 130
    with polarizable ligates, 214
    properties of, 143
    reagents for preparation of, 135
    surface concentration of silanols, 141, 142
Hydrocarbonaceous ligates, 128, 141, 161
    carbon number of, 155–157
    ligates, wetting of, 157
    solvation of, 144
    surface concentration of, 142, 144
    surface configuration, fur, 156
        picket fence, 156
        stack, 156

Hydrocarbonaceous molecular fur, 126
Hydrocarbonaceous surface area, 143, 153, 157
Hydrocarbons, solubility in water, 273
Hydrogen bonding, 229
γ-Hydrooctafluorobutyltrichlorosilane, 162
Hydroorganic eluents, 122, 179
    polar group selectivity of, 183
    preferential binding of less polar component by nonpolar stationary phase, 225
    selectivity, 184–185
    water–acetonitrile mixtures, 168
    water–ethanol mixtures, 171
    water–methanol mixtures, 168
    water–n-propanol mixtures, 170
    water–tetrahydrofuran mixtures, 169
Hydroorganic mixtures, 181, 216
    dependence of activity coefficients on the composition of, 225, 227, 228
    molar volume of, 170
    ternary, 191
Hydrophobic affinity chromatography, 120, 129; *see also* Hydrophobic chromatography
Hydrophobic bond, 202
Hydrophobic chromatography, 130
Hydrophobic effect, 201
Hydrophobic index system, 274
Hydrophobic interactions, 120, 130, 141
Hydrophobic surface, 128
Hydrophobic surface area, 246, 247
Hydrophobicity, biodynamic, 276
Hydrothermal treatment, 134
4-Hydroxymandelic acid, 197
Hydroxymethylnitrofurantoin, 286
Hydroxyphenols, 299
Hydroxyphenylacetic acids, 299
4-Hydroxyphenylacetic acid, 197, 221, 278
4-Hydroxypropranol, 284
5-Hydroxytryptamide, 295
25-Hydroxyvitamin D, 283
25-Hydroxyvitamin D_2, 283
25-Hydroxyvitamin D_3, 283
Hypoxanthine, arabinosides of, 287

I

Idoxuridine, 297
Imazalil, 296

Immobilized enzymes, 124
IMP-GMP:pyrophosphate phosphoribosyl-
    transferase, 291
Indeno(1,2,3-cd)pyrene, 293
Inert tracers, elution time of, 266
Inlet pressure of the column, 16, 17
Inosine, 286, 291
Inosinic acid, 295
Insulin, 262, 263, 290
Interactive chromatography, 151
Interligate space, 157, 158
Intermolecular potential, calculation of,
    206-8
Interstitial space, 8
Intraparticulate diffusion, 149, 159, 242
Intraparticulate space, 157
Intrinsic retention mechanism, 271
Iodoamino acids, 263, 289
Ion association chromatography, 242
Ion-exchange chromatography, 120, 242
Ionic charges, effect on retention in RPC,
    224
Ion-pair chromatography, 67, 119, 122,
    191, 240, 242, 280
    comparison to ion-exchange chromatog-
        raphy, 242
    effect of added salt, 256, 260
    effect of chain length of hetaeron, 256
    effect of hetaeron concentration, 256
    effect of organic solvent concentration in
        the eluent, 257-259
    pH control, 261
    effect of propanol concentration in the
        eluent, 252
    effect of temperature, 256
Ion-pair formation, 122
Ion-pair formation constant, 246, 254
Ion-pair stability constant, effect of or-
        ganic solvent concentration 259-260
Ion-pairing, 244-247, 254, 258
Ionic strength, effect on the retention of
        ionogenic eluites in RPC, 223-225
Ionization of eluite, effect on retention in
        RPC, 228, 238
Ionization potentials, 206
Ionized eluites, dependence of retention on
        the dielectric constant of the eluent,
        217-219
Ionized substances, 209
Ionogenic substances, 190, 221

Irregular retention behavior
    due to conformation changes, 179
    due to silanophilic interactions, 179
Irreversible adsorption, 118
Isocratic elution, 80, 190
Isogentisin, 294
Isohydric eluents, 80, 84
Isokinetic sample introduction, 49
Isomerization, in the mobile phase, 199
Isooctane, 72
Isopropanol, 166, 175, 176
Isopropyl benzoate, 182, 183
L-2-Isopropyl-dien-Zn(II), 262
Isopycnic slurry, 146; see also Balanced-
    density slurry
Isotherm, 237

K

Karl-Fischer titration, 80, 82
α-Keto acid 2,4-dinitrophenylhydrazones,
    265
α-Ketoglutaric acid, 283, 286
Ketones, 275, 298
Ketoprofen, 286
Kieselguhr, 117
Kihara potential, 206
Kinetic resistances, 227
Kinetics of adsorption-desorption, 10,
    127, 159
Kinetics, of secondary equilibria, 159
Kovats index, 280
Kozeny-Carman equation, 6

L

β-Lactams, 275
Langmuir isotherm, 58, 237
LEAC, see Linear elution adsorption chro-
    matography
DL-Leucine, resolution of, 222, 262
Leucine encephalin, 290
LFER, see Linear free energy relationships
LH-releasing hormone, 263, 290
Ligate, carbon number of, 153
Ligates, solute binding to, 213
Limiting retention factors, 239
Linear elution adsorption chromatography,
    58
Linear elution chromatography, 159
Linear free energy relationships, 274
Linuron, 292

Liquid chromatography, 123
  interactive, 133
  size exclusion, 133
Liquid ion-exchangers, 118–119
Liquid–liquid chromatography, 6, 123, 211, 218
Liquid–solid chromatography, 6, 123
LLC, *see* Liquid–liquid chromatography
Loading capacity of column, 59, 61, 68, 86
  determination of, 59
  magnitude of, 59–60
Loading of column with liquid stationary phase
  effect on retention, 81–87
  procedure, 85–87
London parameter, 206
Longitudinal diffusion, 167
Longitudinal molecular diffusion, 8
LSC, *see* Liquid–solid chromatography
LSD, 264
Luminal, 76
Lypressin, 290
Lysine, 289

M

Macroreticular polymeric sorbents, 127; *see also* Porous polymers
  chromatographic efficiency of, 164–165
  preparation of, 164
  stability of columns packed with, 164
Madelung constant, 210
Mandelic acid, 278
Masking agents for surface silanols, 127
Mass transfer resistance, intraparticulate, 151
Maxwell demon, 234
MCS, *see* Moisture control system
Mechanism of retention, in RPC, 224
Mefloquine, 286
α-Melanotropin, 290
Melphalan, 282
2-Mercaptoethanol, 296
Mercaptopurine, 263
Mestranol, 297
Metabolites, analysis by RPC, 283
Metadrenaline, 251
Metal complexes, 230, 298
  of crown ethers, 278
  of diethyldithiocarbamic acid, 299

of nitrosonaphtholsulfonic acids, 278
of nucleotides, 278
Metanephrine, 223
Metformin, 286
Methadone, 297
Methamphetamine, 264
Methanol, 72, 166, 167, 173–176, 180, 190, 275
Methanol–water, 182–184
Methanol–water mixtures
  dependence of viscosity on composition, 193
  dependence of viscosity on temperature, 193
  properties of, 168
  thermodynamic properties as a function of composition, 216
Methiocarb-3,5-dimethyl-4-methylthiophen-ylmethylcarbamate, 297
Methionine, 289
Methionine encephalin, 290
Methionine sulfone, 289
Methionine sulfoxide, 289
Methocarbamol, 264
Methotrexate, 265, 282
(−)-2-Methoxyl-α-1-naphthaleneacetic acid, 289
3-Methoxytyramine, 251
Methylamine, 292
Methyl anthranilate, 295
Methylated bases, 287
Methylated purines, 287
Methyl benzoate, 182–184
Methyl(2-bicycloheptyl)dichlorosilane, 135
N-Methyldiazepam, 282
Methylene chloride, 72, 76
Methylene dichloride, 173
3-O-Methyldopamine, 223
Methylene group selectivity, 180, 181
Methyleugenol, 198
Methyl orange, 267
Methyl parathion, 291, 292
Methyl red, 267
α-Methylstyrene, 42
Methyltrichlorosilane, 135
Metronidazole, 281, 23
Micelle formation, 236, 250, 255
Micelles of hetaeron
  formation of, 251
  partitioning of eluite into, 251, 252

Microparticulate bonded phases, 279
Microparticulate packing, 132
Microparticulate stationary phases, 123,
    145
Misonidazole, 281, 283
Mixed solvents
    free volume, 216
    surface tension correction factor, 215
Model discrimination, in ion-pair chroma-
    tography, 245–252
Modified enzymes, 124
Modulation of selectivity, 229
Modulators, 59
    aliphatic alcohols, 82
    concentration of, in chromatography on
        polar solvents, 78–80
    effect on selectivity, 83
    equilibrium distribution of as a function
        of temperature, 85, 95, 96
    less polar solvents, 83
    for programming the activity of the sta-
        tionary phase, 97
    water, 78, 80, 83
Modulus, 247; see also Retention modulus
Moisture content of eluents, 80, 83
    control of, 80–82
    determination of, 80
Moisture control, 106
Moisture control system, for eluent, 81–83
Molar surface area, 171
Molar volume, 266
Molar volume changes, 267, 268
Molecular associations in the mobile
        phase, evaluation of stability con-
        stants, 276
Molecular contact area, 148, 152, 153, 155
Molecular shape, effect on contact area
        upon binding, 218
Molecular size of the eluite, 217
Molecular surface area, in adsorption and
        partition, 226
    correlation with retention factor, 226
    effect on retention in RPC, 228
Molecular volume, 206
Molybdoheteropoly acids, 299
Monochlorotrialkylsilanes, 139
Monoglycosides, 294
Monolayer, formation by nonpolar eluent
        component on stationary phase in

RPC, 225
    of adsorbed eluent, 102
Monomeric bonded phases, 142, 176
Monoprotic acids, retention of, in RPC,
    238
MTX, see Methotrexate
Multicomponent sample, 5
Mycotoxins, 294

N

NADH, see Nicotinamide adenine dinucle-
    otide, reduced
Naphthalene, 83, 94, 105, 107, 154, 182,
    183
Naphthaleneacetic acid, 296
Narrow bore columns, 53, 54
Natural products, analysis by RPC, 293
Neohesperidin dihydrochalcone, 295
Netilmicin, 282
Niacin, 260, 265, 296
Niacinamide, 265
Nicotinamide adenine dinucleotide, re-
    duced, 287
Nitroanilines, 74
Nitrobenzaldehydes, 182, 183
Nitrobenzene, 74, 182–184
Nitrofurantoin, 286
Nitrofurazolidine, 298
Nitrofurazone, 296
Nitroglycerin, 297
Nitromethane, 72
Nitrophenol, 182–184, 298
Nitrophenols, 182, 183, 265
4-Nitrophenyl phosphate, 298
3-Nitrosalicylic acid, 230
Nitrosoamines, 299
Nitrosonaphtholsulfonic acids, stability
        constants of metal complexes, 278
Nitrotoluene, 182, 183
Nonane, 173
Nonanoic acid, 260
Nonionic surfactants, 298
Noradrenaline, 251, 284
Norephrine, 264
Norepinephrine, 284
Norethistereone, 297
DL-Norleucine, resolution of, 262
Normetanephrine, 223, 251

Norsesquiterpenes, 294
Nortriptyline, 297
DL-Norvaline, resolution of, 262
Nucleic acids, 119
    constituents, 286
    separation by using trialkylmethylam-
        monium chloride on polychlorotri-
        fluoroethylene resin, 287–289
Nucleosides, 286, 287, 295
Nucleotides, 263, 279, 286, 295
    methylated derivatives, 287
    minor forms, 287
    stability constants of metal complexes,
        278

O

Octadecyl ligates, 151
Octadecyl silica, 115, 121, 128, 141, 150,
        151, 154, 160, 161, 171, 174, 181,
        188, 215, 220, 225, 229, 251, 268,
        272, 273
Octadecyldimethylchlorosilane, 135, 136
Octadecylmethyldichlorosilane, 135
Octadecyltrichlorosilane, 135
Octane, 173
Octanoic acid, 260
Octanol, 178, 182, 183
Octanol-coated octadecyl silica, for evalua-
        tion of partition coefficient, 275
Octanol–water partition coefficient, 274
    correlation with retention factor in RPC,
        273–276
    evaluation by using octanol-coated octa-
        decyl silica, 275
    measurement by HPLC, 276
Octyldimethyl silica, 143
Octyldimethylchlorosilane, 135, 136
Octylmethyl silica, 143
Octylmethyldichlorosilane, 135
Octyl silica, 121, 128, 131, 137, 141, 143,
        151, 174, 184, 225
Octyltrichlorosilane, 135
Ogives, 132
Oligoalanines, 219
Onsager reaction field, 208
Opioid peptides, 263
Opium alkaloids, 260
Optical isomers, 231

Optimization of chromatographic system,
        15
Optimization of eluent by using TLC data,
        89
Optimization, at minimum pressure drop,
        29–33
Organic solvent concentration, effect on
        ion-pair stability constant, 259–260
Organic solvent in the eluent, effect on in-
        trinsic mechanism in RPC, 273
Organic solvent concentration in the
        eluent, effect on retention in ion-
        pair chromatography, 258
Organomercury complexes, 296
Oxazepam, 282
Oxipurinol, 287
Oxygen, in eluent, 188
Oxytocin, 290

P

Packing of column, 11
Packing structure, effect on column effi-
        ciency, 19
    effect on pressure drop, 19
Paracetamol, 260, 264
Parameters of chromatographic system,
        column, 16
    operational, 16
Paranephrine, 223
Paraoxon, 292
Paraquat, 296
Parathion, 292
Particle diameter, 6, 16, 23, 27
Particle size, 60–61
    hydraulic radius, 7
Particle size distribution, 7, 9, 23, 125,
        131, 132
Partition, comparison to adsorption, 225
Partition coefficients, 211, 273
    calculation of, 218
    in water and n-octanol, 218
Partitioning model, in RPC, 224
Partitioning of polar molecules between
        eluent and polar adsorbents, 85
Peak asymmetry, 3, 159
Peak capacity, 51, 52, 68, 123, 184
Peak, variance of, 8
Peak width, 3

Pellicular adsorbents, commercial prod-
    ucts, 64
  properties of, 60
Pellicular column materials, 132
Pellicular configuration, 133
Pellicular ion-exchangers, 120
Penicillins, 275, 281, 297
Pentane, 72
Pentanol, 177, 182, 183
Peptides, 286, 289
  opioid, 263
  separation by ion-pair chromatography,
    290
  structure–retention relationships, 290
Perchloric acid, 188
Perfluorinated hydrocarbons, 73
Perfluoroheptyl silica, 162
Permeability coefficient, 167
Perturbation function for potential, 208
Perylene, 105
Pesticides, chromatography on polar sor-
    bents, 104, 105
Pharmaceutical preparations, analysis by
    HPLC, 296
Pharmacokinetics, 280
Phase ratio, 4, 148, 149, 152, 153, 155, 157,
    211, 232, 266, 269, 270
Phenanthrene, 105, 107, 154
Phenformin, 264, 283
Phenobarbital, 282
Phenolic compounds from plants, 293
α-Phenoxyethylpenicillins, 281
Phenylacetic acid, 272, 278
Phenylalanine, 222
Phenyldimethylchlorosilane, 135
2,3-Phenylenepyrene, 105
Phenylephrine, 264
Phenylethyl alcohol, 182, 183
Phenylethylamine, 260, 263
β-Phenylethyltrichlorosilane, 135
Phenylmethyldichlorosilane, 135
Phenylpropanolamine, 264
Phenyl silica, 158
Phenyltrichlorosilane, 135
Phenytoin, 282
pH control
  in ion-pair chromatography, 261
  of retention in RPC, 238
pH dependence of retention in RPC

amino acids, 222
  weak acids, 221
Phosphate buffer, 189
  effect on column efficiency, 188
  trialkylammonium, 188
  with weak amines, 190
Phosphoglycerate kinase, 290
Phospholipids, 283
Phosphoric acid, 162, 188, 290
o-Phthalaldehyde, 284, 292
o-Phthalic acid, 278
Physicochemical measurements by RPC,
    266–279
Physiological samples, analysis of, by
    RPC, 280
Picket fence configuration, 157
Pigments, 295
Pilocarpine, 297
Piprozolin, 283
Pirimiphos methyl, 297
Plate height, 167
  accuracy of measurements, 12
  dependence on flow velocity, 10
  minimum, 9, 13, 18
  reduced, 10
Plate number, 3, 4, 7, 27, 52, 53
Plates per second, 30–33
PLB, see Porous layer beads
Polar group selectivity, 181–183
Polar solvent, selective uptake from eluent
    by polar adsorbent, 85
Polarizability, 208
Poly(ethylene glycol) derivatives, 177, 298
Polychlorinated biphenyl, chromatography
    on polar sorbents, 105
Polychlorobiphenyls, 5
Polycyclic aromatic hydrocarbons, chro-
    matography on polar sorbents, 104,
    105
Polyethylene powder, 118
Polyglutamates, see Pteroyl-oligo-γ-gluta-
    mates
Polymeric bonded phases, 144, 149, 176
  mass transfer resistance in, 159
Polynuclear aromatic hydrocarbons, 293
  chlorinated derivatives, 293
  chromatography on polar sorbents, 104,
    105
  tritiated, 279

Polystyrene, macroreticular, 239
Ponceau 4R, 249
Ponceau MX, 249
Pore
  size, 149, 150
  size distribution, 127, 133, 149, 154
  structure, 133
  volume, 149
Porosity, 6
Porous carbon, 128
Porous layer beads, 9, 60; *see also* Pellicular sorbents
Porous polymers, 127, 128; *see also* Macroreticular sorbents
  cross-linked polystyrene derivatives, 164
Postcolumn reactors, 289
Precolumn, 85, 161; *see also* Guard column
Prednisolone, 298
Pressure drop, 6, 18
  calculation of, 27
  criticial minimum value, 33–35
  dependence on column length, 28
  dependence on particle diameter, 28, 30–33
  minimum, 26
Pressure programming, 93–95
Primidone, 282
Progesterone, 74
Programming of elution conditions, 92
Proline, as hetaeron, 290
Prominal, 76
1-Propanol, 72, 166, 174
2-Propanol, 174
1-Propanol–water mixture, properties of, 170
Propranolol, 284, 286
Propyl chloride, 72
Prostaglandin synthetase, 201, 222, 291
Proteins, 286, 289
Protonic equilibria, 122, 188, 230, 238
Pteroyl-oligo-γ-glutamates, 220
  evaluation of pK_a values by RPC, 278
  pH dependence of retention factor in RPC, 240
Purines, 286, 287, 294
Putrescine, 294
Pyrazolone, 230
Pyrene, 83, 105, 107, 154

Pyridine, 72
Pyridoxine, 265
Pyrimidines, 286
Pyrocarbon, 163
Δ'-Pyrroline-5-carboxylic acid, 291
Pyruvic acid, 283, 286

Q

QSAR, 274; *see also* Quantitative structure–activity relationships
Quantitative structure–activity relationships, 226
Quartz, 65
Quaterphenyl, 66
Quinine, 264
Quinoxalones, 286
Quinquephenyl, 66

R

RPC, *see* Reversed-phase chromatography
Reactive organosilanes, 135
Reduced plate height, 10
  optimum value, 26
Reduced velocity, 10
Reduction term, in solvophobic theory, 217
Reequilibration of column, 100
  after gradient elution, 84
  use of reverse gradient, 84
Refractive index, 208, 209
Regular retention behavior, 176–178
Rejuvenation of bonded phases, 140
Relative retention, 3, 4, 149, 165
  dependence on surface tension of eluent, 216
Reserpine, 283
Resistance to mass transfer, 8
Resolution, 3, 16, 58, 93
  definition of for asymmetric peaks, 3
  dependence on column and operating parameters, 4, 16
Resolution in programmed elution, 92
Response time, 40
  of detector, 40, 44
  maximum permissible, 45
Retention, dependence on eluent composition in chromatography with polar eluents, 103

Retention enthalpy, 269
Retention entropy, 269
Retention factor, 148, 149, 151, 152, 211,
    249, 252, 266; *see also* Capacity factor
  for complex exchange with no displace-
      ment, 233–234
  for complex formation in the mobile
      phase, 232
  correlation with octanol–water partition
      coefficient, 226, 273–276
  dependence on carbon number for ho-
      mologous series in RPC, 219
  dependence on the degree of ionization,
      224
  dependence on hetaeron concentration,
      236, 237
  dependence on molecular surface area in
      RPC, 218
  dependence on structural units for ho-
      mologous series in RPC, 219
  of deuterated and tritiated eluites, 219
  for dynamic complex exchange, 233
  for dynamic ion exchange, 233
  effect of column pressure, 267, 268
  effect of eluite ionization in RPC, 221–
      224
  in hetaeric chromatography of diprotic
      acids and zwitterions, 241
  limiting values for ionized and neutral
      forms of species, 238
  multiple complexes, 277
  pH dependence of, 238
Retention index, Kovats index, 280
Retention mechanism, effect of eluite ioni-
      zation, 234–240
  intrinsic, 199
  in ion-pair chromatography, 240–260
  multiple, 200
  with secondary equilibria, 231–237
  test of identity, 199
Retention modulus, 245–247
  dependence on carbon number of he-
      taeron, 247
Retention time, 3, 8, 18, 58
  adjusted, 3
  dimensionless (retention factor), 4, 5
  of nonsorbed solute, 3
  of unsorbed tracer, 148
Retention volume, 51, 58
Retinoic acid, 283

Retinoids, 283
Retinol, 261
Retinyl fatty acid esters, 261
Reverse gradient, in column reequilibra-
    tion, 84
Reversed-phase chromatography, 71, 92
  column, 121
  comparison to gas chromatography, 115
  definition, 120
  evolution of, 117–120
  mixed-aqueous (MARP), 121
  nonaqueous (NARP), 121
  operating conditions, 121
  origin of name, 114–115
  plain-aqueous (PARP), 121
  scope of, 116
Ribo-oligomers, 288
Riboflavin, 265
Ribonuclease-S peptide, 290
Rosin, 298
Rotenoids, 297
Rotenone, 297
Rubratoxin, 294

S

Saccharin, 295
Safrole, 198
Salicylazosulfapyridine, 281
Salicylic acid, 221, 278, 282, 286, 297
Salicyluric acid, 282, 286
Salting-out chromatography, 120
Sample concentration, at peak maximum,
    51, 52
Sample introduction, contribution to band
    spreading, 48
  duration of, 50
  isokinetic, 49
Sample loading, 59
Sample size, 48, 49, 51, 52, 59
Sample, solubility of, 71
Sampling valve, for short column, 41
Saturator precolumn, 85
Schaeffer's acid, 249
Secobarbital, 264
Secondary chemical equilibria, 230, 280;
    see also Secondary equilibria
  with diprotic acids and zwitterions, equi-
      librium constants and retention, 241
  with micelle formation, 236

Secondary equilibria, 122, 199; *see also*
　　Secondary chemical equilibria
　　retention models with, 231–238
Secondary solvent effects, 76, 103
Selectivity, 180, 229
　　dependence on eluent composition, 180
　　dependence on the ligate, 151
　　effect of mobile phase composition in
　　　RPC, 229
　　effect of modulator in chromatography
　　　with polar sorbents, 84
　　effect of secondary equilibria, 229
　　temperature dependence of in RPC, 192
Selectivity in RPC, correlation with solu-
　　bility, 225
DL-Serine, resolution of, 262
Sexiphenyl, 66
Short alkyl silica, 248
Silanization, 126, 128, 140, 151, 161
Silanizing agents, 134–136, 140, 142, 143,
　　151
Silanol functions, 61
　　accessible, 65
　　reaction of, 65
　　surface density of, 65
Silanol groups, 122, 133, 158, 220
Silanols
　　accessibility of, 144, 145
　　conversion of, 142, 143
　　masking of, 127
　　pK_a, 133
　　shielding by hydrocarbonaceous ligates,
　　　152
　　surface concentration of, 141, 142
　　surface concentration of ligates, 133
Silanophilic interactions, 128, 141, 179,
　　201, 220
Silica, 60, 151
　　acid treatment of, 134
　　argentation of, 68
　　catalytic activity of, 67
　　chemical modification of the surface,
　　　67–68
　　chromatographic properties as a result of
　　　the history of, 67
　　coating with a liquid stationary phase,
　　　68
　　commercial stationary phases, 62–64
　　comparison of retention behavior to that
　　　on alumina, 69

　　dehydration of, 61
　　drying of, 134
　　hydrophobic, 61
　　hydrophobic, rehydration of, 65
　　hydrothermal treatment of, 134
　　naked, 68
　　pore diameter of, 61
　　pore size of, 68
　　　effect on retention behavior, 66
　　pore volume of, 61
　　preparation of, 65
　　pretreatment of, 142
　　retention by, 66
　　retention on different types of, 65
　　silanols at the surface of, 142
　　specific surface area of, 61, 66
　　support for bonded phases, 107
　　as support in liquid–liquid chromatogra-
　　　phy, 68
　　surface coverage of, 144
　　surface properties, 142
　　thermal treatment of, 65
　　wide-pore, 151
Silica gel, 116, 124–126, 131, 133, 141
　　hydrolysis, 125
　　microparticulate, 125, 141
　　particle size distribution, 131
　　pore dimensions, 131
　　reagents for surface treatment of, 135
　　specific surface area, 131, 141
　　surface characteristics of, 141
　　surface properties, 125
　　surface treatment, 125
　　topology of surface, 136
Silicone oil, in reversed-phase paper chro-
　　matography, 117
Siloxane bridges, 128, 137
Silymarin, 294
Sisomicin, 282
Size exclusion, 133, 149, 151
Slurry packing, 145, 146
Soap chromatography, 122, 242, 260; *see
　　also* Ion-pair chromatography
Solubility parameters, 186
Solubilization chromatography, 120
Solution kinetics, effect on column effi-
　　ciency in RPC, 227
Solvent demixing, 89
Solvent effects, 122
　　on selectivity, 191

Solvent-generated (dynamic) ion-exchange chromatography, 242
Solvent gradient, formation of, 100
  linear, 99
  shape of, 98
Solvent properties, 165–166
Solvent strength parameter, 175, 176
Solvent strength, 73; *see also* Eluent strength
Solvents, UV cut-off values, 70
Solvents, miscibility, 75
Solvophobic effect, 201, 203
Solvophobic interactions, 152, 201
Solvophobic ion chromatography, 242
Solvophobic theory, 141, 148, 152, 155, 158, 202, 203, 226, 228, 246
Somatostatin, 263, 290
Sorbents, polymeric, 127
Sorption isotherm, 159
Sorption kinetics, effect on column efficiency in RPC, 227
Speed of separation, optimization of, 35–38
Spermidine, 294
Spermine, 294
Stability constant, of complex in mobile phase, 232
Stability constant of complexes, measurement by chomatography, 276
Stack configuration, 157
Standard deviations, 3
Stationary phases, bonded hydrocarbonaceous, 121
  microparticulate, 121, 123
  modulator activity, programming of, 97
  ideal for RPC, 126
  organosilica type, 126
  polar, 57
Stationary phase surface, solute binding to ligates, 213
Steroids, 76, 77, 105
Sterols, 283
Styrene, 42
  oligomers, 99
Sulbenicillin, 286
Sulfa drugs, 264
Sulfadiazine, 281
Sulfamerazine, 281
Sulfamethazine, 281

Sulfamethoxazol, 281
Sulfanilic acid, 230, 249
Sulfanitran, 298
Sulfapyridine, 281
Sulfinpyrazone, 282
Sulfisoxazole, 284
Sulfobetaine surfactants, 298
Sulfonamides, 281, 297
Sumicidin, 297
Sunset Yellow, 249
Superficial velocity, 167
Surface area, molar, 171
Surface carbon equivalent, 142
Surface coverage, 144
Surface modification, 59
Surface properties, 125
Surface tension, 165, 166, 169, 217
  correction factor for, 205, 215
Surfactant chromatography, 242

T

Tailing of peaks, 58, 67, 82
  reduction by amines, 67
  reduction by treatment with buffers, 67
Tailing reducers, 118
Tartrazine, 230, 249, 265
Temperature of column, 192
  effect on modulator distribution, 85, 95, 96
Temperature control, 95, 194
Temperature profile, effect of viscous dissipation, 267
Temperature programming, 93, 95, 96
  for generation of elution gradient by using a precolumn, 85, 95, 96
Terphenyl, 66, 74
Testosterone, 74
Tetraalkyl ammonium, 240
Tetrabutylammonium hydroxide, 230
Tetrachlorobenzene, 153
Tetrachloroethylene, 83, 105
Tetracyclines, 281, 297
Tetrahydrofuran, 72, 166, 173, 176, 191, 275
Tetrahydrofuran–water mixtures, 169, 183, 184
1-(2-Tetrahydrofuryl)-5-fluorouracil, 282
Tetramethylbenzene, 172

Theobromine, 287
Theophylline, 279, 287
Thermal treatment of silica, 65
Thermodynamic equilibrium constant, 6, 148
Thiamine, 265
Thin-layer chromatography, comparison to column chromatography, 88–90
  use for selection of HPLC system, 88–90
Thiopental, 282
DL-Threonine, resolution of, 262
Time of analysis, dependence on column parameters, 22
  fixed, 21, 22
  minimum of, 37–38
Time constant, *see* Response time
TLC, *see* Thin-layer chromatography
Tolbutamide, 283
Toluene, 94, 182, 183
Tortuousity coefficient, 8
Trace analysis, 292
Tri-*n*-butylchlorosilane, 135
Triamcinolone acetonide, 298
Tribenzylchlorosilane, 135
Tribochemistry, 126
Trichloroalkylsilanes, 137
3,4,4'-Trichlorocarbanilide, 281
Trichlorooctadecylsilane, 136
Tricyclic antidepressants, 264
Tridecylamine, 230
Tridymite, 65
Triethanolamine, 289
Triethylamine, 67, 161
Triethylammonium phosphate, 290
Trifluoroacetic acid, 262, 290
Triglycerides, 295
Trimethoprim, 297
Trimethylbenzene, 172
Trimethylchlorosilane, 135, 139
Trimethyl silica, 143
Triphenylchlorosilane, 135
Triphenylene, 105
Tritiated substances, 219, 279
Tryptamine, 260, 263
DL-Tryptophan, resolution of, 262
Tryptophase, 291
TSH-releasing hormone, 290
Tylosin B, 297
Tyramine, 223, 260, 263

U

Ubiquinone, 283
Unretained eluite, 266, 270, 271
Urinary aromatic acids, 285, 286
Urine analysis, 284
UV cut-off values of solvents, 70

V

DL-Valine, 222
van der Waals energy, 201, 212
van der Waals interactions, 145, 148, 218, 219
  calculation of 206–8, 212
  for solute binding in RPC, 212
  in solution, 206–208, 212
Vanillin, 295
Vanillmandelic acid, 223, 278, 284
van't Hoff relationship, 268
van't Hoff plots, 195, 269
  linear, 196
  nonlinear, 200, 201
Variance, 8
Velocity, of mobile phase, 6
Velocity, reduced, 10
Viscosity, 6, 71, 165, 167, 169, 187, 192
Viscosity of eluent, dependence on composition, 192, 193
  effect on efficiency, 24–26
  temperature dependence of, 192–195
Viscous dissipation, 267; *see also* Viscous friction
Viscous friction, 53
Vitamin A, 283, 295–297
Vitamin B, 295; *see also* B vitamins
Vitamin D, 261, 295
Vitamin D_2, 265, 283, 296, 297
Vitamin D_3, 265, 296
Vitamin E, 296, 297
Vitamin K_1 2,3-epoxide, 284
Vitamin K_1, 284
Vitamins, 294, 296
Vulcanized rubber as stationary phase, 117

W

Wall effect, 49
Warfarin, 284, 297
Water, 72, 166, 167

as a modulator in chromatography on
  polar adsorbents, 78, 80
purification of, 188
Water–acetone mixtures, 180
Water–acetonitrile mixtures, 177, 180
  effect of composition on, 182
  polar group selectivity, 182
Water–methanol mixtures, 171, 176, 275
  effect of composition on, 182
  polar group selectivity, 182
Water structure, 203–204
  iceberg model, 203
Water–octanol partition coefficient, correlation with retention factor, 226
Weak acids, evaluation of pK_a values by
  RPC, 278
  pH dependence of retention in RPC, 221
Wilke–Chang equation, 29, 167

X

Xanthines, 287, 295
Xanthone glycosides, 294
Xanthosine, 286
Xerogels, 61
Xylenes, 173, 182, 183
Xylenol, 74

Z

Zearalenone, 294
Zimelidines, 264
Zwitterions, 210, 239, 261
  equilibrium constants and retention factors in hetaeric chromatography,
  241